ANNALS

OF THE

ASTROPHYSICAL OBSERVATORY

OF THE

SMITHSONIAN INSTITUTION.

VOLUME I.

By S. P. LANGLEY, Director,
Aided by C. G. ABBOT.

[Reprint (1902) from Senate Document No. 20, 57th Cong., 1st Sess.]

WASHINGTON:
GOVERNMENT PRINTING OFFICE.
1900.

PREFACE.

This book is the result of a research originally due to a discovery, made in the year 1881 with the then newly invented bolometer, in the clear air of an altitude of over 12,000 feet, of solar heat in a then unknown spectral region now called the "lower infra-red spectrum." The bolometer has since been used to explore and to map the region in question, through the long succeeding interval, in the latter part of which it has reached an accuracy and a sensitiveness greater than I could once have hoped for.

This map is now (June 18, 1900), after years of constant work, finally published in the present form; not because this edition is final, but because the long labor must come to some term, and because I desire to see its results published while I may hope to see them made useful.

In my early work I was led to notice not only the change of the infra-red absorption spectrum at different hours of the day, but at different seasons of the year, without my observations in the last respect having attained a precision which seemed to justify me in publishing them. Of late, improvements on the early methods seem to be at last bringing more conclusive evidence that there are distinguishable effects in the different seasons of the year upon the absorption of the solar heat by the earth's atmosphere, or perhaps it might be said, evidence that the absorption of the earth's atmosphere is directly associated with the seasonal changes of spring, summer, autumn, and winter.

While we are far from looking forward to foretelling by such means the remoter changes of weather which affect the harvests, or to results of such importance as the power of such a prevision would indicate, still it is hardly too much to say that we appear to begin to move in that direction, and it seems to me that my own early hopes of making the study of the solar energy not simply an interesting scientific pursuit, but one of material usefulness, may one day be justified.

S. P. LANGLEY.

JUNE 19, 1900.

SUMMARY OF THE WORK.

There are here presented tables and charts of the invisible infra-red solar spectrum extending to 5.3 μ and mapping nearly 750 lines, of which most are below 1.1 μ and are nearly all new.

There has been determined in the case of each line the relative absorption of solar energy which has produced it, while its place has been fixed from the comparison of a large number of independent bolometric observations with a probable error in prismatic deviation of less than one second of arc, and for most of the lines with a probable error in wave-length of less than 5 Ångström units.

The infra-red region is shown to be the seat of the principal telluric absorptions of the solar energy, and a discrimination of these absorptions into those which are relatively accidental and those which are found to be annual or peculiar to each season is indicated.

CONTENTS.

	Page.
INTRODUCTION	1
FOUNDATION AND EARLY ANNALS OF THE ASTROPHYSICAL OBSERVATORY	1
Part I.—THE ABSORPTION LINES IN THE INFRA-RED SPECTRUM OF THE SUN	5
Chapter I.—Historical account of contributory researches	7
Early investigations	7
The writer's early work at Allegheny	10
Recent investigations of infra-red spectra by others than the writer	19
Chapter II.—The present research	22
Account of the progress of the investigations at the Smithsonian Astrophysical Observatory	22
Condition of the research July 1, 1892	23
Progress of the research in the year ending July 1, 1893	27
Progress of the research in the year ending July 1, 1894	30
Progress of the research in the year ending July 1, 1895	35
Progress of the research in the year ending July 1, 1896	38
Progress of the research in the year ending July 1, 1897	39
Chapter III.—Description of the observatory buildings and the apparatus employed in the research—adjustments of apparatus	40
The observatory	40
Apparatus	43
Constants of the electrical system	43
Constants of the optical system	45
Siderostat	45
Slit	46
Spectrometer and accessories	46
Bolometer and accessories	47
Driving clock	56
Plate carrier	58
Galvanometer and accessories	59
Battery	63
Comparator	64
Adjustment of apparatus	65
Siderostat	65
Slit	65
Mirror system	65
Prism	66
Cylindric lens	67
Chapter IV.—Methods of procedure in preparing and comparing bolographs	69
The bolographic processes	69
1. Preparing the curves	69
2. Methods of identification of absorption lines	71
Conversion of bolometric curves to linear spectra	73
Comparison and measurement of bolographs to discriminate and fix the position of "real" inflections	74
Chapter V.—Limitations of the research—sources of error now existing	76
Limitations of the research	76
1. Visual resolving power of the prism	76
2. Slit width	77
3. Collimation	77
4. Bolometric resolution	82
Sources of error	85
I.—Optical apparatus	86
II.—Accidental deflections in record curves	105
III.—Imperfect mechanism	112

XXXII CONTENTS.

Part I.—THE ABSORPTION LINES IN THE INFRA-RED SPECTRUM OF THE SUN—Continued.
 Chapter V.—Limitations of the research—sources of error now existing—Continued.
  Sources of error—Continued.
   IV.—Preparing linear spectra from bolographs... 116
   V.—Comparator measurements 118
   VI.—Inherent in the bolographic method 119
  Summary of results of limitations and errors... 122
   A. Limitations... 122
   B. Errors... 123
 Chapter VI.—The absorption lines in the infra-red solar spectrum; results of bolographic spectrum analysis in 1897-98.. 126
  Illustrations... 127
  Comparison of bolographs... 129
  Notes accompanying bolographs.. 133
  Measurements on bolographs... 137
  Reduction of observations 171
  Tables of final values... 176
  Comparison of the results with the solar spectrum maps of Higgs and Abney and with infra red metallic lines of Snow, Lewis, and others.. 200
 Chapter VII.—The variations of absorption in the infra-red solar spectrum............................. 205
  Typical energy curves... 206
  Irregular variations of absorption.. 207
  Emission and absorption experiments 215
  Summary.. 216
Part II.—SUBSIDIARY RESEARCHES... 217
 Chapter I.—The dispersion of rock salt and fluorite 218
  The relative dispersion of rock salt and fluorite 218
  The dispersion of rock salt.. 222
  Have all rock-salt prisms the same indices of refraction?... 236
 Chapter II.—Miscellaneous observations 238
  1. The accuracy of the bolometer 238
  2. Radiation from terrestrial sources... 240
  Supplementary note 242
 Chapter III.—Construction of a sensitive galvanometer... 244
  Purposes for which it was constructed 244
  Considerations governing the construction 245
  Best construction of coils.. 248
  The needle system 251
  Summary 252
Appendix—
 THE DETERMINATION OF WAVE-LENGTHS IN THE INFRA-RED SPECTRUM OF ROCK SALT 253
 The constants for rock salt in Ketteler's dispersion formula 261
 The minute structure of the absorption band ω_1... 263
Index.. 265

ILLUSTRATIONS.

	Page.
PLATE I.—The Astrophysical Observatory	1
II.—Bolographs of D lines taken in 1893 and 1898, illustrating the decrease in accidental disturbances	33
III.—Plan of Observatory buildings and lot, 1900	40
IV.—Plan of main Observatory building, 1897	41
V.—Section through main Observatory building, 1897, looking west	41
VI.—Spectrometer room, 1896, looking south	42
VII.—Galvanometer room, 1896, looking west	42
VIII.—The siderostat, looking east	45
IX.—Plan view of the spectrometer	46
X.—Side elevation of the spectrometer	46
XI.—The bolometer and special rheostat, form used 1896 to 1898	47
XII.—Details of bolometer and special rheostat, form used 1896 to 1898	47
XII A.—Working drawing for combined bolometer and rheostat	52
XIII.—The driving clock of the spectrobolometer	57
XIV.—The photographic recording apparatus of the spectrobolometer	58
XV.—The galvanometer and its accessories	60
XVI.—The comparator	64
XVII.—Galvanometer and battery records, 1895 to 1898	106
XVIII.—Early bolographs of the infra-red solar spectrum	127
XIX.—The infra-red solar spectrum of a 60° rock-salt prism. Composite line spectrum from bolographs of 1896	127
XX.—The infra-red solar spectrum of a 60° rock-salt prism. Energy curves and line spectrum from bolographs of 1898	127
XX A.—The infra-red solar spectrum of a 60° rock-salt prism. From bolographic observations of 1897–98	176
XXI A. } Bolographs of the infra-red solar spectrum of a 60° glass prism	128
B.	
XXII.—Bolographs of the solar spectrum absorption bands ω_1 and ω_2, showing the increase in absorption in ω_1 with declining sun	129
XXIII.—Bolographs of the visible and infra-red solar spectrum of a 60° rock-salt prism	129
XXIV. } Normal map of the infra-red solar spectrum. From bolographic investigation	200
XXV.	
XXVI A.—The relative dispersion of rock salt and fluorite in the infra-red spectrum	220
XXVI B.—The dispersion curve of fluorite in the infra-red spectrum	222
XXVII.—Diagram of apparatus used to determine the dispersion of rock salt	225
XXVIII.—The dispersion of a 60° rock-salt prism	227
XXIX.—The dispersion curve of rock salt in the visible and infra-red spectrum	235
XXX A. } Bolographs of the infra-red solar spectrum of a 60° rock-salt prism, typifying the yearly variations in absorption in spring, summer, autumn, and winter	207
B.	
XXXI A. } Spectrum energy curves of Welsbach and other heated mantles, from bolographs with a 60° rock-salt prism	242
B.	
XXXII.—The dispersion curve of rock salt. Various observers	258

THE ASTROPHYSICAL OBSERVATORY.

ANNALS OF THE ASTROPHYSICAL OBSERVATORY OF THE SMITHSONIAN INSTITUTION.

INTRODUCTION.

FOUNDATION AND EARLY ANNALS OF THE ASTROPHYSICAL OBSERVATORY.

In the reports of the Secretary of the Smithsonian Institution for the years ending June 30, 1888 and 1889, mention is made of the hope then cherished of erecting and equipping an observatory for astrophysical research; and in the year following, 1890, he is at last "able to say that this object has assumed definite shape in the construction of a temporary shed," "begun on November 20, 1889, and . . . completed about the 1st of March, 1890. This building is of the most inexpensive character, and is simply intended to protect the instruments temporarily, though it is also arranged so that certain preliminary work can be done here. Its position, however, immediately south of the main Smithsonian building, is not well suited to refined physical investigations, on account of its proximity to city streets and its lack of seclusion."

This "temporary shed" still continues to be the main observatory building. It has been slightly modified within, from time to time, to make it more suitable for the uses to which it has been put, and in 1898 the southeastern corner received some enlargement to make a more commodious office. Two smaller buildings have been erected north of the original one. The first, dating from 1893, is a single room without basement, which serves for photographic purposes, and the second, erected in 1898 and provided with a double-walled basement of brick and a single story above of wood, is used for such physical researches as are crowded out of the old building by the bolographic apparatus and accessories, which fully occupy it.

Means for the original construction and for making some preliminary observations (consisting for the most part of determinations of the constants of apparatus) were furnished from the Smithsonian funds, but it was believed that the investigations to be undertaken at the observatory were of such a nature that it would be proper for the General Government to appropriate a small sum annually for its maintenance. This view has been accepted by Congress, and the first such appropriation became available in July, 1891.[1]

---

[1] Sundry civil appropriation act approved March 3, 1891: "For maintenance of Astrophysical Observatory under the direction of the Smithsonian Institution," etc.

In bringing the matter to the attention of Congress statements of the purposes of the observatory were made by the writer, which have been summarized by him in the report of the Secretary of the Smithsonian Institution for the year ending June 30, 1892, as follows:

The general object of astronomy, the oldest of the sciences, was, until a very late period, to study the places and motions of the heavenly bodies, with little special reference to the wants of man in his daily life, other than in the application of the study to the purposes of navigation.

Within the past generation, and almost coincidentally with the discovery of the spectroscope, a new branch of astronomy has arisen, which is sometimes called astrophysics, and whose purpose is distinctly different from that of finding the places of the stars or the moon or the sun, which is the principal end in view at such an observatory as that, for instance, at Greenwich.

The distinct object of astrophysics is, in the case of the sun, for example, not to mark its exact place in the sky, but to find out how it affects the earth and the wants of man on it; how its heat is distributed, and how it in fact affects not only the seasons and the farmer's crops, but the whole system of living things on the earth, for it has lately been proven that in a physical sense it, and almost it alone, literally first creates and then modifies them in almost every possible way.

We have, however, arrived at a knowledge that it does so, without yet knowing in most cases how it does so, and we are sure of the great importance of this last acquisition, while still largely in ignorance how to obtain it. We are, for example, sure that the latter knowledge would form, among other things, a scientific basis for meteorology and enable us to predict the years of good or bad harvests, so far as these depend on natural causes, independent of man, and yet we are still very far from being able to make such a prediction, and we can not do so till we have learned more by such studies as those in question.

Knowledge of the nature of the certain but still imperfectly understood dependence of terrestrial events on solar causes is, then, of the greatest practical consequence, and it is with these large aims of ultimate utility in view, as well as for the abstract interest of scientific investigation, that the Government is asked to recognize such researches as of national importance; for it is to such a knowledge of causes with such practical consequences that this class of investigation aims and tends.

Astrophysics by no means confines its investigation to the sun, though that is the most important subject of its study and one which has been undertaken by nearly every leading government of the civilized world but the United States. France has a great astrophysical observatory at Meudon, and Germany one on an equal scale at Potsdam, while England, Italy, and other countries have also, at the national expense, maintained for many years institutions for the prosecution of astrophysical science.

The researches of the observatory have from its foundation been under the general direction of the present Secretary of the Smithsonian Institution, and the work to be described—that of studying the infra-red spectrum—was begun by him before his connection with the Institution while he was Director of the Allegheny Observatory, and its results were described in contributions to various scientific journals.

It was his intention to personally continue it here, and he has at all times endeavored, within the measure of his ability, to do so; but of late years the administration of the general interests of the Institution has so far occupied his attention that the direction of the details of the operations of the observatory, has been increasingly intrusted to other hands.

There have been associated with the writer as his immediate assistants in the work: Dr. William Hallock, Senior Assistant; Mr. F. L. O. Wadsworth, Senior Assistant; Mr. R. C. Child, Aid Acting in Charge, and Mr. C. G. Abbot at present Aid Acting in Charge, who took charge in January, 1896, and with him Mr. F. E. Fowle, still acting as Junior Assistant.

The following is a brief account of the early annals of the Astrophysical Observatory. In recent years its history has been so far wrapped up in its principal investigation, relative to the infra-red spectrum of the sun, that it necessarily must be told in connection with the account of that research, which will be found beginning with Chapter III. Even the buildings have been altered, and in case of the later construction designed especially to facilitate bolographic research, so that they can best be described as portions of the apparatus appertaining to special research.

The main building of the observatory, illustrated in Plates I, III, IV, and V, was begun November 20, 1889, and completed about March 1, 1890. Its cost was defrayed from funds of the Smithsonian Institution.

The Grubb siderostat, Grunow spectro-bolometer, White galvanometer, and Elliott special resistance box for bolometric purposes were all procured at the cost of the Institution, and installed during the year 1890.

Before the entire completion of the building a research, which may properly be called the first to be connected with the Astrophysical Observatory, was conducted.[1] The purpose of this investigation was a comparison of the light and dark radiations of the luminous tropical insect *Pyrophorus noctilucus* Linn., with those of similar kind emitted from the sun and artificial sources of light. Specimens of this tropical insect were procured through the Bureau of Exchanges of the Smithsonian Institution, and were examined with photometric and bolometric apparatus, partly at Washington and partly at Allegheny observatory, where a portion of the work was transferred owing to the then incomplete installation of apparatus at the Smithsonian observatory.

It was shown that the solar spectrum, when reduced to the same total luminous intensity, appeared to the eye to extend notably farther both toward the red and toward the violet than did the spectrum of the insect. The visible spectrum of the latter was practically confined between wave lengths 0.45μ and 0.65μ. Bolometric experiments gave slight but unquestionable indications of radiation from the insect; but by interposing a sheet of glass opaque to wave lengths greater than 3.0μ it was found that these indications then ceased, so that they were due rather to general animal heat than to any rays associated with the luminosity. Other sources of light when reduced to equal total luminosity, including the solar beam, the Bunsen and Argand burners, and the electric light, were all found to expend incomparably more energy in invisible radiations than *Pyrophorus noctilucus*.

Thus nature produces this cheapest form of light at but an insignificant fraction of the cost of the electric light or the most economic light which has yet been devised.

In June, 1891, Dr. William Hallock came to the observatory as senior assistant, and preliminary work in trying and adjusting apparatus was taken up.

The first appropriation by Congress of $10,000 became available, as has been said, July 1, 1891.

---

[1] "The cheapest form of light," by S. P. Langley and F. W. Very, Am. Jour. Sci., third series, Vol. XL, p. 97, 1890.

March 2, 1892, the bolographic research of mapping the infra-red solar spectrum was begun.

In April, 1892, the position of the siderostat pier of the observatory was determined by the United States Coast and Geodetic Survey as follows:

Latitude 38° 53' 17.3" N.
Longitude 77° 01' 33.6" W. of Greenwich.

In the fall of 1892 a dark room was erected closely adjoining the main building, bolographic work being then well started.

A small number of instrument maker's tools, including a lathe and small planer, were procured in the latter part of 1892, and temporarily used in a shed adjoining the Smithsonian building. Two years later they were removed to a room in the Smithsonian stable, and from time to time the number has been added to, so that of recent years most of the instrument making for the observatory has been done at this shop, where one instrument maker is constantly employed.

Automatic control of the heating supply was introduced in 1896, as experience showed that only by securing a constant temperature could satisfactory bolographic work be done.

Cooling by ammonia gas was introduced in 1897, and this cooling system was also provided with automatic temperature control.

A laboratory consisting of one story and basement was erected in the latter part of the year 1898, so that at present the observatory has three buildings—the spectrobolometric building, the photographic room, and the new laboratory.

Further details of an historical and descriptive character will be found in connection with the several researches to be described, for, as already said, the bolographic investigations have been of such a nature as to almost wholly shape the development of the observatory.

PART I.

THE ABSORPTION LINES IN THE INFRA-RED SPECTRUM OF THE SUN.

Of the solar radiations which reach the earth's surface about three-fourths are invisible. While the visible portion of the spectrum has been explored since the days of Newton and mapped with great minuteness, the great and important invisible portion lying beyond the red has remained comparatively unexplored and unknown. The object of the main investigation here published is to give for the first time the principal details of this region.

Chapter I.

HISTORICAL ACCOUNT OF CONTRIBUTORY RESEARCHES.

It will be proper to briefly refer to some of the more important of the earlier and contemporary researches which are introductory to those shortly to be described, and to mention somewhat particularly the writer's previous bolometric work of which the present is a continuation

These references will be restricted chiefly to investigations on the distribution of heat in the spectrum, the absorption of the atmosphere, and on the infra-red line spectra of the sun and of the elements.

EARLY INVESTIGATIONS.

Sir William Herschel observed in 1800[1] that a thermometer exposed in the solar spectrum continued to exhibit heating effects beyond the limit of visible radiations, and he concluded that at least one-half of the heat of the spectrum was obscure. Though he found these dark rays both refracted by glass and reflected according to the ordinary law, yet he was led to the belief that heat and light exist as separate constituents of the solar radiations.

Leslie, in 1804, using a differential thermometer, found that different substances possessed quite unequal powers both for radiating and for absorbing heat. He found that nonluminous radiations were greatly absorbed by glass, which strengthened the mistaken belief that heat and light were distinct in essence.

The idea that light, heat, and actinism were entities was opposed in 1843 by Melloni, who regarded light as "merely a series of calorific indications sensible to organs of sight, and vice versa the radiations of obscure heat are veritable invisible radiations of light." This was the foresight of a man of genius, but this truth was almost universally rejected when first uttered, and it has won its way slowly to its present acceptance. The researches of Melloni and of Seebeck also showed that the disposition of the heat in the prismatic spectrum from any source depends on the substance of the prism employed, and that this is in part due to the absorption exerted by the substance.

---

[1] W. Herschel, Philosophical Transactions, vol. 90, p. 284, 1800.

Melloni in the very word "thermochrose," which he invented, indicated that there were differences in the quality of heat assimilable to those in the color of light, but it was only in 1840[1] that Sir John Herschel published a thermograph which first showed unequal absorption below the red, and which is given in fig. 1.

FIG. 1.—Sir John Herschel's infra-red spectrum.

The reader may be here reminded that the observations which had preceded, and for a long time those which followed this, were made only by the prism, which gives entirely different divisions of the spectrum, according to the substance of which it is composed, and that there was no means then known of referring these to the wave-length scale, except a formula $n = a + \frac{b}{\lambda^2} + \frac{c}{\lambda^4}$ due to the eminent mathematician, Cauchy, which was accepted for fifty years as representing a natural law, but whose values of the wave-lengths the writer has since shown to be utterly wrong and misleading, outside of the visible spectrum.

This may better explain how it was that Dr. J. W. Draper, who observed, in 1842, by phosphorescence, three wide bands, which were called by him Alpha, Beta, Gamma, was entirely deceived as to their position, so much so that, with the ignorance of the truth prevailing at even so late a period, he seems to have supposed these bands, which are almost at the beginning of the infra-red, to have marked its extreme limit, beyond which no radiation could exist.[2]

Foucault and Fizeau, in 1846, appear to have observed the same bands.

Dr. J. Müller,[3] in 1859, gives a construction by means of which a prismatic spectrum may be transformed into a normal spectrum. He conjectures that the extreme infra-red is at a wave-length of about 1.8μ. In his diagram two-thirds of the heat of the spectrum appears below the visible portion.

Tyndall,[4] in 1866, assigned the position of maximum heat energy to the infra-red, and estimated two-thirds the total solar energy to be distributed in this invisible region.

The publication of Lamansky[5] in 1871 may be considered to represent the extent of the knowledge of this region then attained. Lamansky used the thermopile to map out the energy in the solar prismatic spectrum, and exhibits a plot, here reproduced

---

[1] J. Herschel, Philosophical Transactions, vol. 130, p. 1, 1840.
[2] Speaking of Abney's researches, Dr. Draper remarks, in 1880: "Do we not encounter the objection that this wave-length (1.07) is altogether beyond the theoretical limit of the prismatic spectrum?"
[3] J. Müller, London, Edinburgh, and Dublin Philosophical Magazine, fourth series, vol. 17, p. 233, 1859.
[4] J. Tyndall, Philosophical Transactions, vol. 156, p. 1, 1866.
[5] M. S. Lamansky, Monatsberichte der Königlichen Akademie der Wissenschaften zu Berlin, December, 1871. Translation in London, Edinburgh, and Dublin Philosophical Magazine, fourth series, vol. 43, p. 282, 1872.

(fig. 2), which shows three great gaps in the infra-red radiations. He concurs in the belief (then common) that the infra-red rays are strongly absorbed by the atmosphere.

E. Becquerel, who in 1842,[1] had noticed the effect of the infra-red radiations upon phosphorescent matter previously exposed to the light so improved his apparatus from 1873 to 1876[2] that he was able to map out a considerable portion of the solar spectrum, determining his wave-lengths, however, by extrapolation from Cauchy's dispersion formula, whose indications the writer has since shown to be here untrustworthy.

In 1883[3] H. Becquerel much improved this method, and, using a carbon bisulphide prism and Rutherford grating, roughly mapped out the solar spectrum up to about 14,000 Ångstrom units, noting the variation of the great absorption bands under varying atmospheric conditions.

He also showed that the various phosphorescent substances behave differently in the infra-red, showing absorption bands of their own, as to which care must be taken lest they be mistaken for spectrum bands.

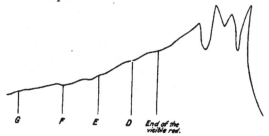

Fig. 2.—Lamansky's curve.

By means of a bromide of silver plate, specially prepared by an elaborate process, Captain Abney, prior to 1880,[4] succeeded in photographing the solar spectrum to a wave-length of 10,000 Angstrom units. Later, with a Rowland concave grating,[5] he photographed the solar spectrum and published a table of absorption lines in the extreme red and infra-red, giving 429 from A down to a wave-length of about 10,000 Angstrom units. These he regarded as accurate to one-tenth of an Ångstrom unit.

In 1881[6] J. Draper improved Becquerel's method in that he obtained a permanent record by placing a sensitive bromide of silver plate on the fluorescent screen. Lommel, in 1890,[7] improved the method further, and published a good spectral chart

---

[1] E. Becquerel, Annales de Chimie et de Physique, third series, vol. 9, p. 314, 1843. See also vol. 22, p. 344, 1848.
[2] E. Becquerel, Académie des Sciences, Comptes rendus, vol. 69, p. 995, 1869; vol. 77, p. 302, 1873; vol. 83, p. 249, 1876. Annales de Chimie et de Physique, fifth series, vol. 10, p. 5, 1877.
[3] H. Becquerel, Annales de Chimie et de Physique, fifth series, vol. 30, p. 5, 1883.
[4] W. Abney, Philosophical Transactions, vol. 171, p. 653, 1880.
[5] W. Abney, Philosophical Transactions, vol. 177, p. 457, 1886.
[6] J. Draper, London, Edinburgh, and Dublin Philosophical Magazine, fifth series, vol. 11, p. 157, 1881.
[7] E. Lommel, Annalen der Physik und Chemie, neue Folge, vol. 40, p. 681, 1890.

made in this manner. He proposed to utilize the process in obtaining photographs of the metallic spectra.

In 1883 [1] E. Pringsheim, using a Crooke's radiometer, studied the distribution of energy in the solar spectrum.

THE WRITER'S EARLY WORK AT ALLEGHENY.

All of the earlier investigators who attempted to measure the relative heat in different parts of the spectrum found themselves unable to obtain results at all commensurate with the labor involved, because the heat of the spectrum was in detail wholly inaccessible to the measuring instruments at their command. It is true that progress was made by supplanting the thermometer by the thermopile, yet only those who have patiently used this instrument in its old form, as the writer has done for many years, can testify how unsatisfactory it was. The immediate need for the acquisition of a more sensitive instrument was felt increasingly by him till, upon attempting to measure the distribution of heat in the diffraction spectrum, a series of painstaking experiments satisfied him that if such a spectrum were ever satisfactorily analyzed by heat-indicating apparatus, something which should be an advance over the thermopile, as the latter had been over the thermometer, must be devised. The thermopile is limited in its action to the current directly generated by a difference of potential, which can never exceed a small and definitely limited amount, and its use in delicate research demands unlimited patience. The writer sought to devise an instrument of (theoretically) unlimited capacity which should act to control the current from an unlimited source of potential, and be at once more sensitive than the thermopile, and be both more exact as a measure, and more exact in pointing, i. e., in work of precision.

After considerable experiment the bolometer was produced, as described at some length at the time,[2] and its behavior, as compared with the thermopile, was from the first exemplary. Since then there has elapsed nearly twenty years, and this instrument has so recommended itself to physicists as now to be used in many forms and for various purposes at research laboratories everywhere.

The principle upon which it is based finds more and more increasing use as a means of heat indication, so that, notwithstanding the extraordinary sensitiveness which has of late been obtained by physical chemists from special mercury thermometers, the Wheatstone's bridge encroaches even in this field. At the same time, however, the principle which is the basis of the thermopile is much used in pyrometry, and has, through the ingenuity of Boys, furnished us the highly sensitive radiomicrometer, and still more recently the thermopile itself has been greatly improved by Rubens, so that the Wheatstone's bridge method does not hold the

---

[1] E. Pringsheim, Annalen der Physik und Chemie, neue Folge, vol. 18, p. 32, 1883.
[2] S. P. Langley, Proceedings of the American Academy of Arts and Sciences, vol. 16, p. 342, 1881.

electrical heat-measuring field alone. The bolometer is, however, so well adapted to use in the spectrum, not alone because of its sensitiveness, but, as has just been observed, also because of its linear form and capacity for being pointed with precision, that it is doubtful if it will be easily displaced.

The first important use to which the bolometer was put was in determining the distribution of energy in the solar heat spectrum, which had hitherto been studied in detail only in that of a prism. In a memoir entitled "Observations du spectre solaire,"[1] not only the prismatic but the normal energy curves were given extending to a wave-length of 2.8 μ. It was shown that the maximum ordinate of the glass prismatic energy curve was at about 1.0 μ, while that of the normal energy curve was in the orange. Nearly three-fourths of the solar energy reaching the earth was found to be of longer wave-length than that at the A line, but the observations showed that the loss in passing through the atmosphere was chiefly confined to the shorter wave-lengths, so that it appeared that an analysis of the solar radiations conducted above the atmosphere might conceivably show a maximum in the violet of the normal spectrum.

This paper was followed in 1883 by "The selective absorption of solar energy,"[2] which describes bolometric measurements made directly upon the grating spectrum and extending from the violet through the visible spectrum and down to a wave-length of 1 μ.

The bolometric indications observed at five different points in the normal spectrum under conditions identical except for the air mass, were compared in order that it might be quantitatively determined what the absorption of the atmosphere was for nearly homogeneous rays of different wave-lengths. These measurements confirmed the earlier conclusion that the maximum ordinate of the normal energy curve was in the orange, and showed that the absorption of the earth's atmosphere increased rapidly with *decreasing* wave-lengths, then a novel statement, for strange as it may now appear, it was even at this late time very generally supposed to increase most in the lower red, though the simple aspect of a sunset might have taught the contrary. Thus it appeared probable, from the present writer's conclusion, that the sun would seem bluish to an eye outside our atmosphere, and of a lavender tint could it be seen without either solar or terrestrial absorption intervening.[3] A reduction of the measurements taken indicated a value of the solar constant as high as 2.84 calories, which was much higher than any previous determination.

Another portion of the investigation may be regarded as the forerunner of the research since undertaken at Washington. This consisted in mapping the infra-red solar prismatic spectrum by means of the bolometer, as far as a wave-length of 2.8 μ.

---

[1] S. P. Langley, Académie des Sciences, Comptes rendus, vol. 95, p. 482, 1882.
[2] S. P. Langley, American Journal of Science, vol. 25, p. 169, 1883.
[3] See "Nature," May 26, 1887.

For this purpose an instrument called the spectrobolometer was devised, which was essentially a great spectrometer with minimum deviation attachment and arms arranged specially for the reception of the bolometer and collimating apparatus.

Four observers were employed frequently in the measurements. One set the circle carrying the bolometer successively on adjacent parts of the spectrum and read it, a second read the galvanometer deflections upon the scale, a third recorded the observations, and a fourth kept the siderostat beam from wandering and called out the appearance of clouds before the sun. The difficulties of the research lay less with the use of the bolometer than with the galvanometer as then constructed, and especially in the tendency of the needle to wander, causing the light spot on the scale to drift with a continuous progress to the north or south when no action on the bolometer took place. This wandering of the needle, here called the "drift," then *averaged* much more than a centimeter in a minute, and was not only the great obstacle to accuracy, but a great consumer of time. In making the spectrum map it was customary to go over 15' of spectrum several times daily, making observations for each separate minute of arc, by opening and closing the slit at each observation, as, if it were left open continuously while the bolometer strip was carried through the spectrum, the movements of the needle caused by spectral heat were confounded with those caused by the "drift." The galvanometer reader had the most trying observations, as the "drift" was sometimes so great that the spot of light did not remain upon the meter-long scale even a minute at a time. It was recognized that the "drift" was largely due to changes of temperature in the apartment, and devices for controlling this were made, but the radical change necessary, that of installing both galvanometer and bolometer in a special apartment within a larger one, itself automatically kept at nearly constant temperature, could not be carried out for want of means.

Under these circumstances more than a thousand readings were required in this part of the investigation to fix the position of a single line, and the work of reducing and plotting the observations was enormous. The result obtained, after some two years of this labor, was to establish nearly twenty deflections in the energy curve (corresponding to absorption lines), some of them indicating regions of almost complete absorption of energy, and showing that the character of the infra-red spectrum was far different from the visual part. The curve illustrating these results is reproduced in fig. 3.

In a research subsidiary to this, entitled "Experimental determination of wave-lengths in the invisible spectrum"[1] the deviation and wave-length curve for the glass prism used was experimentally determined. The sun was employed as a source of heat, and the succession of apparatus was (1) condensing mirror, (2) slit, (3) concave grating, (4) slit, (5) collimating lens, (6) prism, (7) objective lens, (8) bolometer.

The results of this research showed that neither Cauchy's formula (then universally trusted) nor any other dispersion formula then extant would apply even approxi-

---

[1] S. P. Langley, Memoirs of the National Academy of Sciences, vol. 2, p. 149, 1883.

mately in the infra-red when its constants were determined for points in the visible spectrum. It is interesting to note in this connection that Dr. J. W. Draper, in 1880,[1] asks, in speaking of Captain Abney's researches, "Do we not encounter the objection that this wave-length, 1.07 μ (the limit of Captain Abney's map), is altogether beyond the theoretical limit of the prismatic spectrum?"

The research on the selective absorption of solar energy had shown it to be desirable to make observations upon a high elevation, where the air should be both

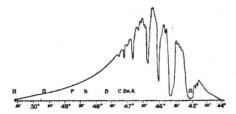

FIG. 3. GLASS PRISMATIC SOLAR SPECTRUM.
ALLEGHENY OBSERVATORY.

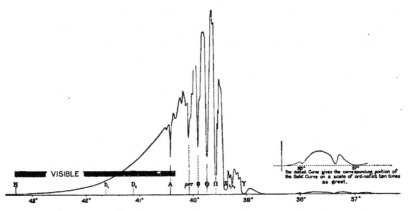

FIG. 4.—ROCK-SALT PRISMATIC SOLAR SPECTRUM.
ALLEGHENY OBSERVATORY.

purer and rarer than at Allegheny. Through private and public aid an expedition was enabled to proceed to Mount Whitney[2] in southern California, in 1883, where a great number of observations with actinometers of various forms and a considerable number with the bolometer and with other instruments were made. The bolometric observations were naturally conducted under great difficulties, as the character of the

[1] J. Draper, Proceedings of the American Academy of Arts and Sciences, vol. 16, p. 233, 1880.
[2] "Researches on Solar Heat. A report of the Mount Whitney Expedition." U. S. A. War Department, Professional Papers of the Signal Service, No. 15, 1884.

In a following memoir upon "Hitherto unrecognized wave-lengths,"[1] reference is made to a former investigation in which the infra-red wave-lengths were determined as far as 2.3 μ, but the research was there discontinued on account of the too feeble energy received from the grating. The determinations of the wave-lengths of the rock-salt prismatic spectrum was now continued to open the field for further researches in solar and terrestrial radiation, by enabling one to husband the available energy by the use of a prism, and still to know the wave-lengths employed; for it will be remembered that even as late as this time the lengths of most of these invisible waves were not known by experiment, but inferred by calculation from almost wholly untrustworthy formulæ.

The sun could not here be used as a source of energy, for though it had been shown that some of its radiation of wave-lengths greater than 2.3 μ is received at the earth, yet this portion was quite too slight to use for this purpose with the means then at command. In the place of the sun was therefore substituted an electric arc, which was caused to play between carbons 1 inch in diameter by the current of a 12-horsepower dynamo. Even with this source the deflections obtained were so small that the measurements could not be extended beyond 5.3 μ, but this was then far beyond where any real measurement had yet been made. The method was to observe a ray or rays which had been both reflected from a grating and refracted by a prism, and to thus observe what wave-length corresponded to a given index of refraction.

As the importance of the result may justify a fuller description of the means, it may be added that the train of apparatus commenced with the arc light mounted immediately before the slit. After considering many devices it was determined to let the rays fall first on a grating and then upon the prism. The rays from this first slit fell then upon a (Rowland) concave grating, which, by the nature of the interference process, necessarily formed them into superposed spectra, lines whose wave-lengths were multiples of one another, lying upon one another, and becoming confused together. Since these lines are of different wave-lengths, however, they are of different refrangibility, and if they are caused to pass through a prism this will separate them.

A second slit was interposed where the grating spectra were formed, and through this slit, as appears from what has been said, only radiations having a multiple of some determined wave-length could pass. For the purposes of the experiment the wave-length of $D_2 = 0.58901\mu$ was chosen. These radiations and their multiples alone passed through the second slit and, diverging, fell upon a rock-salt prism. Out of different possible positions of the prism, that was chosen where the refracting edge was parallel to the lines of the grating, and the prism thus placed refracted the radiations at different angles in azimuth, giving distinct images of them. The linear bolometer was

---

[1] S. P. Langley, American Journal of Science, vol. 32, p. 83, 1886.

moved until it encountered one of these feeble heat images—that due, for instance, to the ray whose wave-length was twice D_2, whose indications were recognized when the bolometer was in a position corresponding to a certain index of refraction of the prism. The bolometer was then moved until it encountered the second heat image, belonging to the next higher order of spectra, and so on through nine in succession, when the wavelength of 5.3μ (nine times that of D_2) was reached. Beyond it the heat was too feeble to be recognized, and in this way the wave-lengths corresponding to refraction in a prism were first definitely measured in this lower infra-red region.

The galvanometer employed was of the four-coil Thomson reflecting type. Its needle system (constructed by Mr. F. W. Very) was composed of hollow steel magnets 1 mm. in diameter, 8 mm. long, and with walls $\frac{1}{20}$ mm. thick. The mirror, 9.5 mm. in diameter, of platinized glass, was of 1 meter radius. Damping was secured by the use of a portion of a dragon-fly wing, a device of the writer, then first used. Suspension was made by means of a silk fiber 33 cm. long, and the time of swing of the needle without damping magnet was from fifteen to thirty seconds.

A cylindrical scale was employed, and the position of the spot could be read to $\frac{1}{10}$ mm.

The sensitiveness of the galvanometer was at that date stated to be such that, with a time of single vibration of twenty seconds, a deflection of 1 mm. on the scale at 1 meter is given by a current of 5×10^{-10} amperes through the coils, which were each of 20 ohms resistance; but the observation of small deflections was rendered very difficult by the drift.[1] Bolometers were then used in connection with this instrument, such that a deflection of 1 mm. corresponded to $\frac{1}{100000}$ degree differential change of temperature of the strips.

It is mentioned that in using the salt prism fourteen months it was necessary to refigure its surface and redetermine its angle thirteen times.

The results obtained (which are given in the appendix to this volume) show the refractive indices of rock salt for wave-lengths extending from $\lambda=0.4\mu$ to $\lambda=5.3\mu$. On comparison with existing formulæ for dispersion none were found which adequately represented the observations. The latter part of the curve showing the relation of n to λ was found to approach a straight line (without becoming one), and an extrapolation was essayed beyond the point where the observations stopped, though it was remarked that all such extrapolations are attended with uncertainty, a remark too soon justified. By this means it was shown that if the curve continued to approach a

---

[1] If we admit that the deflection is inversely as the square of the time of swing, this gives for a ten-second single vibration a millimeter deflection for a current of 2×10^{-9}. That at present (1899) is (for ten-second swing) 5×10^{-12}, showing a sensitiveness four hundred times as great. With the present galvanometer the needle is so very light (2.5 mgr.) that the tremor becomes too prejudicial with a ten-second swing, and being thus restricted to a five-second or shorter swing, not more than fifty times the sensitiveness of the Allegheny instrument (which could be employed with a twenty-second swing) is actually available, but the saving of time is also of great importance.

straight line, as appeared probable, the maximum of the spectrum of melting ice would be at a wave-length of at least 10μ, and that the lowest heat recognized thus far would have a wave-length of something like 30μ.

It must be here remarked that the later observations of other physicists, while they agree well with the observed values determined up to 5.3μ, yet make the results of extrapolation probably so far erroneous, that while the progress of the curve from 5.3μ still, perhaps, requires further investigation, it seems certain that the values of wave-lengths obtained in the above-mentioned extrapolation are too high. The latest values are those obtained by Paschen and Rubens, and some account of their researches will be found in the appendix. In Plate XXXII, illustrating this work, the values of the writer, of Paschen, and of Rubens, have been plotted and a representative curve has been drawn to show the mean result of the three determinations.

In a memoir upon "The solar and lunar spectrum,"[1] which appeared in 1886, the solar spectrum was followed far below previous investigations. The arrangement of apparatus in this research is chiefly distinguished by employing two complete spectroscopes, with rock-salt collimating lenses and prisms, of which the first was called the "sifting train," and was used in order to guard the more effectually against stray radiations from the upper spectrum, which would introduce important errors where so little energy existed. Using the extrapolation already discussed, it was found that from a little beyond the absorption band, known as Y, until a wave-length of 10μ(?) is reached no energy could be observed, and beyond this the amount was but trifling. In an appendix there is given a table of deviations observed in the rock-salt spectrum.

Following this, the bolometer was used in a research on the "Temperature of the moon,"[2] which does not in the main have a direct bearing on the present one, but in which some points of interest appear. It is mentioned, in describing the apparatus, that a battery of 15 gravity cells was employed in the bolometer circuit to diminish the irregularities in the current. This multiplication of the number of cells is due to Mr. Very, and has been found very useful in subsequent practice. In Appendix III of this paper is given a reduction of observations made from time to time during the course of two years to determine the change of deviation of a 60° rock-salt prism with the temperature. It is remarked that the then only available way to estimate the temperature of the prism was by observations of the temperature of the air, which could not have failed to be a fruitful source of discrepancy in the results. The table of values obtained is reprinted on a subsequent page of the present memoir.

---
[1] S. P. Langley, Memoirs of the National Academy of Sciences, vol. 4, p. 159, 1886.
[2] S. P. Langley and F. W. Very, Memoirs of the National Academy of Sciences, vol. 4, p. 107, 1887.

RECENT INVESTIGATIONS OF INFRA-RED[1] SPECTRA BY OTHERS THAN THE WRITER.

Closing here the summary of infra-red researches at Allegheny, we will now turn to a consideration of the more recent determinations of the origin of some of the infra-red metallic lines. One of the interesting phases of this portion of the infra-red investigation is the finding of lines already predicted from the lines now known in the visual and ultraviolet portions of the spectrum. Balmer[2] showed that the wave-lengths of a certain series of lines in the hydrogen spectrum could be represented with surprising accuracy by the empirical formula—

$$\lambda = h \frac{n^2}{n^2 - 4}$$

n representing the successive ordinal numbers above 2 and h a constant. Later, Kayser and Runge[3] showed that in the spectra of nearly all the elements there are series of lines whose positions may be expressed by empirical formulæ of the form

$$\frac{1}{\lambda} = A + Bn^{-2} + Cn^{-4} +$$

where $\lambda =$ the wave-length, and thus $\frac{1}{\lambda}$ is a number proportional to the frequency of oscillation, and n represents the successive ordinal numbers. They also found that in cases where series of pairs occur the difference of the values of $\frac{1}{\lambda}$ between the members of any one of the pairs is a constant for that element and series.

In 1892[4] B. W. Snow, in order to test Kayser and Runge's formula, investigated the infra-red spectra of the alkalies. He used a flint-glass prism, calibrated by means of interference bands, and a bolometer capable of indicating a difference of temperature of $\frac{1}{180000}°$ C. He determined the position of his interference bands to a wave-length of 26,680 Ångstrom units. His spectra were obtained from the salts of the metals inserted as a core in the carbons of an arc lamp.

---

[1] For an account of dispersion and wave-length investigations with rock salt, see Appendix, as well as Chapter I of Part II.
[2] J. Balmer, Annalen der Physik und Chemie, neue Folge, vol. 25, p. 80, 1885.
[3] H. Kayser and C. Runge, Abhandlungen der Königlichen Akademie der Wissenschaften zu Berlin, III, 1888–1892.
[4] B. W. Snow, Annalen der Physik und Chemie, neue Folge, vol. 47, p. 208, 1892.

The following table contains the lines observed by Snow in the infra-red, in which λ represents the wave-length in thousandths of a millimeter (μ), and I the relative intensities of energy as indicated by the galvanometer deflections:

TABLE 1.—*Infra-red metallic lines.* (Snow.)

Lithium.		Sodium.		Rubidium.		Cæsium.	
Wave-length. λ	Intensity. I	Wave-length. λ	Intensity. I	Wave-length. λ	Intensity. I	Wave-length. λ	Intensity. I
μ		μ		μ		μ	
0.811	296	0.770	22	0.775	414	0.775	75
1.800 (?)	?	0.818	660	0.791	443	0.790	107
		0.855	18	0.821	42	0.833	297
Potassium.		0.930	8	0.845	50	0.865	151
		0.995	10	0.878	60	0.882	345
0.768	1,443	1.075	13	0.913	11	0.900	155
0.840	18	1.132	419	0.945	10	0.995	182
0.885	13	1.245	30	0.997	151	1.150	9
0.950	23	1.800 (?)	(?)	1.063	11	1.205	7
1.086	108			1.090	13	1.323	81
1.155	395			1.153	26	1.420 (?)	38
1.220	205			1.224	13	1.450	52
1.470	70			1.318	198	1.520	20
1.500 (?)	50			1.478	102	1.575	8
				1.520	71		

The lines which had been predicted were for—

Lithium	0.819μ			
Sodium	1.150	1.148μ		
Potassium	1.266	1.257	1.253μ	1.244μ
Rubidium	1.718	1.653		
Cæsium	0.922	0.877	0.867	0.828

In the case of lithium and sodium the agreement with the empirical formula seems to be fair, but with the others Snow thinks no conclusion can be drawn.

In 1895[1] E. P. Lewis undertook the determination of the infra-red metallic spectra. He used a radiomicrometer in connection with a Rowland grating. The grating was ruled with 10,000 lines to the inch, and was of 14 feet focus. He used it as in Rowland's well-known method. Its first spectrum was unusually bright and was visible beyond the usual limit in the infra-red. He employed an arc light with its carbons prepared in the same manner as those used by Snow.

The following table gives a summary of the lines found by Lewis, together with a comparison of the intensities of the lines, the D lines being taken as 60.

---

[1] E. P. Lewis, Astrophysical Journal. vol. 2, pp. 1 and 106, 1895.

TABLE 2.—*Infra-red metallic lines.* (Lewis.)

Element.	Wave-length.	Intensity.
Calcium	7,663.7	30
Silver	7,688.4	25
Lithium	8,126.3	40
Sodium	8,183.7	60
Sodium	8,194.2	60
Silver	8,274.0	25
Calcium	8,541.9	50
Calcium	8,662.0	50
Strontium	10,326.8	40
Strontium	10,915.6	45
Sodium	11,381.1	35
Sodium	11,403.9	35
Thallium	11,511.7	40

The following lines, predicted by Kayser and Runge, were not found:

Mercury	9497		
Magnesium	13007	13041	13111
Calcium	11801	11874	12020
Strontium	13022	13345	14086

In closing this extremely succinct account of researches bearing upon the present investigation the writer may mention that researches of his own on the diathermancy of lampblack and hard rubber have showed lampblack nearly (not wholly) without color here. Reference should also be made to an article of E. F. Nichols,[1] treating of the diathermancy of plate glass, hard rubber, quartz, lampblack, cobalt glass, alcohol, and various solutions. The result of chief interest in the present research is that lampblack absorbs nonselectively throughout the region studied (to 3μ).

In connection with these studies of selective absorption the writer calls attention to his own observations made in 1881 on the coefficients of reflection from two surfaces of silver.[2] The following table embodies these observations.

Coefficients of reflection from two surfaces of silver.

Wave lengths μ	0.35	0.38	0.40	0.45	0.50	0.55	0.60	0.65	0.70	0.75
Percentage reflected from two surfaces	37	54	63	73	79	82	84.5	86	87.5	88.5
Reduction factor (reciprocal)	2.70	1.85	1.59	1.37	1.27	1.22	1.18	1.16	1.14	1.13

[1] E. F. Nichols, Physical Review, vol. 1, p. 1, 1893.
[2] S. P. Langley, American Journal of Science, Vol. XXXVI, Nov., 1888, p. 9.

Chapter II.

THE PRESENT RESEARCH.

ACCOUNT OF THE PROGRESS OF THE INVESTIGATION AT THE SMITHSONIAN ASTROPHYSICAL OBSERVATORY.

From the beginning of regular operations at the observatory in June, 1891, till the 1st of March, 1892, efforts were chiefly directed to getting the apparatus in satisfactory condition for observations. Much time was spent on the improvement of galvanometers, in testing bolometers and prisms, and in the determination of their constants.

At length, on March 2, 1892, a "rehearsal" occurred, in which the procedure followed in the bolometric investigations of the infra-red solar spectrum at Allegheny, already referred to on a previous page, was gone through with for the first time at the observatory. A second rehearsal occurred on the following day, and on reviewing it an entry was made by the writer March 4, 1892, in the record book in use by Dr. William Hallock, from which the following quotation is taken:

I think your yesterday's spectral maps were quite successful for a first attempt—indeed, notably so, and give evidence of the goodness both of the system and of the instrumental means. The salient defect of the latter is in the "drift" of the galvanometer, which, though reduced to limits which are insignificant compared to those which it had when I first began the study, is still a barrier to the best work.

My idea (if drift could be eliminated) would be to have a vertical strip of sensitive paper rolled perpetually upward by a clockwork in the focus of the galvanometer mirror. The sides of this paper are marked in degrees and minutes, corresponding to divisions of the spectrometer circle, whose arm is moved by the same clockwork (through electric or other intermediary), so that when the circle is turned through n minutes of arc, the paper is moved upward linearly by a quantity corresponding to the same angular measure. A light is reflected from the mirror onto the paper, on which are traced the movements of the mirror due to the varying heat of the spectrum and to passing inequalities of the sky transmission. (The mirror movement has to be dampened so that there is no sensible swing.) The whole spectrum could be thus traversed in five minutes or less, as many as twelve curves could be taken in an hour, and a composite photograph would eliminate the accidental disturbances.

All this implies that "drift," if not eliminated, is to be greatly reduced. Please consider this "drift," as well as the little movements of the needle due to changes in the apparatus itself, under these three heads:

(a) Changes due to alterations in the galvanometer.
(b) Changes due to alterations in the bolometer.
(c) Changes due to alterations in the battery, and all other sources.

It seems quite certain that these are due largely to temperature.

This is the first enunciation here of a plan long since devised, but not executed, at Allegheny. The writer had indeed prepared to experiment in the synchronous movement of the circle and sensitive film while at Allegheny, but with such impoverished mechanical aid that nothing had then been accomplished by this means. The plan was, as will be seen later, adhered to in its general conception, but constantly modified in detail by practice.

A more thorough examination of the spectrum by the same method used at Allegheny was made on March 19, 1892, and upon looking over the results as plotted the present writer made the following comment in the record book, March 20, 1892:

> This has served its immediate object by enabling us to identify Ω and one or two other landmarks, but by comparison with the prismatic energy curve in "Mount Whitney Observations" it is plain that we have missed the greater part of the lines, and even some bands 5' wide.
>
> The main object in making this Mount Whitney chart was to get the total amount of energy in the spectrum, i. e., the area between the curve and the axis of X, and the mapping of the lines was subordinate to this.
>
> *Our object hereafter is to map the lines.*

In carrying out the purposes thus outlined, steps were immediately taken to procure suitable clockwork for synchronously moving the circle and sensitive film, and systematic observations were begun to determine the source of the drift and its remedy.

The clockwork had been installed and arrangements provided for the photography of the spot of light reflected from the galvanometer mirror, and many preliminary observations and tests made by July 1, 1892. The following description of the arrangements for the work then existing will serve as a basis for comparison with the conditions attained at later stages:

CONDITION OF THE RESEARCH, JULY 1, 1892.

1. OPTICAL SYSTEM.

The great single mirror Foucault siderostat, made under my general design by Grubb, of Dublin, was to send a beam of sunlight due south, and was supplied by him with his extremely elaborate electrical control for right ascension only, designed to accelerate or retard the clock movement. It was very ingenious, but very intricate, and so slow as to be for this special service inefficient. What was needed was some means of directly taking hold, so to speak, from the observer's seat, of the movement, either in right ascension or declination; which implies an external power which will not so much modify the clock movement as will temporarily replace it. I had experimented on such a control with cords at Allegheny, and had found them very inconvenient. Connecting rods are still more so, and I decided to introduce what was then, I think, a novelty, the employment of an electric motor, which was to enable the observer to alter the movement either in right ascension or declination as promptly and as effectively as though he could make the change himself directly at the siderostat. The beam of light thus controlled passed through a slit of 50 mm. height placed upon a wooden stand outside the building. Passing from thence the cone of rays traversed a glass collimating lens of 7.5 meters focus, and the parallel beam resulting fell, after passing a suitable diaphragm, upon a prism of white flint glass,[1] of refracting angle 60° 7′ 20″, height 11.6 cm., and of 10.5 cm. face. The deviations at 20° C. of several prominent visible lines in the spectrum by this prism are given as follows:

H_1 46° 45′ 10″
D_2 44° 10′ 4″
C   43° 47′ 19″
B   43° 38′ 54″

The spectrometer was that now in use and shown in Pls. IX and X, except that the place of the iron base-casting shown was taken by a wooden framework, that the arrangement for minimum deviation was secured by mechanical instead of optical means, and that the focal length of the image-forming mirror was limited by the length of the arm of the instrument which supported it. This concave mirror of 178 cm. focal length brought the spectrum to focus upon the bolometer, the bolometer and the spectrum being both in motion.

From these data may be calculated the following constants:

Height of slit image.. .... 12 mm.[2]
Angular aperture of same for 1 mm. slit.............................. 13″
Angular magnitude of $\frac{1}{10}$ mm. bolometer strip .... 11.5″
Optical resolution of prism at A^3........... .... 1.5″

2. ELECTRICAL SYSTEM.

A four-coil Thompson reflecting galvanometer, by White, of Glasgow, was at first employed. The coils were two in series and two in parallel, and had each 20 ohms resistance. A needle was in use, composed of two systems of six steel tubes specially prepared and magnetized,[4] weighing in all 144 mgrs., supported upon a glass tube weighing 30 mgrs., which carried also a concave platinized glass mirror 9 mm. in diameter, of 10 cm. radius of curvature, and weighing 90 mgrs. Damping was provided by the use of a "dragon-fly" wing, already introduced by the writer, and also by a vane dipping in oil. The time of free swing was twenty-two seconds, but by use of an overhead magnet on a frame of latticework supported from the roof, the time of swing was reduced to ten seconds. Two weak magnets placed behind the galvanometer and moving lengthwise with respect to each other furnished a ready means of adjusting the direction of the mirror without perceptibly altering the sensitiveness. With a time of single swing of ten seconds, with scale at 1 meter, a deflection of 1 mm. on

---

[1] It will be seen that no condensing lens is employed, the advantage of the use simply of the slit and long focus collimator having been recognized by the writer in his previous practice.

[2] A plano-convex glass cylindric lens of about 20 cm. radius of curvature was employed to reduce this height to that of the bolometer strip, which was 8 mm.

[3] See London, Edinburgh, and Dublin Philosophical Magazine, fifth series, vol. 8, p. 261, 1879. See also this book, Chapter IV.

[4] This needle system was made by Mr. F. W. Very.

the scale corresponded to 0.0000000015 (1.5×10^{-9}) amperes in the external circuit, a degree of delicacy considered notable at the time, but which has since been greatly exceeded.

The bolometers employed had $\frac{1}{10}$ or $\frac{2}{10}$ millimeter strips of lampblack smoked platinum exposing a length of 8 mm. and mounted on hard rubber. There were two side strips in these early bolometers, connected in series, and in order to make the instrument symmetrical, so as to avoid "drift," only half the length of the middle strip was exposed. The bolometer case was a hollow cylinder of hard rubber, with appropriate binding posts to connect with the bolometer, provision for the insertion of positive eyepiece for focusing, and metal diaphragms of suitable aperture to admit radiation to the central strip alone.

To complete the bridge a large specially constructed adjustable resistance box, situated near the galvanometer and at about 7 meters distance from the bolometer, was employed. Connection with the bolometer was through flexible copper wires, and a second pair of wires—starting one at the bolometer, the other at the resistance box—led down to the battery, which consisted of a number of gravity cells connected in series, while the current was graduated by means of suitable resistance to 0.1 ampere or less.

3. Early Photographic Arrangements.

A sensitive film, 60 cm. long and 18 cm. wide, wound upon a drum, received the record of the galvanometer needle by means of light from an incandescent electric lamp which was focused upon a ground glass placed behind a pin-hole diaphragm, acting as the source of light for the galvanometer mirror. Much difficulty was experienced in obtaining a satisfactory photographic trace, a difficulty not finally overcome till later by the introduction of the sunlight in place of the electric light just described.

The film and its driving arrangements were contained in a light-tight box, provided with a wide slit covered by shutter when no exposure was taking place. Dependence was made upon the horizontal swing of the galvanometer needle for rectangular axes to the bolographic curve.

4. Driving Clocks.

The considerable distance between the spectrometer and galvanometer made the use of a single clock difficult owing to local conditions, and after many expedients to avoid it, finally imposed the necessity of a line of shafting, occupying the center of the observatory to the great inconvenience of its work. But this was not till, in spite of the inferiority of a double system, from a theoretical point of view, two clocks had been essayed—one driving the circle through a tangent screw, the other the drum with its photographic film—both being electrically controlled and synchronised from a time clock. Change gears were provided for a wide range of speeds.

5. SURROUNDINGS.

No provision had then been made for isolating any part of the spectrobolometric apparatus in special chambers of the observatory, and consequently it was throughout subject to temperature changes often amounting to 15° C. in twenty-four hours. The building had been provided with special sliding shutters, both for the windows and the skylights, and darkness was obtained during the progress of observations by closing them.

6. EARLY RESULTS.

In Pl. XVIII is shown a portion of the bolograph I of June 14, 1892, extending from A to below $\rho\ \sigma\ \tau$. The following notes are copied from the record of that date made by Dr. Hallock:

> State of sky, hazy.
> Height and aperture of slit, 50 mm. by 0.5 mm.
> Angle of same at bolometer, 13".
> Prism used, "Great Hilger."
> Galvanometer, White.
> Time of single vibration.(not stated, but for June 17, 11 seconds).
> Bolometer VI, 0.1 mm. strips, subtends 11.5".
> Current of battery (not stated, but June 13, 0.127 ampere).
> Speeds, 1' arc of spectrum in one minute of time corresponding to 1 cm. of film.
> Observers, William Hallock and C. T. Child.
> Remarks by Dr. Hallock: "Drift very slight. The setting on A changes several minutes during a day. Does the wooden arm warp?"
> Grade of bolograph (not taking state of sky into account), "A."

This bolograph is, as its grade indicates, one of the best at that time produced. The chief defects in the arrangements were: (1) The galvanometer was not sufficiently sensitive, so that the slit was necessarily wide and the spectrum impure. (2) The "drift" was nevertheless frequently uncontrollable and always troublesome, and accidental deflections, due to various causes, were not inconsiderable. (3) Inaccuracy in the motion of the circle, drum, and minimum deviation mechanism, taken in connection with the warping of wooden supports to the optical apparatus, and the shrinking of the photographic films, made discrepancies between successive energy curves in the same region far too great. (4) Change of temperature, which may be said to be the father of evils in such research, was of great magnitude, and owing to the temporary character of the building very difficult indeed to overcome.

The major part of the work of the observatory from this time on has been devoted to the tasks of making the recording apparatus more sensitive, the disturbance of the record less serious, and the position of deflections corresponding to absorption lines less subject to error. It has been only by the most patient pursuit of these aims that it has become possible to obtain satisfying evidence for the reality of the more minute absorption deflections now given as real. It will be interesting to note successive advances along these lines.

PROGRESS OF THE RESEARCH DURING THE YEAR ENDING JULY 1, 1893.

The observatory was unfortunate in losing the benefit of Dr. Hallock's services, and during about two months in the autumn of 1892, after the close of his connection with it, this part of the work was discontinued.

Upon again taking it up the improvement of the galvanometer was first sought. Orders were placed for the special winding of new coils in accordance with the theoretical considerations of Maxwell, and in the meantime attempts were made to render the galvanometer in use more sensitive by reastaticising and strengthening the magnets. In this success was so far attained that the constant was reduced to 0.00000000075 (7.5×10^{-10}). A storage battery was introduced in the place of the gravity battery, as experience showed its discharge to be far more uniform than that of the latter. Additional diaphragms were placed in the bolometer case to more effectually prevent the circulation of air currents. By winding coils of tubing about the bolometer case and passing through them a continuous water current, a provisional attempt was made to introduce constant temperature conditions at that part of the circuit. A bolometer case of metal with more adequate provision for a continuous water current was ordered. An experimental plant was introduced for the partial cooling and drying of the air of the observatory in summer with ice, but this proved unsatisfactory.

In the interest alike of accuracy and of the appearance of the photographic curve, the size of the light spot on the sensitive plate was limited by a metal slit placed in front of the film at right angles to the direction of motion of the drum. The recording apparatus was still further improved by bringing a beam of sunlight from the siderostat mirror, which, passing through a vertical slit in the place of the pin-hole diaphragm, replaced the electric light before used. This change did away at the same time with the proximity of a considerable electric current, a potent source of disturbance to the galvanometer. Glass plates provisionally superseded films near the close of the year, and the necessary change from circular to rectilinear motion was temporarily provided for.

The driving mechanism was changed by doing away with one clock and causing the other, by means of suitable steel shafting, to move both circle and drum, thus assuring closer synchronism, though at great local inconvenience. In place of the tangent screw at the circle there was procured a very accurate worm and wheel segment.

For the mechanical arrangement for minimum deviation was substituted an optical arm.

Several months were largely devoted to experiments and theoretical considerations relative to the availability in the research of the grating in connection with a "lifting prism" for removing the higher-order spectra. It was found inadvisable to replace the prism by this arrangement, as the great loss of energy under the most favorable conditions would make it necessary to use a slit of prohibitive width with any galvanometer likely to be obtained.

In order to extend the scope of the investigation a rock-salt prism[1] superseded the glass prism toward the close of the year. Collimation was sometimes omitted, sometimes provided by means of a rock-salt lens of 11 cm. aperture and of 9 meters focal length.

The photographic work of the observatory had been carried on in the photographic department of the National Museum, but had now become so considerable as to make advisable the erection of the small house marked B in Pl. III. This house was finished in June, 1893.

From the very beginning of the work of mapping the infra-red spectrum it had been intended to produce a line spectrum, so as to make the results graphically comparable with those in the visible spectrum, and the production of these line spectra (called frequently "cylindrics" because of the use of a cylindric lens in making them) kept pace with the taking of bolographs.[2] The processes used for this purpose are so distinct from those employed in producing the curves that it has been thought best to give a connected account of them in a place by itself, and this will be found at another page.

RECAPITULATION FOR THE YEAR 1893.

1. CHANGES IN THE OPTICAL SYSTEM.

Salt had replaced glass to enable observations to be extended beyond a wavelength of $2\,\mu$. A silvered plane mirror, fixed parallel to the base of the prism, which I had employed only for separating the grating spectra, had, at the suggestion of Mr. Wadsworth, most advantageously replaced the mechanical minimum deviation arrangement, and given a longer optical arm.

2. CHANGES IN THE ELECTRICAL SYSTEM.

The galvanometer constant had been reduced to 0.00000000075 and a new galvanometer (described later) almost prepared for use. Storage cells had been substi-

---

[1] I have already mentioned (American Journal of Science, third series, vol. 30, p. 477, 1883) that when I commenced these researches I was assured by the European optician of most reputed skill in working rock salt that no prism of that substance had ever been made which had shown a single Fraunhofer line and that none ever could be made. Very shortly, nevertheless, through the skill of Mr. J. A. Brashear, I obtained prisms as perfectly figured as any of glass could be and which showed hundreds of lines. A great difficulty was experienced, however, in obtaining the salt at once large enough and clear enough for my purpose. After searching everywhere in Europe and in this country, I found the best at the Royal salt mines of Baden. More recently, however, several exceptionally fine blocks have been obtained through the courtesy of the Russian Government. These prisms and lenses are of exceptional size and have as good definition as can be desired, and, so far as I know, all possess the same index of refraction. I am not prepared to state that this is absolutely true, but it is at least probable that a 60° prism of salt, used at a standard temperature, may be used as a standard of reference in preference to one of any other substance—certainly to any of glass. (See comparison, at a subsequent page, of the indices of refraction in three 60° prisms—two of Russian salt, the other of German salt.) It remains to add that with a dry room of uniform temperature like that in which the rock salt is at present used the hygroscopic qualities of the substance are scarcely prejudicial, the surfaces remaining practically perfect for at least several consecutive months.

[2] I think the use of the cylindric lens for this purpose was suggested to me by Prof. Edward C. Pickering, of Harvard, who had already employed it in such eminent success in his stellar spectral work.

tuted for a primary battery. Various measures had somewhat reduced the disturbances to the galvanometer from accidental causes, but the "drift" was still most formidable.

3. PHOTOGRAPHIC IMPROVEMENTS.

A beam of sunlight reflected from the siderostat had replaced the electric lamp and a slit was now in use at the recording box. Most important of all was the provisional substitution of glass plates moved by rack and pinion for films moved on a drum. For general photographic purposes a dark room had been built within the observatory lot.

4. ALTERATIONS IN THE MECHANISM.

Under the local conditions it was so inconvenient to drive the mechanism at the circle and that at the photographic plate from a single clock that the attempt had been made to use two independent ones electrically controlled. The results were, however, so unsatisfactory that it was decided to incur the necessary inconvenience, and a line of shafting was introduced, transmitting the motion from a single clock. A worm and wheel motion replaced the tangent screw at the circle.

5. CHANGES IN THE SURROUNDINGS.

In this respect very little had been accomplished beyond a local provision for constant temperature at the bolometer by the use of coils through which water flowed, surrounding the bolometer case.

6. RESULTS OBTAINED.

In Pl. XVIII is shown the bolograph No. II of June 24, 1893, which may be regarded as typical of the results thus far attained. The following data are taken from the records of that date (the noting of the passage of each wagon on the nearest road is significant):

Station, Washington.
Date, June 24, 1893.
State of sky at 11 a. m., fair, blue.
Source of heat, sun.
Height and aperture of slit: 50 mm. by 0.5 mm.
Prism used, rock salt, S. P. L. T.

Galvanometer, White, No. II.
Time single vibration, 12 seconds.
Bolometer, No. 5 (0.2 mm.), $11\frac{1}{2}''$.
Current of battery, 0.12 ampere.
Observer, F. L. O. W.

Angle subtended by slit at bolometer, $11\frac{1}{2}''$.
Remarks: When started the drift was nearly zero.
Object: Whole spectrum from A down.
Circle speed: 1° in 30 minutes. Drum 1 revolution in 60 minutes.[1]
A line, 339° 49''.
Start, 339° 47''.
At 1^h 6^m 45^{sec} Exposed plate.
        7^m 45^{sec} Opened slit.
        12^m Wagon passed.

---

[1] These speeds are such that 1 cm. of plate corresponds to 2 minutes of arc in the spectrum and to 1 minute of time. "Drum" in the notes seems to have been retained from association, though the film had been replaced by a glass plate.

There is a very considerable amount of vibration of a period of one-fourth or one-fifth of a second. Amplitude of vibration about one-half millimeter.

17^m Set beam south.
20^m Plate box fell over (caused deflection of 3 cm.).
$1^h\ 30^m$ About one-half plate gone.
Carriage passed.
Spot began vibrating.
$1^h\ 55^m$ Stopped.
Circle at end, 340° 36′ 55″.
Sky very good all through run.
Instruments also working very well.
Grade of bolograph, 1 to 2b.[1]

PROGRESS OF THE RESEARCH IN THE YEAR ENDING JULY 1, 1894.

1. Optical System.

Acting in the interest of greater accuracy, iron castings were procured, making complete metal supports for the slit end of the spectrometer, and iron castings with wooden tops were obtained to hold the bolometer and concave mirror. The great salt prism and the 10 cm. slit now in use were received in March, 1894.

After December 6, 1893, collimation was discontinued, and the optical arrangements at the close of the period under consideration were as shown in the diagram (fig. 6).

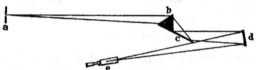

Fig. 6.—Diagram of the spectrobolometer as used from 1893 to 1896.

a is a bilateral motion slit 10 cm. in height.
b is a rock-salt prism 13 cm. face, 19 cm. height.
c is a plane mirror.
d is a concave mirror of 312 cm. principal focus.
$ad = 1{,}280$ cm.
$de = 412$ cm.

2. Electrical Apparatus.

The new galvanometer, made with 20 ohm coils according to specifications prepared at the observatory, was first used July 18, 1893. Unfortunately, no record of its sensitiveness under the conditions of use[2] other than that furnished by comparison of

---

[1] The "grade" is intended to signify the state of sky and instrumental conditions prevailing. Numbers refer to sky, 1 being best (and very unusual), while letters indicate the instrumental behavior, including freedom of record from drift and accidental inflections, as well as the condition of optical apparatus. With increasing instrumental efficiency the highest degree of excellence (which is indicated by the letter A) has constantly advanced.

[2] Each coil was tested separately on May 4, 1893, with a needle weighing 27 mgs. and only moderately damped. For a time of single swing of four seconds each coil gave about equal deflections, and for a deflection of 1 mm. it required a current in one coil of 0.00000030097. Whence for four coils connected in series and under a ten-seconds time of single swing, it was calculated that the constant would be 0.00000000004 (4 by 10^{-11}).

bolographs was preserved. It was tested September 24, 1895 (the coils being connected in two parallel series), with the same needle and same damping used in 1893 and 1894, and for a time of single swing of four seconds required for a deflection of 1 mm. on a scale at 1 meter a current in the external circuit of 0.0000000011, which, making the supposition that the deflection is proportional to the square of the time of swing, gives for a ten-second swing a constant of 0.00000000018 (1.8 by 10^{-10}).

The needle system first prepared, of 27 mgs. weight, proved too unstable for photographic purposes, and another (that used in the test above mentioned), of 56 mgs. weight, was made (by Mr. Wadsworth) of twenty pieces of glass-hard sewing needles, from $2\frac{1}{2}$ to 4 mm. long, arranged in circular outline in four groups of five each, symmetrically placed about the glass stem, two at each end. It had midway between the magnet systems a concave platinized glass mirror of 3 mm. diameter, weighing 4 mgs. When first used no damping other than that from four brass rods in the center of the coils was employed. This, however, proved entirely insufficient, and a piece of dragonfly wing, 9 mm. wide and 5 mm. high, was fixed to the back of the mirror. A square cast-iron tank with mercury had been for some time in use as a galvanometer support, but being found intolerable for magnetic reasons with this new galvanometer, it was replaced by four flat stones, set up with rubber blocks between at the corners, which system of support continued unchanged till 1896.

The drift became, July 22, 1893, "with the new galvanometer and new ($\frac{1}{10}$ mm.) bolometer, very troublesome indeed." The drift was largely reduced by the introduction of a new metal bolometer case with provision for water circulation as now in use. In January, 1894, a new pattern for bolometers, substantially that now employed, was proposed by Mr. Wadsworth, in which the hard-rubber back was replaced by metal and mica, but this frame did not come into use till later. Bolometers were first made at the observatory shop in this year.

3. MECHANISM.

Alterations of considerable value were made in the gearing of the driving clock in March, 1894, and its position was changed so that the shafts leading to the spectrometer and to the plate carrier ran at right angles north and west, respectively, as now. At nearly the same time a new plate carrier with rack and pinion motion was prepared at the shop. Considerable improvement was also made in the siderostat.

4. PHOTOGRAPHIC OPERATIONS.

With the exception of the provision of the improved plate carrier above mentioned no essential change was made in the registering apparatus. Glass plates, 60 cm. long and 20 cm. wide, were employed throughout the year and have continued in use ever since.

S. Doc. 20——5

5. SURROUNDINGS.

Very marked improvement in this respect was made in January, 1894, by the long-deferred construction of a double-walled wooden chamber about the spectrometer, and the separation of the galvanometer, battery and balancing resistance box from the rest of the observatory by means of a tight partition. The constant temperature arrangement for the bolometer has been already mentioned.

6. RESULTS.

In August, 1893, a research was begun on temperature and radiation, which was intended to have consisted in forming and bolometrically examining the spectra of bodies emitting radiations at various well-determined temperatures, from bodies under ordinary conditions to incandescent ones at the highest temperatures measurable. Little progress was, however, made before the work had to be temporarily abandoned.

With a view to compare the appearance of the bolometric-cylindric spectrum with that directly photographed, a spectroscope of the Littrow type was fitted up and some photographs prepared of the visible spectrum from a glass prism.

In the latter part of the year preparation was made for a preliminary communication of the results then attained. In that communication, which, though presented in the autumn of 1894, was illustrative of the state of the work in the spring of that year and should be mentioned in connection with the period of which we are now treating, a summary of the writer's investigations in the infra-red solar spectrum prior to those at the Astrophysical Observatory was first presented. These were shown to have given us an approximate idea of the form of the energy curve representing the rock-salt solar spectrum, and the position of about a dozen inflections below $1.5\ \mu$, representing unknown absorption bands. However meager these early results, it may be remarked that they represented years of painstaking labor.

The new method for carrying on these researches, which had been step by step elaborated in the Astrophysical Observatory, was then described, and it was shown that in accuracy of result and in rapidity and ease of working it eclipsed all former efforts in this direction. In illustration of this great advance, three curves, representing a half day's work, were shown superposed, which gave evidence of nearly a hundred absorption bands, where less than twenty had been known before.

At this time the amount of disturbance introduced into the record by ground and other tremors was inadequately appreciated, and an undue confidence in the number of authentic lines discriminable was introduced. The "cylindric" of the region given in a communication to the British Association for the Advancement of Science was, it was then stated, exhibited only in illustration, and not to be treated as a criterion of the final results to be obtained by the composite process. Indeed, the test of "reality" to which the lines appearing on this plate was submitted was subsequently found

ASTROPHYSICAL OBSERVATORY ANNALS, VOLUME I. PLATE II.

BOLOGRAPHS OF THE D LINES TAKEN IN 1893 AND 1898,
ILLUSTRATING THE DECREASE IN ACCIDENTAL DISTURBANCES.

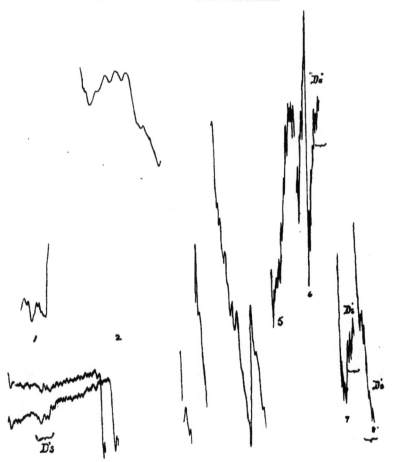

insufficient, not only because the error in the position of deflections in the bolographs had not at that time been sufficiently investigated, but especially because the number of accidental deflections was not duly recognized. The latter criticism applied equally to the curve and cylindric of the D lines.

The bolograph of the D lines is here reproduced (Pl. II) to give an idea of the drift and accidental deflections which were then frequently encountered, and the following are the notes taken at the time:

>Date, November 28, 1893.
>Source of heat, sun.
>State of sky at 12 M., fair, blue, with clouds.
>Height and aperture of slit, 50 mm. by 0.3 mm.
>Angle of same at bolometer, 6.8''.
>Prism, rock salt, No. II, S. P. L.
>Galvanometer, new, No. 2.
>Time single vibration, 4 to 5 seconds.
>Bolometer, No. 6 ($\frac{1}{10}$ mm.), subtends 5.8''.
>Setting on D_1 339° 28' 0''.
>Current of battery, 0.065 ampere.
>Observers, F. L. O. W. and R. C. C.

(Speeds for Nos. 5, 6, 7, and 8, of which No. 6 was that used in illustration of the Oxford report.)

>Circle speed,[1] 1° in 120 minutes.
>Drum (plate) speed, 1 cm. in 4 minutes (whence 1 cm. = 2').
>No. 6. Start 339° 27' 35''.
>   Start at $2^h\ 13^m\ 50^s$.
>   (Shutter open) 14.
>   15   35 D_2 entered bolometer strip.
>   15   46 D_2 passed off bolometer strip.
>   15   58 D_1 entered bolometer strip.
>   16   12 D_1 passed off bolometer strip.
>   (Continual vibration of the D's.)

Mr. Wadsworth remarked at the time: "This gave fairly good record of D lines."

With the present experience in regard to the effects of accidental disturbances it would seem doubtful if the deflections in the curve No. 6 were really due to the D lines and not to fortuitous causes.[2] To illustrate the advance made before July 1, 1894, through the introduction of the improvements already mentioned, the curve No. VI of May 10, 1894, Pl. XVIII, fig. 3, is shown as the best representative then attained, so far as instrumental working is concerned, of curves taken at the standard

---

[1] At that time the spectrum speed was habitually recorded as "circle speed," which is really half that thus indicated.

[2] The curves may be compared with those taken March 1, 1898, and presented in the lower left-hand corner of Pl. II, in which the D lines are undoubtedly indicated. These latter curves were taken with the great glass prism and at a speed such that 1 cm. of plate = 1 minute of time.

speed 1 cm. of plate to 1' arc of spectrum to 1 minute of time. The following are the notes accompanying this bolograph:

Station, Washington, D. C.
Date, May 10, 1894.
State of sky, fair blue, grade 1½.
Source of heat, sun.
Height and aperture of slit, 100 mm. by 0.5 mm.
    Angle of same at bolometer, 8''.
    Prism used, rock salt "R. B. I."
    Galvanometer, new, No. 2.
    Time of single vibration, 3 seconds.
        Object: Detail bolograph, A$-\Omega$.
        Speeds: Circle[1] 1° in 60 minutes. Plate 1 cm. in 1 minute.
        A line 159° 37' 15''.

Bolometer, No. 5 (0.2 mm.), subtends 11¼''.
Current of battery, 0.074 ampere.
Recorder, A. K.
Observer, F. L. O. W.
Remarks: Northeast wind.

No. VI. Circle at start, 159° 36' 30''.   Temperature at 12^h 56^m = 22.5°.

Started,	1^h 6^m 00^s 10		Set beam,	1^h 50^m 00^s	
Wagon,	10	40	R. C. C. opened side door,	2	2 00
Set beam,	19	00	Set beam,	3	30
Wagon,	22	30	Stopped,	2	6 10
Wagon,	23	15			20
Wagon,	29	30	Circle at end,	160° 6' 38''	
Set beam,	31	10			
Set beam,	40	00	Temperature,	23.5°	
Wagon,	45	15	Grade[2] of bolograph, 1½a.		

DEGREE OF ACCURACY ATTAINED IN 1894.

It is of interest to note that the degree of accuracy and the sources of error which existed in the early part of 1894, were apparently compatible with a variation of much more than a millimeter in the absolute position of any line. The following abstract is from a discussion of sources of error made in February of that year:

1. CHANGES OF TEMPERATURE.

Effect on prism.—The effect of a change of temperature in the prism is to change the deviation of a line, $-11''$ at A, and $-14''$ at Ω for every degree change in temperature.

If, then, in running from A to Ω the temperature changes 1°, the lines at Ω will be displaced 14'' (viz, with above ratio between circle and plate speed, which we shall hereafter assume 2½ mm). Now, the actual conditions of running with the new 'house' around the spectrobolometer are such that the change of temperature during an hour's run is from 0.1° to 0.2° only, and hence the error from this source will be from 0.23 to 0.45 mm., at a maximum not more than one-half the prescribed error.

This change is so slight that changes in other parts of the apparatus in consequence of it are insignificant,[3] particularly since the apparatus is now mounted, with the exception of the concave mirror, on massive iron supports.

With the slit, the other essential part of the prismatic train, the case is different. At the distance at which this is placed an error of 6'' corresponds to a movement of a little less than 0.4 mm. linearly. The cast-iron tripod on which the slit rests is about 104 cm. high in the center; legs about 1.5 meters long; angle of legs 45°.

---

[1] Should be interpreted "spectrum speed."

[2] 1½ refers to condition of sky, 1 being best. a Refers to instrumental conditions, a being best. Grading by F. L. O. W.

[3] It was afterwards found that this was not the case, but that a displacement of 10'' of arc for 1° change of temperature was not unusual, owing largely to the use of wood in mounting the concave mirror.

Hence the lateral displacement by an expansion of one of the legs will be, roughly,

$$etl \cos 45° = 1,000 \times 0.00001 \times t = (\text{say}) \tfrac{1}{2}.$$

$$t = \frac{0.50}{0.01} = 50°.$$

Hence, to produce a displacement of one-half millimeter the change of temperature in one leg would have to be 50°, and this, though quite possible through the day [?], is not possible or probable in the course of an hour.

2. Stability of Parts of Apparatus.

All the parts of the apparatus within the "house" are without doubt free from any serious disturbance of an order of five seconds. In regard to the slit this is not true. On windy or even on moderately breezy days the spectrum, when received with a fixed eyepiece, is seen to be in continual lateral vibration, amounting sometimes to more than the width of the bolometer strip, 5″. This motion is due partly to vibration of the slit itself, but largely to the changes in atmospheric refraction going on in the column of air between the slit and the prism. Here is, then, a source of error of an order of magnitude of that which must be eliminated, and the way to do it is evidently to inclose the whole column of air from slit to lens by means of a suitable tube.

3. The Errors due to the Clockwork.

These errors are the only important ones left to consider, and in the present state of the apparatus are the most important of all. To obtain an idea of the accuracy demanded, consider what an error of six seconds, viz, 1 mm., means mechanically.

At the tangent screw 6″ of spectrum corresponds to 3″ of motion. The tangent arm is 11 inches long. Hence 3″ corresponds to a movement of $\frac{3 \times 11}{206000} = \frac{1}{6200}$ inches, say $\frac{1}{6000}$ of an inch, a quantity measurable even only under the best mechanical conditions. It is unreasonable to expect any tangent screw driving such a mass of metal as this is required to drive should do any better or even as well as this under average conditions of running.

The tangent screw is of such pitch that one revolution = 20′ of circle = 40′ of spectrum; 6″ therefore means $\frac{1}{400}$ of a turn, or for 100 teeth one-fourth of a tooth. The teeth of the gear are no doubt accurate to this degree.

The tangent motion shaft turns twenty times as fast as the tangent screw. Its motion, therefore, must be correct to one-twentieth of a revolution, and this at present is not the case.

Changes in gearing and position of the clock were recommended, which were actually accomplished in March, 1894, with the exception of the provision of a larger circle driving shaft, which was delayed several months. Beyond these changes and the provision of a single-walled tin tube extending from the slip to about 2 meters inside the observatory, which somewhat diminished the "boiling," no changes of importance were made between the time of this letter and July 1, 1894. It may, therefore, be taken as setting forth, so far as was known, the degree of accuracy to be obtained with the apparatus of the time.

PROGRESS OF THE RESEARCH IN THE YEAR ENDING JULY 1, 1895.

During this period there was taken a large number of bolographs, of which many of the best were translated by the "cylindric" process into line spectra with the purpose of preparing a final composite for publication. But while this was in progress some important instrumental changes were made with a view to render the apparatus more efficient.

1. OPTICAL ARRANGEMENTS.

No change was made during the year in the arrangement of optical apparatus. The only new departure was the introduction of the practice of refiguring the salt surfaces at the observatory. This custom was begun after a comparatively trying series of experiences with the great salt prism, which fogged several times in succession before being used at all, after its reception from a distant optician to whom it was sent to be repolished. A local optical instrument maker, Mr. Kahler, was instructed in the art of polishing salt by Mr. Brashear, and since that time he has continued to refigure the salt prisms and cylindric lenses as required.

2. ELECTRICAL ARRANGEMENTS.

A new galvanometer needle was made in duplicate of its predecessor after an unsatisfactory attempt to permanently improve the astaticism of the latter. These needles were of about equal sensitiveness, and neither retained for very long a suitable degree of astaticism, probably because of the excessive temperature changes they were subjected to.

3. PHOTOGRAPHIC OPERATIONS.

The general arrangements for recording the galvanometer deflections were not changed, but improvements were made in the temporary plate carrier, and a new one was ordered in which the plate was to be moved by a screw instead of the rack and pinion mechanism thus far used.

4. MECHANISM.

Improvements were made in the driving clock by introducing a larger shaft to actuate the circle, and by making some changes in the governor. But as tests showed the motion of the plate still not sufficiently regular, an entirely new driving clock (by Warner & Swasey) was procured and installed in April, 1895. This instrument will be illustrated and described later, as it still remains in use and is eminently satisfactory.

Considerable improvements were made in the various parts of the siderostat, which increased the regularity of its motion.

5. VARIOUS CHANGES.

The galvanometer pier was rebuilt in March, 1895, to secure greater steadiness, and at the same time the pier was prepared on which the driving clock is now placed. Very considerable improvement in the steadiness of the needle resulted from the rebuilding of the galvanometer pier.

A partially successful attempt to secure better control of the "drift" was made by inserting a micrometer valve in the water supply for the jacket of the bolometer case. The use of the furnace in the basement was discontinued, owing to its bad effects in producing drift and to the dust arising, and steam coils, supplied with steam from the main building of the Smithsonian Institution, were introduced.

6. RESULTS OBTAINED DURING THE YEAR 1895.

It was thought at the beginning of this period that the work had then reached such a stage that with the exception of the mechanical arrangements, which were, as we have seen, greatly improved, no part of the apparatus required changes of importance, but that from a large number of bolographs it would be possible to deduce final results. But after the new galvanometer pier was made many deflections which had been in doubt disappeared, and it was at first thought that the absence of detail might have resulted from imperfect definition. However, the prism and mirror surfaces were improved and carefully adjusted without effect, and it became certain that this lack was due to the elimination of spurious detail, and must be attributed to the improvement rather than to the degeneracy of instrumental conditions.

Experiments made about this time limited the confidence heretofore reposed in the composite process employed with the "cylindrics" or linear spectra, a process which it was shown could not be certainly depended on to eliminate all the smaller false deflections without an inordinate number of plates, so that unless it should be determined to limit the research to the establishment of the position of about eighty lines, it became more important than ever to reduce those occasional deflections in the record, which are liable to be confounded with the smaller "real" ones.

In Pl. XVIII, fig. 4, is shown a portion of one of the best bolographs of the year, and the following is taken from the record accompanying it:

Station, Washington, D. C.
Date, June 8, 1895.
Wet bulb, 64° in inner chamber.
Dry bulb, 71° at 8 a. m.
State of sky at 9 a. m., milky blue.
Source of heat, sun.

Galvanometer, No. 2.
Time single vibration, 3½ to 4 seconds.
Bolometer, No. 11, (0.1 mm.), subtends 5.7''.
Current of battery, 0.076 ampere.
Observers, R. C. C. and F. E. F.

    Height and aperture of slit, 100 mm. by 0.40 mm.
    Angle of same at bolometer, 6.4''.
    Prism used, rock salt "R. B. I."
        Object: Bolograph, A—Ω.
        Speeds: 1° spectrum = 60 minutes. 1 cm. plate in 1 minute.
          A line at $9^h\ 10^m$. 160° 9' 50''.
          Scale, 1 cm. = 1' = 1 minute of time.
            No. IV. Temperature at $12^h\ 0^m = 24\ .0°$.

Circle at start,	160° 8' 20''	Set beam east,	$2^h\ 00^m\ 56^s$
Start,	$1^h\ 18^m\ 00^s$	F. E. F. in galvanometer room,	1 10
Sun in light clouds,	19 10	F. E. F. out of galvanometer room,	12 20
Set beam east,	1 29 00	Set beam east,	12 50
Wagon,	32 15	Wagon,	2 14 40
Set beam east,	39 15	F. E. F. in galvanometer room,	15 20
Wagon,	47 10	Close,	{ 18 00
Set beam east,	50 20		10
F. E. F. in galvanometer room,	40	Circle at end,	160° 37' 20''
F. E. F. out galvanometer room,	12 20	Temperature at end,	25.9°
		Grade of bolograph[1] 2 (a–b).	

---

[1] 2 refers to the condition of sky, 1 being best. (a–b) refers to instrumental conditions, a being best. Grading by F. E. F.

PROGRESS OF THE RESEARCH IN THE YEAR ENDING JULY 1, 1896.

In the fall of 1895 a very considerable number of bolographs were taken under conditions not materially different from those of June, 1895. These curves confirmed the conclusion that the accidental deflections were more numerous and more considerable in size than had at first been supposed, and showed that very considerable improvement must be made in these conditions as well as in the control of the "drift." The following changes were made in the effort to eliminate accidental deflections from the curves and errors in position of the "real" ones, so that the evidence in favor of the latter should become more conclusive by reason both of their greater prominence and of their greater fixedness of position.

1. The number of battery cells employed was increased from four to sixty (storage cells). It had been observed from the time of the early work at Allegheny that it was advantageous to employ a large number of cells in order to limit the force of local changes in any cell by thus making the average of many. The use of sixty cells is an extension of this method.

2. A specially constructed rheostat with water jacket, designed by Mr. Abbot, but with the advice and assistance of Mr. R. C. Child and Mr. W. Gaertner, in which there are two coils, each of the same resistance as the bolometer strips, and which is provided with a slide wire operated from outside the water jacket, was obtained to balance the bolometer. This piece of apparatus having been placed close to the bolometer, the circuit became very compact, and the disturbances arising from differences of temperature in the distant resistance box were avoided.

3. To reduce the effect of ground vibrations at the galvanometer, a system of support similar to that employed by Julius[1] was constructed at the suggestion of Mr. Abbot with some improvements of his own.

4. The double-walled room which surrounds the spectrobolometer was extended to include also the galvanometer.

5. A system of automatic temperature control was introduced and heat supplied at night as well as day through the winter from a steam boiler in a distant building.

6. A double-walled tube was extended from the slit to the spectrometer room to reduce the irregular refraction by air currents.

7. A new galvanometer needle of greater sensitiveness and also more perfect astaticism was constructed.

It may be added that it was proposed near the end of this period to return to the practice of collimating the beam, and that improvements of the circle-motion mechanism were projected. Both these changes have since been made with advantage.

---

[1] W. H. Julius, Annalen der Physik und Chemie, neue Folge, vol. 56, p. 161, 1895.

NOTE.—This device was adopted before the publication of his more recent communication (Zeitschrift für Instrumentenkunde, XVI, 267, 1896), in which he describes improvements, but owing to the size of the galvanometer and the slight height at command in the galvanometer room it is doubtful if the later form could be used with advantage here.

During the latter half of the year only a comparatively small number of bolographs were made, attention being given to perfecting the new apparatus, and to devising and executing quantitive tests of the accuracy of the process at all its stages.

PROGRESS OF THE RESEARCH IN THE YEAR ENDING JULY 1, 1897.

With the improvements made in the early part of 1896, the bolographs taken in the fall and early winter months showed wonderful improvement in freedom from drift and accidental deflections. Besides this, the optical definition of the spectroscope was very materially improved by the introduction of a system of collimation by two spherical concave mirrors, so that the results of these few months of observation far outvalued all that had been previously obtained in years for the purpose of discovering the position of absorption lines in the infra-red solar spectrum. Hence it was felt desirable to prepare for the immediate publication of the results indicated by these bolographs.

In this work the composite "cylindric" method for eliminating false lines and retaining the true was replaced by a method of comparision and measurement, as experience had shown that the former method was not entirely trustworthy except with the employment of an inordinate number of bolographs, and even then likely to eliminate some true as well as false lines. The composite process was, however, retained in a slightly modified form to prepare a line spectrum for purposes of illustration, to exhibit the lines resulting from the comparison. As the method of comparison and measurement is still in use it will be described in detail later.

The first six months of the year 1897 were, therefore, almost wholly devoted to preparing for the press an account of the work and of the results then obtained. These results gave with great accuracy the position of 222 lines in the infra-red spectrum between wave-lengths $0.76\,\mu$ and $5.3\,\mu$, and included besides several minor contributions of value.

Unforeseen complications connected with the relations of the Astrophysical Observatory to the Government prevented the funds for the publication of this report from being available, so that its issue was unfortunately delayed. However, the delay was not wholly unproductive of good result, for improvements made within the next year in the apparatus of the observatory, including the construction of a new and extraordinarily sensitive galvanometer, and the substitution of cylindric for spherical mirrors for collimation, enabled the repetition of the observations with results far richer than before.

We will here close the historical account of the tedious stages of improvement by which the apparatus and processes have been brought practically to their present condition, and will proceed in the next chapter to describe the arrangements now existing.

Chapter III.

DESCRIPTION OF THE OBSERVATORY BUILDINGS AND THE APPARATUS EMPLOYED IN THE RESEARCH.—ADJUSTMENTS OF APPARATUS.

Recalling again the account given in the Introduction of the construction and equipment of the observatory in a manner intended from the first to be only provisional, the actual conditions under which the work has been carried on will be gathered from what follows. It must be remarked at the outset that the greatest difficulty has been encountered because of the nearness of streets where considerable traffic, including electric cars, goes on. In comparison with this other obstacles are not formidable, though among them may be cited the fact that it is almost impossible to heat a temporary wooden building of this character with uniformity.

THE OBSERVATORY.[1]

Pl. III shows a plan of the observatory buildings and lot. A is the main building, containing the spectroscopic and bolometric apparatus within the double-walled chamber. The photographic work is carried on in the smaller house B, which is provided with a large camera, b, and with suitable sinks $c\ c$, and cupboards and shelving d. D is a small shelter in which is kept a Newtonian telescope of 50 cm. aperture and but 1 meter focus. The great Grubb siderostat, with its movable covering, is shown conventionally at C. A platform, E, connects the two buildings, and all is inclosed by a high board fence surrounding the lot.

The main building is shown in greater detail in Pls. IV and V, which are similarly lettered, in plan and longitudinal section in elevation. It is, as we have constantly to remember, not a permanent installation, but one originally intended only as a provisional shelter for the instruments. To moderate the intense heat of summer, as well as to facilitate heating in winter, the roof is sheathed within, leaving an air space as shown. The walls are tightly double boarded. Shutters are provided at all the windows, so as to allow the whole building to be used as a dark room during pho-

---
[1] The descriptions which follow treat the observatory as it was up to September, 1898, when the new laboratory and extension to the office were added, as shown in the frontispiece and in Pl. III. These new parts have been unused in the research here described, and they are reserved to figure in some later publication.

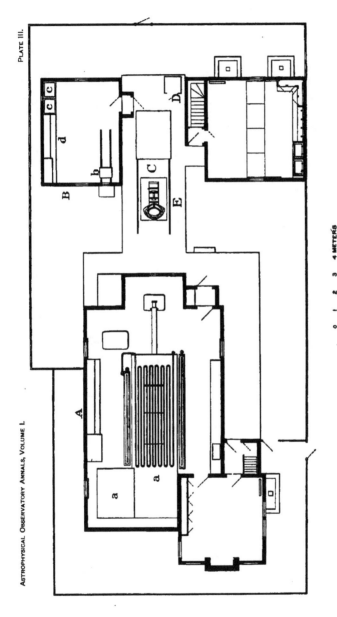

PLAN OF OBSERVATORY BUILDINGS AND LOT, 1900.

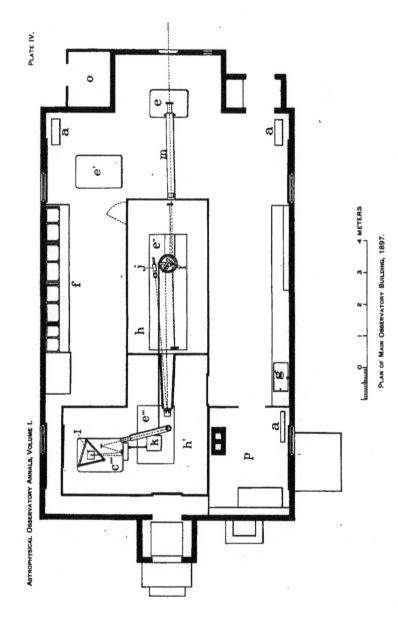

PLAN OF MAIN OBSERVATORY BUILDING, 1897.

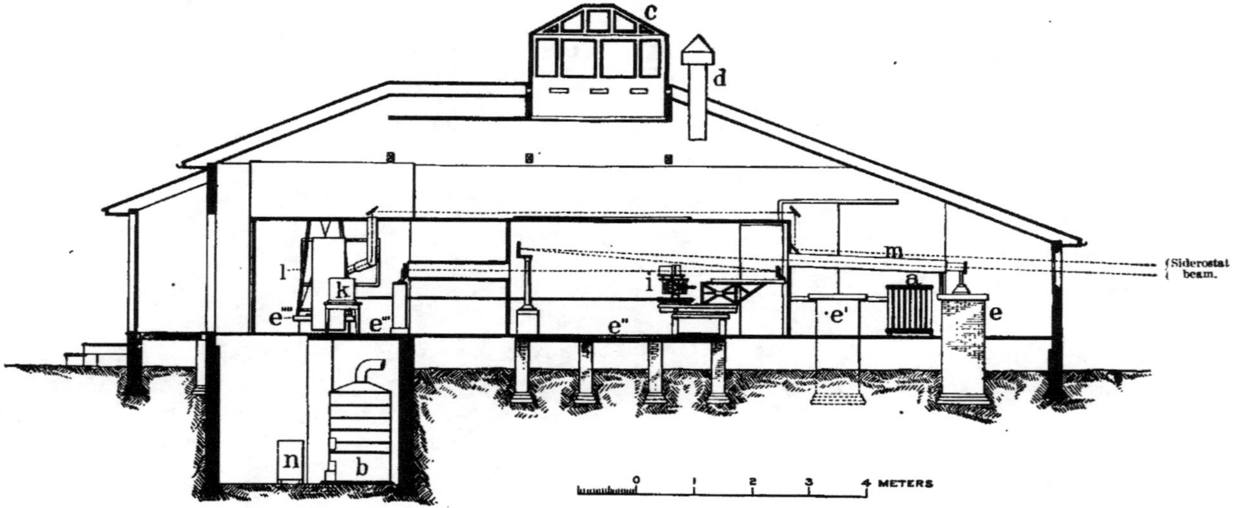

ASTROPHYSICAL OBSERVATORY ANNALS, VOLUME I.

PLATE V.

SECTION THROUGH MAIN OBSERVATORY BUILDING, 1897. LOOKING WEST.

tographic operations. Heating is at present accomplished through the steam radiators and supplied from the distant Smithsonian building, but formerly through the hot-air furnace *b* in the cellar, which proved to be such a source of dust as well as of local heating that it has been advantageously replaced by steam heating. It had at all times been plain enough that next to the difficulty introduced by the neighborhood of traffic and electricity, the great difficulty of such a research conducted with the galvanometer was due to irregular movements of the needle, themselves due in turn to causes which were nearly all reducible ultimately to changes of temperature. Attempts to furnish a chamber of constant temperature had been made as long ago as the time of the writer's previous researches in Allegheny, but the experience had then and subsequently seemed to show that nothing but a complete installation of a chamber of constant temperature would do, and this involved some very serious considerations, that of the expense being one of them. It was not until 1897 that the writer was able to completely carry out his long deferred purpose of establishing such a chamber, which was done by building internal rooms in the temporary observatory building, and by bringing in the De la Vergne system of cooling by ammonia, from engines established at some distance in the main building. This was in 1897.

The primary object of the heating and cooling apparatus is to maintain a constant temperature of 20° C. in the inner chambers of the observatory containing the spectrobolometric apparatus. This has been found so absolutely essential to a successful prosecution of the research that the greatest pains and much expense have been incurred in securing it. The cooling coils, supplied by day with cool compressed ammonia from apparatus in the Smithsonian building, are chiefly contained in two long tanks of brine running above and on either side of the inner chamber of the observatory. These brine tanks are inclosed in boxes open at the top and provided with vents at the bottom, through which the cold air falls by its own weight, and thus produces an automatic circulation round the spectrobolometer chamber. At night the cooling is kept up by the ice in the tanks. For the very hottest weather some auxiliary cooling coils exposed to the direct circulation of the air are added to those in the two brine tanks.

But the system of heating and cooling would be wholly incomplete without some automatic regulation. The writer had tried this also at Allegheny by the use of mercurial thermometers with electrical attachments, but with only partial success, due as much to the lack of time to experiment as to lack of means. Since then commercial demands have introduced systems into certain manufactories in need of automatic regulation, and several systems are in employment. After considerable examination the writer, discarding his own early methods, has found it advantageous to adopt one of these systems, the Johnson system of automatic regulation, which controls both the steam coils and the vents to the cooling tanks, and thus retains the temperature of the outer observatory within 1° C. of that desired in the spectrobolometer chamber. The insulated wall of this latter still further reduces the fluctuations of temperature within to so great an extent that often twenty-four hours will pass with an extreme variation as small as 0.1° C. The regulating system has been recently extended to include the steam coils in all the buildings of the observatory.

At c is shown a skylight, which may, if desired, be moved off on a track. The funnel d is provided for the escape of fumes arising from experiments with radiating substances. Shelving and drawers appear at $f\,f$, and a tap at g connects with the city water mains.

Piers shown at e, e', e'', e''', e'''', of which that at e'''' is by far the most stable, have been constructed especially for the support of the sensitive galvanometer. Inner chambers $h\,h'$, with double walls and intervening air space, inclose the spectrometer and bolometer and the clock and galvanometer shown, respectively, at i, j, k, and l. The entrance of the beam from the siderostat is effected through the double-walled tin tube m, and its path within the spectrometer chamber is shown by dotted lines. An auxiliary beam for photographic purposes is shown, passing by the agency of various mirrors into the galvanometer room.

The storage battery of sixty cells, which supplies the bolometer current, is contained in a box in the basement, shown at n. A suitable copper switch connecting the cells all in parallel for use, and in two series of thirty cells each for charging, is placed immediately upon the battery box, and from its terminals go a pair of heavy copper wires to a milliammeter and series resistance in immediate proximity to the bolometer.[1] Another storage battery of twelve cells, for use on lamp circuits, siderostat control, and for other occasional uses, is placed in the little room o, and connections are readily made with a dynamo in the Smithsonian building for the charging of this and the bolometer battery. The room p is used as an office and for the storage of books for frequent reference.

In Pls. I, VI, and VII are shown photographs of the observatory and of the interior of the spectrometer and galvanometer rooms.

Before going on to describe in detail the various pieces of apparatus employed, let us recall (1) that the main object of the present investigations is the mapping out of the absorption lines in the infra-red solar spectrum; (2) that the method of discovery of these absorption lines is to note depressions in a curve whose ordinates correspond to energy of radiation, and whose abscissæ correspond to prismatic deviations; (3) that such energy curves are produced automatically by causing the spectrum to pass steadily over a linear bolometer strip, whose variations in temperature (corresponding to varying energy in the spectrum) produce horizontal motion of a spot of light reflected from a

---

[1] For the most careful work the milliammeter is used only temporarily, and a single platinoid coil of suitable resistance is soldered into the circuit in place of the variable resistance, thus avoiding extra contacts.

SPECTROMETER ROOM, 1896. LOOKING SOUTH.

GALVANOMETER ROOM, 1896. LOOKING WEST.

galvanometer mirror upon a vertically moving photographic plate. With this preliminary remark we may now examine in detail the various parts of the bolographic apparatus, and the more at length because that the success of the undertaking depends on the excellence of devices to secure a pure yet energetic spectrum, sensitive yet not unmanageable bolometric apparatus, and accurate synchronism of the motions of the spectrum and recording photographic plate.

APPARATUS.

Diagrams of the electrical and optical systems are shown in figs. 7 and 8, respectively. The electrical connections are in the main so familiar that it is superfluous to describe them further than to point out the rather unusual device of a coil, e, auxiliary to the slide wire d, for making a fine adjustment between the resistance of the bolometer strips b and b' and the balancing coils c and c'. All the circuits are arranged in form with special care to avoid induced currents, and are protected by a thick layer of

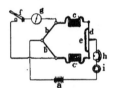

FIG. 7.—Diagram of electric circuit of spectrobolometer. 1896 to 1899.

felt throughout. The source of electromotive force at a is the battery of sixty storage cells connected in parallel. By means of the resistance i the strength of current (as indicated by temporarily interposing the milliammeter h) is adjusted to that value which experience has shown to be the most satisfactory for working under the present conditions. All the apparatus indicated in the diagram, except the battery, is inclosed in the isolated chambers of constant temperature, and for the bolometer and balancing arrangements still other provision for constant temperature is made, as elsewhere described. The battery is in the cold basement and protected by a close box, well wrapped in cloth.

CONSTANTS OF THE ELECTRICAL SYSTEM.

The constants of the electrical system depend on the bolometer in use and are, with the two bolometers which were used in 1897 and 1898, as follows:

Electromotive force of battery, 2 volts.

Current in battery circuit for bolometer No. 20, 0.083 ampere; for No. 21, 0.028 ampere.

Resistances at 20°.

	Bolometer No. 20.	Bolometer No. 21.
	Ohms.	Ohms.
i	0.0	26.0
b	4.23	41.73
b	4.23	41.54
c	43.54	43.54
c	43.28	43.28
d	0.2	0.2
e	8.0	8.0
g	30.0	30.0

In the diagram of the optical system, fig. 8, a represents the siderostat mirror, b the vertical slit, c and d a pair of fixed cylindrical mirrors, of which c is convex and d concave, which are so adjusted as to collimate the beam. The entire height and about one-seventh of the width of the beam from the slit is utilized to fill the curtate face of the prism. The prism e and plane mirror f, mounted as described in another place, disperse the beam and reflect the part which passes the prism in minimum deviation in a direction parallel to that of the beam when incident upon the prism. This portion of the beam is brought to a focus upon the bolometer strip at i by the concave mirror g

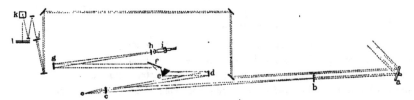

Fig. 8.—Diagram of optical arrangements of spectrobolometer, 1897 to 1899.

A double-convex cylindric lens of salt h with its elements horizontal is interposed at such a point as to cut down the height of the spectrum to that of the bolometer strip.

An auxiliary beam for recording the movements of the galvanometer needle is carried into the galvanometer room by means of several plane mirrors. It passes through the slit j, and by two other plane mirrors is made to fall upon the concave galvanometer mirror at k under such conditions as to cause the image of the slit j to be formed at the photographic plate l, on which the movements of the galvanometer needle are recorded.

The Siderostat, 1896. Looking East.

CONSTANTS OF THE OPTICAL SYSTEM.

The constants of the optical apparatus are as follows:

Mirrors.	Dimensions.				Principal focal length.
	Diameter.	Height.	Width.	Length.	
	Cm.	Cm.	Cm.	Cm.	Cm.
a	46				
c		16	8		−58
d		18	12		+543
f		17.5		25	
g	22.5				468
k	.25				90

Prisms.—(1) Rock salt "R. B. I." Refracting angle (1896–97), 59° 55′ 9″; (1897–98) 59° 53′ 30″; (1898–99) 59° 56′ 15″. Faces 13 cm. wide, 19 cm. high. (2) Flint glass. Refracting angle, 59° 59′ 50″. Faces 13.3 cm. wide, 16 cm. high.

Lens.—Rock salt cylindric, double convex, diameter 6 cm., thickness 1.5 cm., radii of curvature equal and about 40 cm.

Slit.—Double motion. Height 10 cm. Width elsewhere specified.

Distances:

$bc = 790$ cm. $oo = 54$ cm.
$gi = 468$ cm. $od = 541$ cm.
$kl = 120$ cm. $hi = 25$ cm.

The center of the beam from the siderostat after collimation is 110 cm. above the surface of the pier on which the spectrobolometer rests.

SIDEROSTAT.

An illustrated description of this substantial and accurate piece of apparatus as it was originally designed, from general indications given by the writer, was furnished by Sir Howard Grubb in a paper read before the Royal Dublin Society and reprinted in Engineering, May 16, 1890. It is worthy of the maker's well-earned reputation, but in the course of its use many changes dictated by experience of local needs have been made from the writer's first design. It is a well-made instrument, and as now used is shown in the photograph, Pl. VIII.

As will be seen, it is of the Foucault type, and very solidly built. The original governor has been replaced by one constructed by Messrs. Warner & Swasey. The original clockwork (still in use) is of an extremely massive construction, driven by a weight of nearly a thousand pounds, which rises and falls in a well underneath the instrument. The provision for movement in altitude and azimuth is rarely used, but that in right ascension and declination is in constant employment, being controlled by motors actuated by a keyboard beside the observer within. Thus the solar beam

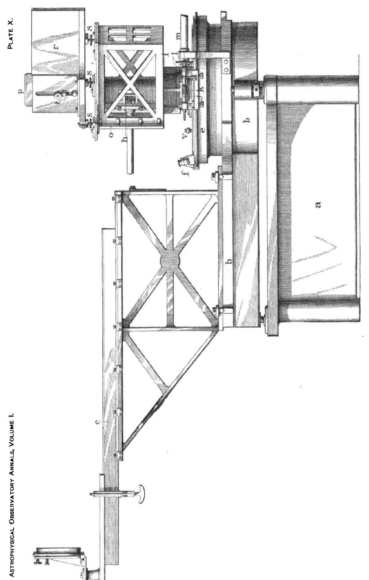

SIDE ELEVATION OF THE SPECTROMETER.

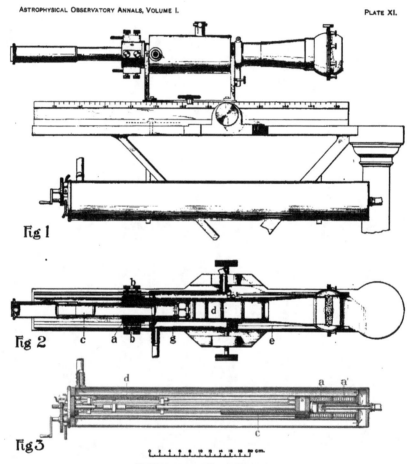

THE BOLOMETER AND SPECIAL RHEOSTAT.
Form used 1896 to 1898.

Turning, now, to the detail illustrations, fig. 1, Pl. XII, shows bolometer No. 20 itself in perspective. The strips a a', each of about 4 ohms resistance and about 12 mm. in length, are soldered to the copper pieces b b' b''. The pieces b' and b'' have provision by means of a screw and slot for slightly changing their distances from b, thus enabling the approximate equalizing of the resistances of the strips a and a'. A circular brass plate, c, covered on its front face with a sheet of mica, serves as the frame for the bolometer. Through the brass plate, and insulated from it by hard rubber, pass three hollow copper cylinders, upon the ends of which are tightly screwed the pieces b b' b''. The whole is supported by four copper posts, which make connection with the copper binding posts shown in fig. 1, Pl. XI, on the outside of the bolometer case. Wedges are inserted in the ends of the copper posts after affixing the bolometer, to insure good contacts, and it will be seen that from the binding posts to the bolometer strips there is a complete copper connection.

Preparation of Bolometers.

Bolometers are usually made at the observatory, and it is thought that an account of the method employed in preparing and fastening the strips will not be without interest.

Having made use, experimentally, of iron, steel, and carbon without satisfactory result, platinum, which was early employed by the writer, has been definitely adopted as the best material. Thin platinum sheets have been used in preparing strips for many bolometers, but the difficulty in cutting them out with satisfactory regularity in form and resistance is a drawback to this means, especially in preparing fine linear bolometers. Various methods have been proposed for electrically depositing the strips, but none have been successfully used here.

Linear bolometers.—For linear bolometers, the early method of cutting the strips from platinum foil has been followed by the easier method of construction which makes use of silver-covered platinum wire ("Wollaston wire") as the raw material, for bolometers of from 0.05 mm. to 0.20 mm. width of strips. This wire is obtained of an exterior diameter 0.004 inch (0.1 mm.), inclosing a platinum core 0.0005 inch (0.0125 mm.) in diameter. A portion long enough for both strips is flattened between steel flats to such a width as will give the desired width of platinum. The silver is then dissolved off by nitric acid, the remaining platinum ribbon carefully washed, and all is ready to insert the strips in the bolometer frame. This is accomplished with ordinary solder, using rosin as a flux. The resistances of the two strips are made equal within 1 per cent by repeated measurement and resoldering, after which they are carefully smoked over burning camphor. When a little skill is attained a bolometer may thus be filled and balanced in three hours.

These bolometers have resistances of only about 4 ohms for each strip. Bolometers of higher resistance, requiring proportionally smaller wire, are of course more difficult to construct satisfactorily, but the above method has been satisfactorily followed here for them.

Area bolometers.—For bolometers of considerable surface, designed for use with radiations of small intensity, sheet platinum is advantageously used. The general plan of their construction is the same as that described by the writer in an early article on the bolometer[1]—that is, of making two gridiron-shaped rows of strips arranged on two frames in such a manner that when the frames are superposed the strips are all connected in series and break joints, appearing, as seen from the front, to form a continuous surface.

Two alterations have been made in details of construction to avoid drift, sometimes caused by the warping of the vulcanite frames and the "creeping" or slow increase of the galvanometer deflection due to the ends of the strips being out of sight, and hence receiving heat only by slow conduction. To obviate these two defects the frames are now made of brass, with a number of suitably placed copper pieces insulated from the brass by ivory and mica. To these copper pieces the ends are soldered, so that all the electrical resistance of the central area is in the surface exposed to radiation.

It has always been observed here that the difficulties from drift and accidental deflections increase largely with decreasing thickness of the platinum strips. Recent experience indicates that a thickness of much less than 0.002 mm. is unadvisable. Inasmuch as linear strips should not be less than five times as wide as they are thick, this sets a fairly definite limit to the resistance of linear bolometers.

BOLOMETER CASE.

A horizontal section of the bolometer case appears in fig. 2, Pl. XI. The bolometer itself, affixed by the four insulated copper posts to the hollow cylindrical support *a* and connected by insulated copper wires to the copper binding posts *b b*, has already been described. A suitable positive eyepiece, *c*, for focusing upon the spectrum and bolometer strip, slides within the hollow support *a*. In front of the bolometer is placed a series of blackened brass diaphragms, *d*, with central holes of progressively increasing aperture, which serves the triple purpose of preventing circulation of air currents, of cutting off stray radiation from the central strip, and of hiding the side strip. Until recently it was found by the writer that this series of diaphragms, originally employed by him, was sufficient to prevent prejudicial air currents, especially as a strong draft of air blowing at one end of the case would hardly affect a candle flame at the other. But as the bolometer as well as the galvanometer was perfected by successive improvements, smaller and smaller defects in the record came under notice. It was later noticed that the galvanometer always indicated some dis-

---

[1] "The bolometer and radiant energy," Proceedings of the American Academy of Arts and Sciences, Vol. XVI, p. 342, 1881.

turbing cause at work in the circuit when the wind blew outside the observatory, even though at so moderate a rate as to barely raise a drooping flag on a neighboring building. The opening and shutting of the door to the observatory was also clearly noticeable in the new and perfected record. The source of the disturbance was traced to the bolometer itself, and it was found that when the case was made air-tight the disturbance ceased. Accordingly, a glass plate was cemented in to close the eyepiece end of the case, while still allowing the use of the eyepiece, and the salt cylindric lens was set in a ball and socket mounting, which forms an air-tight continuation of the case. It is now found necessary that the case occupied by the bolometer shall be air-tight to a pressure difference of a third of an atmosphere, in order to entirely overcome the prejudicial effect of the wind.[1] The water jacket, with its outlets, is shown at g.

RHEOSTAT.

Section views of the special rheostat are shown in fig. 3, Pl. XI, and figs. 2 and 3, Pl. XII. Like the bolometer, this instrument is provided with a surrounding water jacket, and provision is also made for the circulation of the water within the coils, as shown by the arrows, fig. 3, Pl. XI. The resistances and mechanical arrangements of the slide wire are all firmly framed together, and upon loosening four screws at the front can be entirely removed from the surrounding jacket.

The coils a a' are of well insulated platinoid wire, wound double to avoid inductive effects, and are of about the same resistance as the bolometer strips employed. Several spools are provided with coils of different resistances for use with different bolometers.

The slide wire is shown in section in fig. 3, Pl. XII, and a portion of its length in fig. 2 of the same plate. The wire is marked b in each figure. It is obscured by other details in fig. 3, Pl. XI, but it runs from the coils to the front of the instrument. At c, fig. 3, Pl. XI, are shown two heavy copper wires, which lead from the coils without break to the binding posts of the bolometer, but are of sufficient length to allow of moderate movement of the latter. Direct contact is made between the four ends of the balancing coils and the ends of the slide wire and of these two connecting wires, respectively, by suitable clamping devices. It is a little difficult to remove and insert the coils on account of the peculiar disposition of their ends, but extra connections are thus avoided.

An auxiliary coil is shown at d, fig. 3, Pl. XI, of which the ends make contact with the slide wire at variable points on each side of the contact between the latter and the battery terminal, and by its shunting action this auxiliary coil enables a finer adjustment to be made.

---

[1] In considering the effect of local tremors from these outside disturbing conditions, it will be borne in mind that they exactly resemble the minor tremors introduced from proper solar causes, and as it is from an interpretation of these latter tremors that the infra-red spectra are here determined, the elimination of the former class—that is, of local tremors—is a matter of the most vital concern.

The devices employed for moving the three contacts upon the slide wire are clearly shown in fig. 3, Pl. XII. A central screw, e, extends through the whole length of the slide wire and is moved from without by means of a crank either directly or from the shaft f, according as a slow or rapid adjustment is desired. Three sliding pieces g g' g'' carry the pieces which make contact for the auxiliary coil and for the battery terminal, respectively, with the slide wire. The sliders g and g' have half nuts h and h' actuated by springs which cause them to engage with the central screw e when not prevented by the eccentric k, and thus provide for the motion of the sliders g and g'. No screw is provided for the slider g'', which is pushed by either g or g', according as it is wished to move it to the right or to the left. According as the crank l at the end of the eccentric k is thrown to the right or to the left or remains in the central position, so is the slider g, or g', or neither, allowed to remain stationary on turning the screw e. At m and m' are shown the devices which, through moving the eccentric n, clamp and unclamp the auxiliary balancing coil d from the slider. By means of pieces of ivory the sliders g and g' are entirely insulated from the slide wire. The central slider g'' has outside its central brass core a block of hard rubber, o, into which are screwed the two pieces of the copper clamping device p. This clamp is worked by a third eccentric of copper, q, by means of a third crank, r, into the center of which is fastened, by means of set screws, the end of the battery wire.

A reference to the diagram, fig. 7, page 43, will make the working of this piece of apparatus clear. It may be stated of it in general that the sliding contacts work satisfactorily as thus constructed and that the use of the apparatus as described is a decided improvement over that of the ordinary resistance box.

A New Form of Apparatus in which Bolometer and Rheostat are Combined.

In 1898 a design was prepared at the observatory by Mr. C. G. Abbot[1] which brought all the features of the bolometric apparatus just described into a compact form. The advantages of this are obvious when they are enumerated. They consist in avoiding extra electrical connections and lead wires, in allowing the bolometer to be moved without disturbing its electrical balance, in improving the appearance of the apparatus, and making it more easy to manage by the less expert. The apparatus was constructed as designed and has been used in the research described in Chapter I, of Part II, and has justified all expectations in its performance.

---

[1] The bolometer case, as now constructed, is the product of years of slow elimination of undesirable conditions rather than of the sudden introduction of any new device; but as it is shown in Plate 12 A it represents an actual working case into which several pieces that were formerly distinct have been introduced, and not the least important of these changes is the introduction in the very case of the bolometer itself of what used to occupy a large distant resistance box, now practically dispensed with. The form of the mouth with the rock-salt cylindric lens cover will also be noticed, and also many minor changes for which the reader may be referred to the drawing. These are derived from the suggestions of much past experience, and it is difficult to say in every case to whom they belong, but so many more of them are due to Mr. C. G. Abbot than to any other that it is only just to connect his name in this way with the instrument as at present employed.

Pl. XII A is a reproduction of the working drawing of the instrument, and shows incidentally a number of frames for bolometers of different types. Before going on to describe the arrangements in detail it may be said that it was desired to secure in a single air-tight water-jacketed case (not much larger than that of the bolometer case alone, as given in Pl. XI) the bolometer, its two balancing coils, and the adjusting slide wire, with a mechanical device for operating the adjustment, and to provide the bolometer also with a cylindric lens in a ball and socket mounting and an eyepiece for focusing the spectrum upon it. All these matters were easily accomplished with the exception of providing the adjustable slide wire. It was impossible to use the device shown in Pl. XII, for this would have made the case too long and would besides have prevented the use of an eyepiece. Accordingly, a double spiral slide wire was employed, and a contact made to follow this in all its convolutions, and to rigidly connect the two spirals wherever desired, thus practically removing from the circuit all that portion of slide wire which lay beyond the contact. In order to vary the sensitiveness of the adjustment to suit different resistance coils a suitable portion of the wire of one of the coils is shunted across the terminals of the double slide wire, so that the latter becomes merely a variable shunt to modify the resistance of one of the balancing coils.

DESCRIPTION OF COMBINED BOLOMETER AND RHEOSTAT.

Figs. 1, 2, 3 show the support to the instrument, fig. 4 a longitudinal section, figs. 5, 6, 7, 8, 9, 10, 11, 17, 18, and 19 details, and figs. 12, 13, 14, 15, and 16 are views of three types of bolometer frame.

The support.—Fig. 1 is an end view from the left and shows the form of the Y and a split ring, a, which is clamped by screws to the bolometer case at b (fig. 4), and is provided with an adjustment for the rotation of the bolometer case in the Ys. Fig. 2 is the side elevation and shows an auxiliary support to the water jacket for use in the absence of the front Y. Fig. 3, an end view from the right, shows adjusting screws to act upon a lug, c (shown also in fig. 4), and designed to rotate the bolometer without moving its water jacket. The front Y is split so that it may be removed at pleasure while the water jacket remains supported. Grooves are shown in figs. 1 and 3, which fit the track of the stand on which the bolometer is to rest.

The bolometer case.—It is designed to have the chamber occupied by the bolometer air-tight and at constant temperature. This is accomplished by inclosing it in a water jacket, $d\,d$, closed at the front end (left hand in the drawing) by a circular lens, E, and at the back end by a cylinder sliding air-tight inside the water jacket and having an air-tight end at f. The bolometer is indicated at g. The cylinder just mentioned has five projecting copper posts, h, which support the bolometer and its balancing resistance coils (carried on the frame i, figs. 4 and 7), and at the same time these posts are the conductors for the passage of electric currents. Within the cylinder is a slide wire

Asti

wound in a double spiral, j, and having an adjustable clamp, k, which is caused by a suitable mechanism operated from without to follow the double slide wire, and at any point to connect to adjacent turns, thus allowing a fine adjustment to be made between the bolometer resistances and those of the two coils wound on the bobbin i. The lens E is mounted in a ball and socket mounting so as to allow its adjustment in all directions. At the front of the case is a positive eyepiece for viewing the bolometer strip.

In more detail the apparatus is as follows: The water jacket $d\ d$ is provided with two tubulures, $l\ l$, of which one is shown dotted in fig. 4, the other full in fig. 10. These tubulures are horizontal and situated on opposite sides of the water jacket and with their centers 2 cm. from its opposite ends. The inner wall of the water jacket is supported at 190 mm. from the front by a plate, m, which has two concentric rows of apertures, through the inner of which the water flows round within the coil bobbins. A conical tube, n, completes the inner wall of the water jacket. At its front end the conical tube is made cylindrical and upon this cylindrical part slides, air-tight, the socket of the ball and socket mounting to the lens.

This mounting is shown also in cross section from the right in fig. 5.[1] (All the figures, from No. 5 to No. 11, are views from the right and have the part corresponding to the top in fig. 4, on the left.) The ball of the mounting has a shoulder in its cavity, against which the lens rests, and at the center o is turned out a groove into which melted wax is poured through the opening p to render the lens air-tight in the ball.

The socket is in two parts with flanges screwed together by thumbscrews, and having an opening cut away for the insertion of a handle, q, to the ball, by means of which the position of the lens is adjusted.

Two series of diaphragms are provided for the conical tube n, and are separated at intervals of 4 cm. by short conical tubes, r. A rod, s, passing through the diaphragms, keeps their apertures in alignment.

The bolometer and coils are, as already said, supported upon five copper posts, h. Fig. 6 shows five copper segments, $t'\ t''\ t'''\ t''''\ t'''''$, which are screwed by insulated screws to the brass piece u, and are insulated from u by mica and silk and from each other by air spaces. These copper segments have holes for the insertion of the posts h. There are concentric slots passing through these holes, and all the segments except t''' have second slots outside these. The second slots are for the insertion of the wires coming from the coils through the holes in the end of the bobbin (fig. 7). Two radial screws countersunk in each segment are designed to close up both slots simultaneously and thus to clamp the coil wires and the posts h in good electrical contact.

The coil bobbin i has a projecting cylinder, v, which slips into the inside of the ring u and thus secures the coil bobbin.

---

[1] An air-tight cock was inserted at the side of the conical tube n as an afterthought. It aids in testing the tightness of the apparatus and enables the bolometer to be used in vacuo.

The bolometer is supported on three copper posts, w, having split ends. These posts are 90° apart on a circle 28 mm. in diameter, and the posts themselves are 25 mm. in diameter.

At the front end of the posts h they pass through the brass plate x and are insulated from it, and the holes are made air-tight. Copper segments, y' y'' y''' y'''' y''''', fig. 8, are soldered to the front ends of the copper posts and are insulated from the brass plate x by mica and silk ribbon and screwed to it by insulated screws. Projecting flanges, z, extend forward from the segments 13 mm., and flanges, z', forming a split tube are extended forward clear to the end of the case (138 mm.) from all the segments but y'. The four segments z' are fastened by small bolts at their front ends to the binding posts, as shown in fig. 11. The ends of the double spiral slide wire are soldered to the flanges z of the segments y' and y''. (Note all soldering is done without other flux than rosin.)

The brass plate x has short cylindrical portions projecting forward and forming its inside and outside. Within the inner flange is inserted a glass plate, f, and the outer flange is soldered into a recess of an interiorly screw-threaded cylinder, a', which slides air-tight in the water jacket. Upon the outside of the inner flange of the plate x is a brass tube, which extends to the front end of the case and within which the eyepiece fits. The front end of this tube has holes provided for the free passage of the heads of the small bolts shown in fig. 11. Between this tube and the split tube z' of fig. 8 is wound a layer of silk insulation, and outside the split tube z' is an ebonite tube, b', on which is coiled (in very shallow grooves) the double spiral of No. 16 (B & S gauge) platinoid wire j. The loop of the spiral is soldered to a projection of a copper ring, c', also shown in fig. 19. This projection, being mortised into the ebonite tube, prevents the spiral from unwinding. The pitch of the spiral is 8 mm. to a diameter of 52 mm.

Contact is made between adjacent loops of the spiral by means of a clamp like a double monkey wrench carried on a short exteriorly screw-threaded cylinder, d'. The cylinder, d' has interior grooves in which fit projections of the cylinder e', shown in perspective in figs. 17 and 18. The cylinder d' also has a copper lug, f', which projects down between adjacent turns of the spiral j. It will be seen that as the cylinder e' is turned the cylinder d' is caused to move along and the lug f' will follow the space between the wires of the spiral. It therefore only requires the locking device consisting of two steel jaws, g', sliding in guides on the lug f' and worked by a right and left hand steel screw, h' (which is held by a flange within the lug f', but with some end play), in order to connect the adjoining wires of the spiral when desired. The screw h' is turned by means of a square rod, i', along which it slides. Stops are provided on the lug f' to prevent screwing back the jaws g' so far as to touch the wires on their outsides.

The cylinder e' has a bevel gear wheel, j', at its outer end, which is moved by the small bevel k' (fig. 10). To clamp the cylinder e' in the desired position an eccentric pushes up a pin against the flange l', at the same time pulling forward a bolt whose head is within a groove in the flange l'. This device is shown at m'. The flange l' is secured by being screwed onto the piece j' and keyed.

The square rod i' has bearings in interiorly projecting parts of the cylinder e', as shown in fig. 17 and in fig. 4, and has nearly 2 mm. end play. In front the rod i' bears a gear wheel, n', engaging with an annular gear wheel cut on the wheel o'. The wheel o' turns on the outside of the hub of wheel j', but it is ordinarily constrained to rotate with it, by means of an adjustable friction disk fixed to the flange l'. When l' is locked by means of the eccentric m', o' may be turned by means of the small bevel wheel p' (fig. 10), which engages with bevel teeth cut on the outside of the wheel o'.

In order then to work the sliding locking device at k, the crank r' (fig. 10) is moved out so as to turn the arbor of the wheel p', thus turning o', n', i', and eventually the double-threaded screw h', and unscrewing the jaws g'. Then the crank r' is again returned to the position shown in fig. 10, and after loosing m' it may turn the bevels k' and j' and the cylinders e' and d', thus moving along the contact lug f' to the desired position. The eccentric m' is now again thrown over, locking this latter train, and the jaws g' are again screwed up.

The hub of the wheel j' rotates on a steel bushing, s', projecting back from the brass piece s''. The piece s'' rests on the front Y, and is secured by eight radial screws to the end of the interiorly screw-threaded cylinder a'. The four copper segments u' having the binding screws (fig. 11) are screwed to the piece s'' by insulated screws, but are insulated from s'' by a sheet of mica. These binding screws are connected with the four sections of the split tube z' (fig. 8) by copper bolts whose heads are insulated by air spaces from the tube within which the eyepiece slides.

Bolometer frames.—It has already been observed that the essence of the bolometer lies in the immediate arrangements for electrically balancing the thin strips, which themselves rest on portions that for lack of a better word we will designate as "bolometer frames," and which expose them to the radiations they are to register. Figs. 12 to 16 show bolometer frames.

The very earliest bolometers were made with a single frame. Even in the earlier construction, however, two frames were employed with every bolometer, one having certain holes the other the split pins which passed through them, and held the frames together. As at present constructed, only the linear bolometers employ but a single frame.

Figs. 12 and 13 show a frame for an area bolometer 10 mm. square. The bases are of brass with raised rims and having a central aperture 10 by 22 mm. Each base is covered on the inside by a sheet of mica glued upon it for insulating the connectors. These have each insulating cylinders of vulcanite. Copper cylinders just fitting the posts w are screwed into two of the vulcanite bushings of fig. 13 and into the third of fig. 12. Thus, upon folding the two figures together like a watch case, the three copper cylinders pass clear through the combined frame. The electric current entering at a'' (fig. 12) passes through the thin insulated copper piece and divides. One portion from the upper corner of this piece crosses (on the bolometer strips not shown) over to the small insulated piece l'' and thence similarly back and forth to f''; thence through the split pin to the corresponding cylinder d'' of fig. 13; thence to g'', through the pin and cylinder at e'', and finally leaves at k'' (fig. 12). The other branch of the current at a'' passes back and forth to c'', thence to f'', and finally leaves through the copper piece h'' at the point i''. All the small copper pieces like l'' are fastened with a single flush insulated pin. The conductors like h'' are to be soldered to their cylinders or pins. No screw clamp electrical connections are tolerated. The strips are soldered in at the observatory and are each 1 mm. wide.

Figs. 14 and 15 show similar halves of a frame for bolometer 10 by 5 mm.

Fig. 16 shows the single frame required for the ordinary linear bolometer. In this the side copper piece has a sliding adjustment of about 1 mm., as shown, the copper piece being screwed down by a nut upon a shoulder of the copper cylinder, but when the strip has been inserted this joint also is soldered. The other two copper pieces are soldered to their cylinders. Rosin only is used for flux in soldering.

DRIVING CLOCK.

The earliest use of the bolometer in the spectrum was effected by letting different portions of the spectrum successively fall upon it, while at least three persons were needed to note the effect, one person being seated at the circle which carried the prism, and reading aloud the degrees and minutes which indicated the part of the spectral heat which fell on the bolometer strips, another person reading the simultaneous deflection of the galvanometer, and a third recording both. There were, in fact, usually more than three employed, and the process of measuring the position and intensity of an invisible absorption band in the infra-red was excessively slow, owing not only to the difficulties of the method of reading, but still more to the fact that the galvanometer needle was in incessant movement upon the scale, even when the same portion of spectral heat was falling upon the strips, this motion being due to causes extraneous to the observation properly in view. This incessant wandering of the needle, known technically as "drift," was the reason why an automatic method of registry was not earlier introduced, but its desirability was from the first apparent.

Some sixteen years since, when the fluctuations of the "drift," though not eliminated, had been reduced, the first attempt at automatic registry was made at Allegheny by preparing a sheet of sensitive paper which moved vertically in the focus of the galvanometer mirror, the vertical movement being rendered synchronous with the movement of the circle of the spectrobolometer; but then, and for a considerable

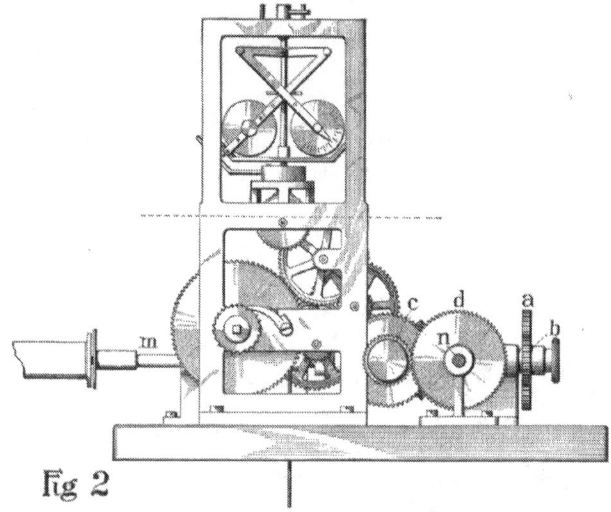

Fig 2

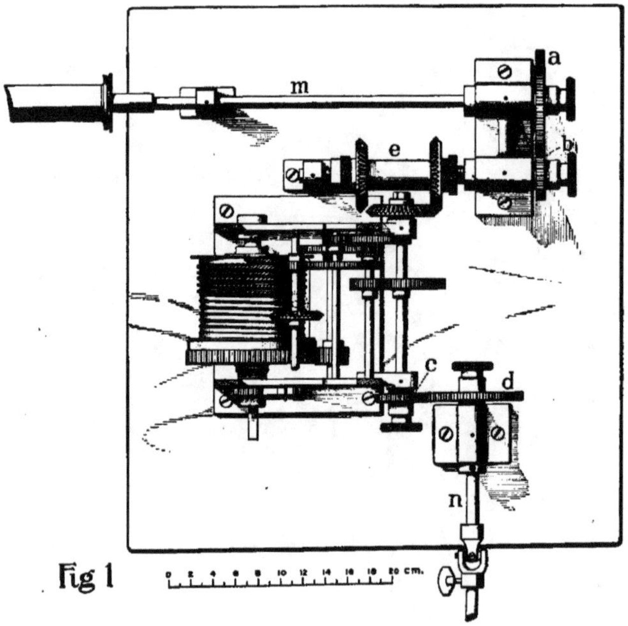

Fig 1

THE DRIVING CLOCK OF THE SPECTROBOLOMETER.

time, the difficulties of any exact registry continued to be insuperable. Subsequently at Washington, the first attempts at a more perfect control were made, by employing a cylinder, like a chronograph cylinder, carrying the sheet of prepared paper, the movement of the cylinder and of the distant circle of the spectrobolometer being independently controlled by two similar clocks which were synchronized by electricity. Even these attempts, however, were crowned with only very partial success, but they led to better things. As the other portions of the apparatus were improved and the accidental deflections grew less, these methods of synchronizing were definitely abandoned, though with great reluctance, in favor of a radical measure which had always been postponed on account of its inconvenience, that of connecting the circle mechanism and that vertically moving record (now a plate of glass) directly by a shaft through the observatory, which insured absolute synchronism, but only at the expense of great inconvenience, the disposition of the floor of an observatory not being adapted to such an arrangement as that of a machine shop might be.

I have given here but the most imperfect sketch of the process of a development which, like that of the bolometer, went on for many years and which finally resulted in the apparatus now described. The method will be understood from what has preceded, and has been fully described elsewhere.

Briefly, it is essential that every movement of the great circle carrying the prism and sending a ray into the bolometer shall be accompanied by an absolutely synchronous motion of the plate which receives the record of the galvanometer swing, this swing being actuated by the heat of the spectrum in the part under examination. If there be no irregular motion of the galvanometer due to external causes (e. g. to the drift or to tremors due to electric influence or to the jarring of the ground without), then we shall have with perfect synchronism a perfect record on this plate of whatever is transpiring, for there will be once for all established a definite relationship between the movement of the circle and the movement of the plate. If the circle, for instance, moves through one minute of arc in one minute of time, the plate during that one minute may move through one centimeter, while, by the use of change wheels in the clock, another definite relationship may be set up. This is the essential character of the mechanism which is controlled by the driving clock about to be described.

The driving clock, by Warner & Swasey, which moves the spectrometer and recording plate synchronously, is shown in plan and elevation in figs. 1 and 2, Pl. XIII.

As will be seen on inspection of the illustrations, the shafts m and n, which lead, respectively, to the spectrometer and plate-carrier mechanisms—and which appear with the same lettering in the illustrations of these instruments—are connected by a continuous system of gearing entirely independent of the clock proper. The function of the clock is simply to drive one of the shafts of this connecting system at the proper rate of speed; but a failure on its part to regulate the speed of this motion would not in the slightest degree affect the relative movements of the plate and the circle.

The gears a, b, c, and d can be readily removed and replaced by others of different sizes. The speeds thus at command are as follows:

TABLE 3.—*Plate and spectrum speeds.*

Spectrum speed (=twice circle speed).	Plate speed.
1' of arc in $\frac{1}{4}$ minute time.	1 cm. plate in $\frac{1}{4}$ minute time.
1' of arc in $\frac{1}{3}$ minute time.	1 cm. plate in $\frac{1}{3}$ minute time.
1' of arc in $\frac{1}{2}$ minute time.	1 cm. plate in $\frac{1}{2}$ minute time.
1' of arc in 1 minute time.	1 cm. plate in 1 minute time.
1' of arc in 2 minutes time.	1 cm. plate in 2 minutes time.

As these speeds of circle and plate are independent, it follows that there are twenty-five possible arrangements of the circle and plate motions. Of these, the ratio and speed 1' spectrum = 1 minute time = 1 cm. plate has been most used in these researches.

Provision is shown at e in the plan for means of reversing the direction of the circle motion. A similar device is shown in the view of the plate carrier for reversing the motion of the plate. Thus it is possible to take two exactly similar bolographs in opposite directions through the spectrum, and hence to have a means of determining if any appreciable displacement of the deflections is inherent in the bolographic method, and to correct the results for such displacements if they exist.

PLATE CARRIER.

A plan view and rear elevation of the plate-carrying mechanism are shown in Pl. XIV, figs. 1 and 2. This piece of apparatus is supported upright upon the end of the galvanometer pier, and is inclosed in a dark box, which has two slides for the adjustment of the plate. The back of the box is removed in the illustration in order to show the mechanism more clearly.

The plate is held fast between the clamps a a and slides vertically on the guide rod. The slider c has affixed to it a piece, d, carrying a tongue, e, which is pressed by a spring into the groove of the screw f, of 1 cm. pitch. This screw is moved by a reversible arrangement g precisely similar to the one shown more clearly in the illustration of the clock. The shaft n, of which the other end is shown in the illustration of the clock, is provided with a double Hooke's joint to prevent straining of the mechanism.

The parallel-motion slit h at right angles with the motion of the plate furnishes the entrance for the beam of light reflected from the galvanometer mirror.

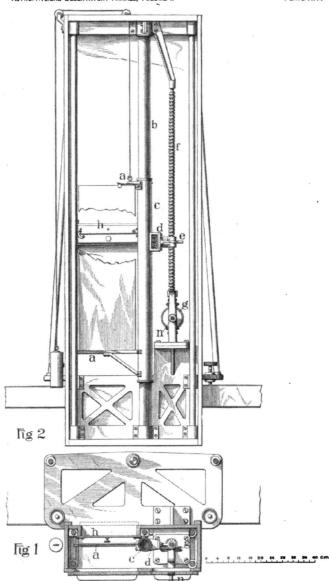

THE PHOTOGRAPHIC RECORDING APPARATUS OF THE SPECTROBOLOMETER.

GALVANOMETER AND ACCESSORIES.

The first galvanometer employed by the writer antedated the use of the bolometer itself, having been one of a very early Thomson pattern made by Elliott. The mirror was about a centimeter in diameter, and this, with its magnets and the suspending stem, would now be considered preposterously heavy, while the mirror was suspended by a raw silk fiber of something like an inch in length, which twisted every way, through day and night, without ascertainable cause, and produced an amount of exasperating irregularities, which only those who remember the use of these early instruments can sympathize with.

My first considerable improvement at Allegheny was due to the introduction of a longer fiber. One of over 12 inches in length was employed, I think at the suggestion of Professor Rowland of Baltimore. The day of Boys' quartz suspension fiber had not yet come, and the young physicist of to-day will never know what he owes to it. The small mirrors of Boys also were still in the future. The needle system in use at Allegheny at the time of the improved four-coiled Thomson galvanometer (by White, of Glasgow, with the kind oversight of Lord Kelvin, then Sir William Thomson) was considered then to be a marvel of lightness when taking into account its elaborate system of magnets. The whole (magnets, mirror, tube, and damper) weighed about 300 milligrams, and it owed this lightness in a considerable measure to the absence of the heavy metal vane, in place of which the writer introduced a wonderfully light and stiff construction which Nature herself furnished in the shape of the wing of the dragon fly.

Up to the time of the early work of the Smithsonian Observatory, in 1891, the arrangements of the galvanometer had been made entirely by the writer. The subsequent improvements in it have been partly due to him, but increasingly to the others who have been associated with him in the observatory in subsequent years, and they have been so continuous that it is difficult to note any single improvement. Perhaps for the most essential ones credit must in fairness be given to Professor Boys, whose quartz threads and especially whose small mirrors are so well known for such purposes that the writer need only express his acknowledgments for their help in putting an end to such troubles as the present generation of electricians engaged in like work no longer knows. In the final improvements here represented the writer has had only a minor share.

While that which the writer brought here from Allegheny in 1888 and completed the installment of himself in the year 1891 has been the basis on which subsequent improvements have been built up, the most considerable changes have been made within the past three years.

The following conditions of sensitiveness of the galvanometer are always to be taken with the consideration that they are more or less illusory, unless we can be sure of such an immunity from tremor and drift as has only come in the last four years.

It is convenient here to note the actual results in progressive years, remembering that those given as being actually attained at present are almost incomparably more certain than the previous ones. The values given as the galvanometer "constants" are the currents required to give a deflection of 1 mm. on a scale at 1 meter with a time of single swing of the needle of 10 seconds. The resistance in each case is about 20 ohms.

Comparative sensitiveness of galvanometers.

Name of galvanometer.	Date.	Galvanometer constant.
Allegheny, No. 2	1881	0.000000240000
Allegheny, No. 3	1885	0.000000002200
Smithsonian, No. 1	1892	0.000000001500
Smithsonian, No. 2	1894	0.000000000200
Smithsonian, No. 2	1896	0.000000000100
Smithsonian, No. 3	1897	0.000000000020
Smithsonian, No. 3, with lighter needle[1]	1897	0.000000000005

[1] This light needle, of 2.5 mgs. weight, was found somewhat too unsteady for bolographic purposes.

It follows from the above statement that the galvanometer which the writer used in his earliest bolometric work at Allegheny was successively improved till the then latest form of needle and suspension used by him in the first galvanometer at Washington was over two hundred times as sensitive, and that successive improvements here have made the one at present used seventy-five times more sensitive still, or that the whole increase has been over ten thousand times.

The present arrangement of apparatus on the galvanometer pier is figured in Pl. XV. Two Thomson reflecting galvanometers have been employed, both of which have been constructed after designs prepared at the observatory. The principal data of the earlier of these forms, known here as "No. 2," are as follows:

The four coils are wound each in five sections, of such shape and of such wire as to exert about equal influence on the needle. The maximum radius is 50 mm., minimum radius $2\frac{1}{2}$ mm., and extreme thickness 40 mm. A copper rod is inserted in the center of each coil for damping purposes, but is supplemented by a damping vane on the needle, as later described. The resistances of the coils are about 21 ohms each, and they are at present connected in two series of two each in parallel, giving a total resistance of about 21 ohms. The case is of brass, with tightly fitting glass sides, and with a brass tube to support the suspension and overhead astaticising magnet.

The astaticising magnet a is provided with slow motions, by means of which it may be raised or lowered or rotated about a vertical axis. Its function is to reduce the time of swing of the astatic needle from about ten seconds to four or five seconds as it is employed, and to direct it so that the beam of light reflected by the mirror falls on

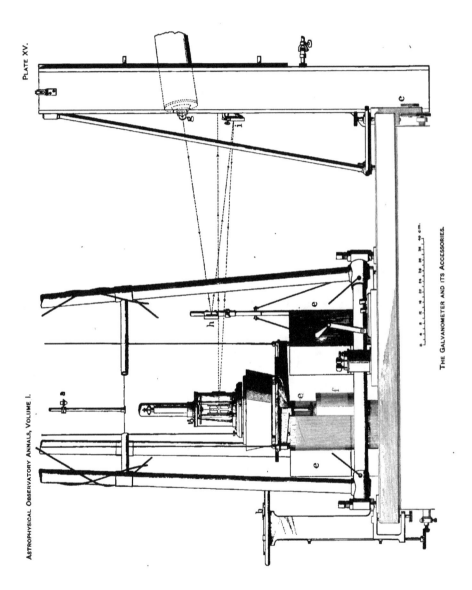

THE GALVANOMETER AND ITS ACCESSORIES.

the scale. A finer adjustment of the spot of light upon the scale is effected by means of two very weak magnets placed at b directly behind the galvanometer and pointing east and west with their unlike poles together. A slow motion worked from c at the east end of the galvanometer pier moves these magnets east and west in opposite directions and thus serves to change the direction of the needle as desired.

GALVANOMETER NEEDLE.

The needle is suspended by means of a quartz fiber 55 cm. long hanging within the glass tube d, which is itself supported within the larger brass tube by means of two rubber rings near the top and bottom. A sketch of the needle is shown in fig. 9,

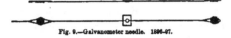

Fig. 9.—Galvanometer needle. 1896-97.

though of course in incorrect proportion, as the stem and other parts are quite too fine for delineation.

Its distance between the centers of the magnet systems is 120 mm. The two systems are of six magnets each, of such length as to make a nearly circular outline 3.5 mm. in diameter. These magnets were made from hardened No. 12 sewing needles, magnetized when in place on the needle by means of an electromagnet. They are fastened with shellac to thin pieces of mica on one side only, and the glass rod and quartz fiber, being fastened similarly on the same side, the moment of inertia is reduced to a minimum. The concave mirror is of silvered glass, 2.5 mm. in diameter, and was constructed by Mr. J. A. Brashear. Under the usual temperature conditions its definition is very good, but if the temperature is allowed to fall below 16° C. the beeswax with which it is fastened to the stem strains the mirror slightly and the focus is appreciably changed. Behind the mirror is fixed with beeswax a piece of "dragon-fly" wing, 7 mm. wide and 4 mm. high, which completes the damping. The weights of the separate parts are as follows:

	Milligrams.
Magnet systems with mica and shellac, each	11.3
Glass stem	6.0
Mirror	2.0
Vane	0.3
Total completed weight	31.3

Several other needles have been prepared, but only one other as sensitive as this one. This latter weighed 40 mgs. and was disposed as follows:

Magnet systems, each of one large magnet 4 mm. long, composed of Stubb's steel wire, and having three additional very small magnets of watch spring used in attaining sufficient astaticism. Each of the large magnets weighed 13.4 mgs. Weight of stem, 4 mgs.; of mirror, 3.8 mgs.

All the other needles which have been compared were found only about two-thirds as sensitive as these two, and included two weighing 56 mgs. each, and one of 19 mgs. weight. Each was composed of twenty magnets, those of the last-mentioned needle being of watch spring 2 mm. long and weighing altogether 15 mgs.

The much more sensitive galvanometer used in 1897, 1898, and 1899, which appears in Pl. XV, is described more at length in Part II.

Mounting of the Galvanometer.

It will be remembered that a great difficulty of the research for which the galvanometer is to be used arises from the unsuitable location of the observatory, surrounded as it is by streets filled with traffic, the jar from which has introduced in previous years such prejudicial irregularities in the galvanometer record that it seemed at times that either the site or the research must be abandoned. It has been a matter of the very first consequence to diminish these accidental vibrations, but this has seemed impossible until a comparatively recent plan proposed by Mr. Abbot has effected an almost unhoped for improvement.

Julius[1] has suggested the advantages of suspending the apparatus by wires in such a manner that the center of gravity may coincide with the point of support. It has been unadvisable here, however, to suspend the apparatus from the ceiling. A support of three heavy brass tubes, each about $2\frac{1}{2}$ inches in diameter and about 10 feet in length, has been built up from the galvanometer pier, and from these the galvanometer was suspended. The result was at first disappointing, as no greater steadiness was apparent than before, but attention being drawn to the sonority of the supporting tubes, the experiment was tried of wrapping them with India rubber until this sonority was deadened, when all other vibrations were found to be proportionally extinguished. The ground vibrations have not been absolutely eliminated, but they have been reduced to a small fraction of their former amount, to the very great improvement of the record. Too much, then, can hardly be said of the advantages of this method of the suspension in the actual case.

The galvanometer rests upon a brass plate suspended by three brass wires, 220 cm. in length and 3 mm. in diameter, from the summit of the tripod, which is well braced by cross wires and rods, and is supported on rubber blocks which rest upon the pier. Rubber blocks also support the galvanometer on the suspended table, and rubber washers are placed at the joints in the wires to further deaden vibrations.

Three oil tanks, $e\ e\ e$, filled with heavy mineral oil, in which dip right-angle vanes, prevent circular vibrations of the galvanometer table. The center of gravity of the suspended parts was experimentally adjusted by means of the lead weight f to secure the most satisfactory results in the steadiness of the needle.

---

[1] W. H. Julius, Annalen der Physik und Chemie, neue Folge, vol. 56, p. 161, 1895.

Still further efforts to reduce earth vibrations by the expedient of floating the galvanometer on mercury were made in December, 1897, and resulted in a slight improvement. Two heavy lead pans were made, one fitting within the other, and both were covered with hard wax, turned and still again coated with plaster of paris. The lower pan rests on the swinging table and the upper, floating in it, carries the galvanometer. Connection with the galvonometer switch is made through mercury cups, so that the galvanometer has now no rigid connection with anything. The demands on the observer's patience, when through some mischance the needle sticks or turns round, are extraordinary. Before practice was had in dislodging it, an hour's work or more was sometimes necessary. It now rarely takes more than five minutes for the practiced hand. An idea of the improvement in steadiness resulting from the introduction of the Julius suspension and of mercury floating may be gained by an examination of Pl. XVII.

At the right of Pl. XV is shown in side view the plate carrier, which has already been described. The course of the solar ray is from the small slit g to the plane mirror h, interposed to adjust the situation of the conjugate focus; thence to the plane mirror i; thence to the galvanometer, and finally to the horizontal slit in the front of the plate carrier, as shown by the dotted lines and arrows.

BATTERY.

To one not experimentally acquainted with the photographic recording of bolometric indications. the degree of immunity from the slightest fluctuations required of the electric current employed is surprising. The disproportionate deflections of the galvanometer, caused by small changes of the battery current, are doubtless to be traced to inequalities in the capacity for the radiation and conduction of heat of the very small bolometer strips. Former experiments of the writer[1] show that such small strips sustain the heating effects of a current of 0.05 ampere (about that here employed) without being raised in temperature more than 1 or 2 degrees. It might, however, readily be that owing to a slight difference in the character of their surfaces a change in the current in one part in a hundred thousand might effect a difference in the temperatures of the two strips of a millionth part of their excess of temperature over their surroundings, and such a change would immediately be recorded by the galvanometer. At all events, the best constructed bolometers show such effects, and it has been necessary to take great pains to secure a constant current.

A battery of sixty storage cells entirely inclosed in a wooden case, which is well wrapped with cloth to avoid sudden temperature fluctuations, supplies the current which by external resistance and by that of the bolometer circuit is reduced to 0.1 ampere. The cells of the battery are placed in three sliding trays, one above the other in the case, so that they may be moved out and refilled or inspected.

---

[1] S. P. Langley, Proceedings of American Academy of Arts and Sciences, vol. 16, p. 345, 1881.

Copper wires are soldered to the lead terminals and pass up to a switch on the top of the case, where they are soldered—the positive in one row, the negative in another—to copper springs or brushes pressing hard on opposite sides of a wooden cylinder. Two copper bars run lengthwise on opposite sides of this cylinder, and thus connect all the cells in parallel for discharge. By turning the cylinder a quarter turn, a system of brass bars makes contact with the brushes and connects the cells in two series of thirty each for charging from the 110-volt dynamo. No. 16 B. & S. gauge copper wire, doubly insulated, runs from the battery (which is, on account of the fumes attending its charging, in the basement) to the milliammeter, series resistance, and bolometric circuit, all in the spectrometer room at constant temperature. Very recently a battery of ten cells of Type I of the so-called "Cupron element" has been employed. This type of cell is a modification of the Lalande cell. The performance of the battery has been so far very satisfactory, and is perhaps better than that of the sixty cells of storage in freedom from minute irregularities; but the difference, if any, is very slight. The Cupron battery has the disadvantage that it is more troublesome to maintain, for in reoxidizing the copper oxide plates they must be removed from the cell and dried, and the disagreeable task of amalgamating the zincs also has to be attended to. This battery has, on the other hand, the advantage that no fumes come from it, so that it has been set up in the spectrobolometer room itself, where the temperature is constant.

COMPARATOR.

Among the recent accessions to the equipment of the observatory is the comparator by Warner & Swasey, illustrated from a photograph in Pl. XVI. This instrument was specially constructed for the purpose of making rapid and accurate measurements in abscissæ and ordinates on glass plates 60 cm. long and 20 cm. wide, and has been found very convenient for the purpose.

The plate to be measured rests at an angle of 45° in a frame with glass bottom, supported through six adjusting screws by the base casting. Upon accurate ways at the front and back of the casting slides a carriage containing the reading microscope. Cross-ways are provided upon this carriage, accurately at right angles to the ways upon which the carriage slides. The motions of the microscope are roughly accomplished by hand, but fine adjustments are perfected by screws parallel to the axes of ordinates and abscissæ, and which engage in half nuts actuated by springs. Readings are made to hundredths of a millimeter by verniers presenting continuous plane surfaces with the scales of ordinates and abscissæ, which are graduated to half millimeters.

By means of four screws the whole surface of the plate is adjusted to be in focus for the reading microscope; but the construction of the latter is such that a pair of cross wires immediately above the plate are also in fair focus. To facilitate measurements, still another long wire crosses between the plate and these cross wires

THE COMPARATOR.

exactly parallel to the cross-motion ways. The wire is of great utility, as it enables the observer to set very closely in abscissæ without looking in the microscope, so that it takes only a few seconds to set exactly upon any point of a curve. As in the measurement of bolographs, to which this instrument is put, it is of no importance to measure ordinates with very great accuracy, it can be readily seen that a pair of cross wires in the microscope play but a very subordinate part compared with this wire. Error from parallax is quite negligible when a small aperture of the eyepiece is used, since the length of the microscope is about 20 cm.

ADJUSTMENT OF APPARATUS.

SIDEROSTAT.

For the present research it is not of paramount importance that the solar beam should be quite motionless, nor consequently that the running of the siderostat should be of the highest uniformity, for the width of the beam and the focal lengths of the mirrors are such that an angular motion east or west of 10' of arc in the position of the beam will hardly affect the deflection of the galvanometer, and will, of course, in no way whatever affect the position of a spectral line. With the present governor, which is similar to that of the circle clock, so great a variation as this need not occur in an hour if the siderostat remains untouched. Yet, for purposes of occasional adjustment in right ascension (and in declination also at certain seasons of the year), reversible electric motors, which give a slow motion in these directions, may be operated from within the observatory through suitable electric circuits. Such adjustments have been made as often as once in twenty minutes, and each such adjustment is recorded, together with the instant of its occurrence, in the notes accompanying the experiment.

SLIT.

As the slit is accurately at right angles to the upper face of its mounting it is sufficient to level the latter to insure that the edge of the slit is in a vertical plane. Provision has been made for motion about a vertical axis, so as to bring the faces of the slit jaws at right angles with the medial plane of the beam, and this adjustment is made by one person, while another observes the definition of the spectroscope and indicates when the seeing is at its best. In setting the slit for widths under 0.30 mm. the slit should be always reduced to the width desired to avoid error from backlash.

MIRROR SYSTEM.

Referring to the diagram, fig. 8, page 44, the mirror c is first turned so as to cause the beam to fall in the proper position on the mirror d when the latter is roughly adjusted in distance from c for making the beam parallel in a horizontal plane.[1] The prism is then turned to the approximate angle of incidence for minimum deviation in the part of the spectrum under investigation, and the mirror c adjusted so as to cause the beam to cover the prism face.

To render the prism mirror vertical a carefully adjusted theodolite is next leveled at the same height as the prism (being placed for this purpose on the northwest pier), and the filament of an electric light is placed in such a position as to form an image tangent to the horizontal cross hair of the theodolite and to be visible at the same time from the prism table. The latter is now turned about so as to cause the prism mirror f to reflect the image of the lamp filament upon the theodolite, and the leveling screws under the prism table are adjusted so as to bring this image again tangent to the theodolite horizontal cross hair. The beam is leveled by turning the screw back of the mirror d until the center of the light spot falling on the opposite wall of the spectrometer house is on a level with the center of the mirror d^2.

To adjust the mirrors exactly for collimating the beam d is moved back and forward upon its track till the image of the slit is made as sharp as possible at the slit itself, when the beam from d will be parallel in its horizontal section. To insure a minimum of deviation (see discussion below) the prism is for the moment removed, and the mirror g and the bolometer are so adjusted that the beam from the mirror d, falling directly upon g, is brought to a focus upon the bolometer strip.

PRISM.

Next, putting the prism in place upon its table, the theodolite and lamp filament are again brought in use to make its faces vertical. The leveling screw under the end of the plane mirror f (see also Pl. IX) and the screw t at the back of the prism are employed in this adjustment. Owing to the disposition of the three leveling screws s the motion of the one indicated does not throw the prism mirror sensibly out of a vertical plane.

Passing next to the consideration of minimum deviation, let $f\ m\ n$ in fig. 10 represent a section of an isosceles prism, with its base parallel to the plane mirror $n\ p$, and let $a\ b\ c\ d\ e$ represent a ray of such refractive index that it emerges in the direction $d\ e$ parallel to its original direction $a\ b$. Under these circumstances, since the angles $p\ d\ e$, $c\ d\ n$, and $m\ o\ b$ are all equal, the angle of incidence of the ray upon the prism will equal the angle of emergence, and the ray will pass in minimum deviation. Draw $f\ g$, $l\ h$, and $k\ h$, respectively, perpendicular to the three faces of the prism. Then, since the triangles $f\ g\ m$, $l\ g\ h$, and $k\ g\ h$ are similar, the angles $l\ h\ g$ and $k\ h\ g$

---

[1] See discussion of collimation at a later page.

[2] With spherical collimating mirrors an easy and accurate way to level the beam was as follows: Having thus made the mirror f vertical, it is set so as to be normal to the beam from the slit, and the slit itself is turned horizontal in its Ys. The mirror d is then slightly rotated by its adjusting screw about its horizontal axis till the slit image falls on the slit, thus insuring that the collimated beam is horizontal. This method is impossible with cylindrical collimating mirrors.

will be equal each to each, and equal to $(90° - \frac{1}{2}\alpha)$ where α is the refracting angle of the prism, or $n f m$.

Applying these considerations to the prism mirror combination of the observatory, it is clear that, since the image-forming mirror is adjusted to bring all rays parallel to the beam from the mirror d (fig. 8, p. 44) to a focus on the bolometer strip, any ray which falls upon the latter will have passed the prism in minimum deviation if the bisector of the refracting angle of the prism is perpendicular to the prism mirror. This condition is secured by bringing the image of the lamp filament reflected from the prism mirror tangent to the vertical cross hair of the theodolite, then turning the circle through $(90° - \frac{1}{2}\alpha)$ and slightly adjusting the prism till the image reflected from its face is also tangent to the cross hair.

It will be later shown that the accuracy of this theodolite adjustment is quite sufficient for the purpose. The theodolite and electric light are used in preference to a larger telescope and the slit, because it is not possible to use the mirror face to reflect an image of the slit when the beam is in adjustment. For measuring the refracting angle of the prism, however, the position of the beam is altered, and a 3-inch telescope is used by the observer.

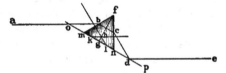

FIG. 10.—Adjustments of prism in fixed arm spectroscope.

Referring again to fig. 10, it may be seen that the minimum deviation of the ray under consideration is equal to twice the angle $m\,o\,b$; and therefore, if the circle be turned from its present position through $(90° + \frac{1}{2} D)$, where D is the angle of minimum deviation, the plane mirror will then be perpendicular to the beam $a\,b$. In order, therefore, to measure angles of minimum deviation, the circle is turned until the prism mirror reflects the beam back on its own path and forms an image of the slit upon the slit. A reading R_0 being then made, the minimum deviation of any ray subsequently falling upon the bolometer strip is given from a circle reading R_1 by the relation:

$$D = 2\left((R_0 - R_1) - 90°\right)$$

A discussion of the probable error of such observations will subsequently be given.

CYLINDRIC LENS.

The height of the spectrum, or rather the length of the spectral lines, depends on the height of the slit, focal lengths of the mirrors $c\,d$ and g (fig. 8, p. 44), or, in the case of the cylindric mirror collimation, only on the height of the slit, focal length of g, and total distance traversed by the beam in passing from the slit to g. These latter conditions are such that the length of the spectral lines would be about 5 cm., or some five times the height of the bolometer strip. A rock-salt cylindric lens with its elements horizontal is therefore interposed in front of the bolometer to reduce the height of the spectrum to that of the bolometer strip. The ball and socket mounting of this lens allows of its rotation about a vertical and two horizontal axes.

At each day of use it is desirable, but usually not indispensable, to perform the adjustment now to be described. The observer stations himself at the eyepiece of the bolometer and focuses upon the bolometer strip. He then brings a visible spectrum line (which we may suppose to be A) into the field, and having removed the cylindric lens from the front of the bolometer case, moves the bolometer back and forth to obtain good definition, and brings the line under observation to the edge of the bolometer strip. He then inserts the cylindric lens again, roughly adjusts it about its horizontal axes for fair definition, and if the line does not now coincide with the bolometer strip, he turns the lens about its vertical axis till it does so, and is then confident that in the infra-red, since the deviation of the lens is less there than it is where he is observing, no lateral shifting will occur. The lens is now turned about a horizontal axis at right angles with the beam till the line appears equally sharp at top and bottom, and finally about a horizontal axis parallel with the beam to secure the best definition. As the lens has appreciable thickness the bolometer will now require a slight movement along its track to bring the strip exactly in the focal plane of the spectral lines.

As a test of the adjustment of the whole apparatus, the spectrum is now turned until the bolometer strip bisects the head of the A line, which is the datum point for the investigation. Under constant temperature conditions circle readings on A usually agree for several weeks within 30'', no change in the permanent adjustments of the mirrors and prism having in the meantime been made.

Chapter IV.

METHODS OF PROCEDURE IN PREPARING AND COMPARING BOLOGRAPHS.

THE BOLOGRAPHIC PROCESSES.

Before entering on the technical description which follows in the next two chapters, the reader may recall the very simple theory of the action of the clock-driven spectrobolometer which is so complex in practice. This was devised by the writer fifteen years ago at Allegheny, and slowly developed to its present condition. In theory it is only needful that the galvanometer mirror should, by means of a ray of light, affect a photographic plate before it, which is moved steadily upward or downward at a rate proportional to the movement of the clockwork actuating the prism carrying the circle, in order that the record of every solar line within the reach of the instrument's optical capacity shall be made. This is all that is necessary in theory and all that is necessary in practice to obtain the principal and most interesting divisions of the infra-red spectrum, and indeed would be for all of those of secondary importance, except for the minute tremors of the ground and the piers on which the instruments rest, tremors so minute as to be negligible until we have developed the method so that all considerable indications of solar or terrestrial absorption of what we may call the second order have been secured; we then reach a point where the minute ground or magnetic tremors, which exist under every condition, are of such an order of magnitude, as compared with those from veritable spectral causes, that extreme precautions become necessary both in preparing and in interpreting the curves to avoid the introduction of error. The nature of these precautions will be noted by the reader in the following description of procedure:

1. PREPARING THE CURVES.

Let us now suppose that the siderostat clock has been started, the beam directed so as to fall correctly on the prism, the slit set at the proper width, both the permanent and daily adjustments of the optical apparatus perfected, the bolometer circuit balanced, the spot of light reflected from the galvanometer mirror made to fall upon the slit in the plate-exposing apparatus, and that darkness prevails in the apparatus chambers, so that all is ready to actually make a bolograph. Two observers are in general employed, though not continuously, and it is possible, though not advantageous, for one to do all that is necessary.

Observer A, by the light of a small incandescent lamp, sets the circle to the proper point (e. g., just above the A line) for the start, reads the two thermometers near the circle, one of which is in air the other being inserted in a great prism of rock salt (of nearly the size of that in optical use but of inferior quality), and reads the milliammeter in the battery circuit, calling out all these observations to B, who records them, along with the width of slit and other data, in the book kept for the purpose A then puts a plate in the exposing apparatus, sets its end in front of the exposing slit, and marks a line parallel to the axis of ordinates across the plate by causing the light spot reflected by the galvanometer mirror to travel back and forth over its width several times.

At a signal from B he starts the clock, and exposes the plate for a run of several minutes (but without letting any radiations fall on the bolometer strip) to give a record of the amount of local disturbance.[1] When this record is closed at another signal from B, the plate is set back to its end and the light spot brought to its edge. A then, as quickly and at the same time as quietly as may be, throws the eccentric at the spectrometer (Pl. X), thus putting the circle in gear.[2] Returning to the galvanometer room, after two or three minutes which are suffered to elapse that the disturbance of temperature about the bolometer may subside, at a signal from B, A starts the clock and exposes the plate, and at a second signal, after ten or fifteen seconds, drops the shutter from the path of the beam, thus exposing the bolometer.

A's part is now done till the end of the run, an hour or other interval later according to the plate speed, except that it is wise, unless he be absolutely sure of the behavior of the apparatus, for him at some critical points in the curve to cautiously reenter the galvanometer room, and observing the spot of light from a distance in a mirror provided for the purpose, to see that it has not approached too near the edge of the plate. If such a state of affairs should be found, A adjusts the spot by turning the milled head at the east end of the galvanometer pier, as already described.

B meanwhile takes care that the beam of the light from the siderostat is properly directed, observing for this purpose the auxiliary beam of light as it falls on the mirror on the north end of the spectrometer chamber. He also notes passing clouds or other unusual phenomena likely to disturb the curve. At the end of the time alloted to the bolograph he again gives two signals to A, who closes the shutter and stops the

---

[1] This very important indication of the amount of minute local disturbance, as compared with the deflections due to minor solar lines, was not introduced as a part of every bolographic record until March, 1896.

[2] A recent alteration of the driving clock, described in Chapter I of Part II, gives the observer entire control and exact knowledge of the position of the circle without leaving the galvanometer room.

clock. Another instrumental record curve of two or three minutes is taken, the plate is marked, and is then ready for developing. In bolographs of the spectrum below Ω it is usual to open the slit to about four times its first width while passing the absorption band X, and again to twice this latter width at Y, so as to give suitable deflections in the regions of comparatively little energy. Observer B makes these changes and records the instant of their occurrence.

The bolographs are examined and graded as described in the note on page 30. Nos. 1, 2, etc., refer to the state of the sky, 1 being very unusual and indicating a cloudless dark-blue sky. Letters a, b, etc., refer to instrumental conditions and are now chiefly concerned with accidental deflections in the record curves. In the earlier work "drift" played a very important part compared with what it does at present, but it is now practically negligible.

2. METHODS OF IDENTIFICATION OF ABSORPTION LINES.

It has already been remarked that the great difficulty of these later years of this research has lain in the unfavorable site in which it has been carried on, and in the consequent introduction of local disturbances which tend to be confused with those (caused by absorption in the solar and terrestrial atmosphere) which we are studying.[1]

If there were no such disturbances, the so-called battery record would be a smooth line, and the smallest inflection of this would denote a genuine case of solar or telluric absorption. Let there be, on the other hand, an extremely minute disturbance of this battery record (let us suppose, for instance, that an accidental deflection of a tenth of a millimeter is as likely to be found as not, but that this accidental deflection never exceeds a half millimeter); then it is equally clear that any deflections exceeding half a millimeter are to be set down with confidence as "real." As a matter of fact, in 1896 the battery record on any but unfavorable occasions had been brought, by the introduction of the suspension of the galvanometer and other improvements, to a condition where the mean deflections were between one-tenth and two-tenths of a millimeter. With the present much more sensitive galvanometer they are somewhat larger, averaging about four-tenths of a millimeter. The magnitude of these accidental deflections, as has been explained, is registered on each plate at the commencement and end of each observation.

There is a familiar empirical rule in use with the application of the doctrine of least squares, under which the computer feels authorized to treat any deviation of five times or more the amount of the probable accidental error as not due to chance. There is then at least a very strong presumption that where the average deviation due

---

[1] Possible disturbances from passing clouds are not here in question. Visible clouds are recognized by the observer. Others, practically invisible, and which yet obstruct heat radiations, have been long ago referred to by the writer. The effect of these is to produce occasional deflections of a somewhat greater magnitude than those of the battery, but which do not appear on every plate, nor, of course, in the same position.

to accident lies within one-tenth of a millimeter, a deviation exceeding five times that amount is not due to chance, and this would be the case if it were observed only on a single plate. When, however, we find this exceptional deviation repeated in the same position and magnitude on six or more plates, we are justified in considering that the presumption amounts to practical certitude and that the deflection is "real," *i. e.*, due to solar or telluric atmospheric disturbance.

We will then, under the given conditions, decide as a practical rule that deflections of five times the amplitude of the average accidental deflection, when identified in position and magnitude on several independent plates, are "real." In pursuance of this rule, deflection of 1 millimeter occurring on the plates of 1896–97 and deflections of 2 millimeters occurring on those of 1897–98 are, when identified on several plates, to be treated as real without examination. Below this magnitude it must be evident that very many real ones exist and are identifiable, but these are placed in Class D, hereafter described. Of deflections falling below this limit we may, then, have all degrees of confidence as to their reality, from a practical certitude, in the case of those nearly as great as this and identified on numerous plates, to a condition of wholly suspended judgment as to those which are of the exact order of the accidental ones, and whose identification on numerous plates is uncertain or impossible. *We are here concerned only with those which do not fall below this limit, and as to whose "reality" there is no question.*

It has already been stated that it was originally intended to automatically produce numerous line spectra corresponding to the bolographic curves and then to combine these spectra into a final line spectrum, from which the false should be eliminated from the true by the well-known process of composite photography. It has been found that this procedure, when applied to the class of smaller deflections here described, demands so great a number of originals that from this cause and other difficulties it can not be used with advantage as a criterion of the genuineness of these smallest deflections, though it is efficient for others. It was, in 1897, therefore, thought best to restrict its use to conveying to the eye an approximate idea of what the appearance of the infrared prismatic spectrum would probably be could it be seen, and the results then attained were thus illustrated.[1] The decision as to the reality of the lines is made to rest upon other evidence.

The actual application of the composite process is as follows: The curves under consideration, which, it will be remembered, consist of a dark line traced photographically on a transparent glass plate, are superposed with well-known corresponding deflections in alignment one under the other, and the observer selects and marks those which seem to him from their size and reappearance in the same position in the curves to be most certainly not due to accident. From this class of deflections, then, where magnitude, position, and verification on independent plates gives practical certitude of reality. the composite is to be composed.

---

[1] On account of the difficulty of the composite process a linear spectrum has been drawn corresponding to the results of 1898, but the composite of 1897 has been reproduced in Pl. XIX.

CONVERSION OF BOLOMETRIC CURVES TO LINEAR SPECTRA.

The curves obtained by the bolometer are in one important respect superior to ordinary spectra, since they show not only the position of light and dark lines, but also the amount of the absorption which produces them, and especially the relative intensity of heat energy at successive parts of the spectrum.

The writer has desired, however, to present in addition, for the convenience of the reader linear spectra, especially if these could be obtained from the curves by some purely automatic process; partly to compare the bolographic indications with those obtained from photographs of the visible spectrum; partly also because linear spectra lend themselves better than the curves to that process of composite photography, by means of which (in theory at least) all accidental deviations due to tremors of the apparatus or like minor causes may be eliminated, while all real ones due to solar and telluric absorption may be preserved.

The difficulty in producing these linear spectra is not a difficulty due to the bolometric process, as might at first appear, so much as to the nature of the special region which we are studying, which is quite unlike the visible spectrum in being composed of great alterations, corresponding to wide regions of light alternating with wide regions of shade, while like detail is found in each region.

The difficulty this introduces is not wholly unlike that which a photographer experiences who is called upon to photograph a tree or some similar object in the fullest sunlight. The tree presents large masses in high light and large masses in deep shade, and on each of these masses are superposed innumerable minor details. If he adjusts his exposure for the high lights, he misses all details in the shadow; if he give a sufficiently long exposure to discriminate details in the shadow, he "burns out" all those in the high lights.

In a similar way, if he had to photograph such spectral regions as that about the H lines, where the energy is feeble and the light, even apart from physiological causes, is weak; if we had such essential dark regions not only in immediate juxtaposition to the most brilliant parts of the spectrum, but alternating with them, we could not detect the minuter lines in these two juxtaposed regions without altering the conditions of light, as we are, indeed, accustomed to do even for the eye in the visible spectrum when passing from the brighter to the darker portion of it.

Many methods which have been tried here to procure the linear spectrum directly from the galvanometer curves have been frustrated by an analogous difficulty. One method consists in "blacking-in" all of the plate below the photographic trace and then photographing this through a combination of a spherical and cylindrical lens,[1] which draws the blacked-in portion out into regions of a greater or lesser shade, according to the linear depth of the blacked-in portion, as is shown in the illustration, fig. 11 (*b*) and (*c*).

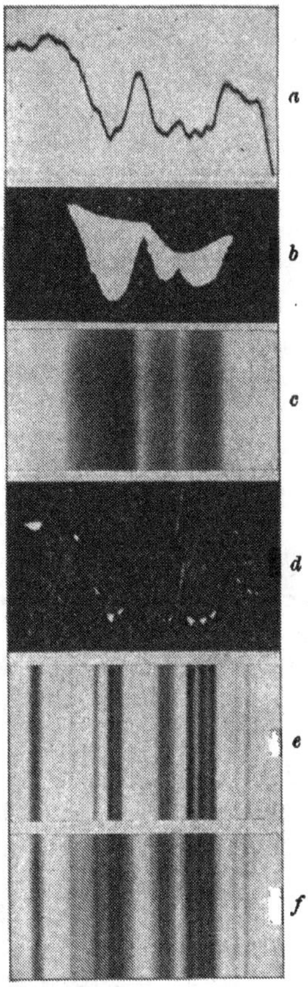

Fig. 11.—Photographic production of line spectra from bolographs.

In the original curve (*a*) the great elevations represent regions of the greatest heat, and the greatest depressions regions of the greatest cold, and if we fix our attention on these great

---

[1] The linear spectra thus produced have been termed "cylindrics," but as the lens introduces a considerable distortion, it has been replaced in recent practice by a horizontal slit, to which no such objection applies.

regions only they can be adequately rendered as in (c) into bright and dark bands, respectively, but the detail is comparatively ill rendered without a special adjustment, which would in turn give a false presentation of the great masses of light and shade.

The difficulty can not perhaps be entirely overcome, but it is otherwise met by fixing the attention primarily upon the smaller inflections of the curves. Thus, in (d) and (e) the same curve is shown with a treatment calculated to develop the minor inflections. The result is a linear spectrum where the lines appear distinctly, but now the great absorption bands are no more adequately indicated than are regions of little energy in ordinary photographs of the visible spectrum. If we combine the two treatments, however, by superposing one of these plates upon the other, we get perhaps as good a result as is attainable, as is shown in (f), and it is by a similar process that the "composite" linear spectrum given in Pl. XIX is obtained.

COMPARISON AND MEASUREMENT OF BOLOGRAPHS TO DISCRIMINATE AND FIX THE POSITION OF "REAL" INFLECTIONS.

The composite process which has just been described is essentially a graphic one, and is not used as an independent method of determining the reality of the lines. Let us next, then, consider the process which is employed for this latter important purpose. We have divided the inflections of the curves into two classes—in the first of which are those which are unquestionably real, and in the second those of which a large part are believed real, but of which a portion is in doubt.

The first of these classes may be subdivided into three groups, which we may call (a), (b), and (c), which are distinguished by the depth of the deflections in the curves, or, in other words, by the relative intensity of the absorption lines they represent.

(a) In (a) appear a number of great deflections which correspond to absorption bands extending through whole minutes of arc of the observed regions in the spectrum, such as those designated here as ψ, Ω, and the like.

(b) A second series contains deflections of less magnitude, but still of such considerable importance as to have a depth of 5 mm. or over, and these constitute class b.

(c) Class c contains deflections identified on all the plates, of a less depth than 5 mm., but of more than five times the average depth of the accidental deflections. This lower limit has been set at 1 mm. for bolographs of 1896–97, and 2 mm. for those of 1897–98.

Within the three divisions, (a), (b), (c), are included those deflections whose "reality" is from their magnitude obvious. While the visual resolving power of the prism is consistent with the definition of more than a thousand lines, the actual number under the enforced restrictions falling in this first great class of bolometrically determined ones is approximately but 300.

Every line included in the tables given later is marked with the letter (a), (b), or (c), whose significance has already been explained, or with the letter (d), which designates a class of deflections which are believed to be real on the evidence which follows, and which refers particularly to the determination of this last class now in question.

The observer examines all the similar bolographs and selects those which are by reason of the conditions of sky and of instrumental behavior most trustworthy, rejecting, for instance, as a rule, any in which the mean amplitude of accidental inflections in the battery records exceeds one-fifth the lower limit of amplitude in class c, as has just been described. These curves, to the number of at least six, but more usually twelve, he superposes with good light from beneath, and so that corresponding deflections are opposite. He then picks out the coincident deflections and numbers them in order, and divides them in the four classes a, b, c, and d, which indicates to which division of the catagory already described they fall. In class d, with which we are at the moment particularly concerned, belong all those which may be regarded from their coincidence in position and appearance as apparently genuine, yet which are very minute. A subscript is often appended to the designation of deflections of class d to indicate on how many of the curves examined the deflections fail to appear.

Having thus selected and classified the deflections he proceeds to measure their positions on each of a number of the best curves with the comparator. In this measurement ordinates are read to millimeters, but abscissæ (corresponding to prismatic deviations) to hundredths of millimeters to insure accuracy in the tenths place. Two independent series of measurements are made on each plate. In the comparison presented in the accompanying tables six similar curves were measured.

Having completed the comparison and measurement the observer's results are reduced as follows: The measurement of abscissa corresponding with each deflection is subtracted from that corresponding to some sharp deflection agreed upon between the observers which appears on each plate and which is thus taken as a reference. The mean of these differences for the two observations on each plate is then taken. A temperature correction, determined as indicated in another place, is next applied (where necessary) to reduce the length of the plate to that for constant temperature. All these corrected means of separate plates are now brought together and reduced, throwing out such deflections as seem from the divergence of their measured positions to be accidental. The other observer having completed the same work independently, the mean results of all are now for the first time brought together and compared, as shown in Table 18, and doubtful deflections omitted.

The reader will note that in the procedure just described there are furnished four checks on the "reality" of deflections. These depend upon, first, magnitude; second, similarity of form; third, coincident position; fourth, the independent judgment of two observers.

Chapter V.

LIMITATIONS OF THE RESEARCH—SOURCES OF ERROR NOW EXISTING.

LIMITATIONS OF THE RESEARCH.

1. VISUAL RESOLVING POWER OF THE PRISM.

Lord Rayleigh,[1] in treating of the visual resolving power of prism spectroscopes, decides that we must consider a bright double line having equal components to be fairly resolved when the brightness midway between the two maxima of the image is about eight-tenths that of the maxima themselves. Under these circumstances he finds that the components will subtend an angle equal to that subtended by the wave-length of light at a distance equal to the horizontal aperture of the prism employed.

Whether this definition can be adopted without modification for bolometric observations will be considered a little later. Assuming its applicability, let us proceed to consider the resolving power of the great salt prism whose face is, allowing for defects in the edge of the prism, fully 120 mm. wide.

The angle of incidence for minimum deviation at the A line is 50° 8′, and for the deviation of the the extreme part of the region studied about 49° 20′. The horizontal aperture is therefore equal to or exceeding

$$120 \cos 50° 8' = 77^{mm}.$$
For A, $\lambda = 0.00076^{mm}.$
For the lower limit investigated $\lambda = 0.0053^{mm}$, approximately.

Calling $\Delta\theta$ the difference in deviation of two lines as close as can be fairly resolved and a the horizontal aperture,

$$\Delta\theta = \frac{\lambda}{a}$$

Whence at A, $\Delta\theta_{(40° 20')} = \dfrac{0.00076}{77} = 2''$ nearly;

at 1.2μ, $\Delta\theta_{(39° 44')} = \dfrac{0.0012}{77} = 3''$ nearly;

and at 5.3μ, $\Delta\theta_{(38° 38')} = \dfrac{0.0053}{77} = 14''$ nearly;

---

[1] Lord Rayleigh, London, Edinburgh and Dublin Philosophical Magazine, fifth series, Vol. VIII, p. 266, 1879.

If we consider the deviation curve of rock salt to be made up of two straight lines within the region under consideration, extending

$$\begin{array}{ll} \text{from} & \theta = 40°\ 20', \lambda = 0.00076 \\ \text{to} & \theta = 39°\ 44', \lambda = 0.0012 \\ \text{and from} & \theta = 39°\ 44', \lambda = 0.0012 \\ \text{to} & \theta = 38°\ 38', \lambda = 0.0053 \end{array}$$

respectively, and that the mean resolution in these regions is 2.5" and 8.5", respectively, we have the number of lines which might, if they existed, be distinguished within the whole region as approximately 1300, of which about 450 would be found in the lower of the two divisions, which is not as yet mapped photographically.

The great glass prism has been employed for auxiliary investigation in the region between 0.76μ and 1.8μ on account of its greater dispersion. Having equal resolution, it might be expected to discover in this region nearly twice as many lines as the salt prism on account of its great dispersion.

Let us now pass from the question of resolution thus considered to other causes of imperfect definition.

2. SLIT WIDTH.

It has been found necessary, under the most favorable circumstances yet secured, to use a slit image of about 1.2" angular width together with a bolometer subtending 1.2" in the region from A ($\theta = 40°\ 20'$, $\lambda = 0.76\mu$) to X ($\theta = 39°\ 20'$, $\lambda = 2.5\mu$) in order to obtain suitable deflections of the galvanometer. From this point to Y ($\lambda = 4.3\mu$) a slit width of about 3.6" has been employed together with a bolometer subtending 3.4", and beyond this wave-length a slit width of 9" was used with the same bolometer.

3. COLLIMATION.

At the time I commenced this research in Allegheny the theory of the collimator appears not to have been universally understood. At any rate, rather by practice and experiment than by any aid through theory, I worked out a system (then I believe novel, though to-day common) whose merit lay partly in its simplicity. If, for instance, a solar image was needed of 3 inches in diameter, I preferred to obtain it by using a lens of so long focus that this diameter would be directly given there, and not by the magnification of the image of a short-focus lens.

Applying this system to the collimation of the spectrobolometer, I have never made use of a condensing mirror or lens before the slit in solar work, for it can be shown that there is no real gain of energy resulting from their employment here. My practice has been to use a collimator of such long focus that the prism or grating was entirely covered by the natural divergence of the solar beam. Thus linearly a comparatively wide slit may be used for a required small angular aperture, and losses of light by diffraction, always attending the use of narrow slits, are avoided. At

time the corresponding width of slit, which he varied. In the accompanying figur⟩ these results are plotted, together with a straight line showing the deflections which would have been observed if they had been throughout proportional to the slit width.

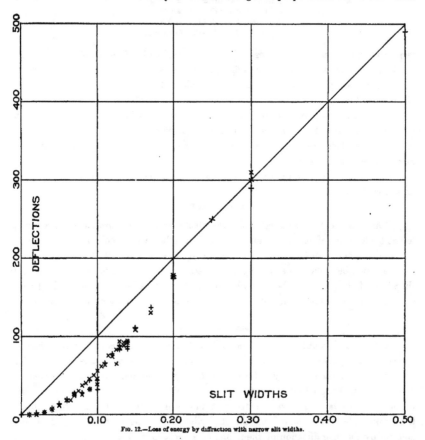

FIG. 12.—Loss of energy by diffraction with narrow slit widths.

Two series of measurements were made upon a very clear day and appear in the table. Another series in the main substantiated the result of these, but having been made on a day when the sky was rapidly growing worse it is not here given.

TABLE 4.—*Observations on Diffraction.*

[Date, January 23, 1897. Observers—at slit, F. E. F.; at galvanometer, C. G. A.]

Series I.			Series II.		
Width of slit.	Number of the observation.	Galvanometer deflection.	Width of slit.	Number of the observation.	Galvanometer deflection.
mm.		*mm.*	*mm.*		*mm.*
0.30	1, 26	290, 290	0.30	1, 32	310, 300
.25	2	230	.25	2	250
.20	3	173	.20	3	178
.17	4	137	.17	4	130
.15	5	111	.15	5	108
.14	7, 18	87, 84	.14	8, 38	94, 92
.13	6, 20	88, 83	.13	6, 37	94, 85
.12	8, 19	77, 75	.12	9	75
.11	9	65	.11	7	64
.10	11, 22	44, 32	.10	10, 31, 36	56, 46, 38
.09	10, 23	44, 32	.09	12, 27, 35	45, 32, 32
.08	13	28	.08	13, 29, 34	37, 26, 26
.07	12	27	.07	14, 28, 33	26, 26, 25
.06	14	19	.06	15	18
.05	15	13	.05	22	12
.04	16	7	.04	23	7
.03	17	2.5	.03	24	2.5
.02	21	1.5	.02	25	0
.01	24	0	.01	26	0
.00	27	0	.135	11, 39	92, 88
.50	25	490	.125	16, 40	65, 83
			.115	17	75
			.105	18	62
			.095	19	50
			.085	19	40
			.075	21	30
			.065	30	19

REMARKS.—The second column gives the order in which the observations were taken. Circle reading at point of observation, 160° 30′ 10″. Circle reading on A line, 160° 6′ 25″, whence the distance from A in angular deviation in the spectrum was 47′ 30″ corresponding to wave length of about 1.8μ.

It will be seen by inspection of the plot that the effects of diffraction first became noticeable with a slit width of 0.20 mm., and become important at a width somewhat less than that which it is customary to use, namely, 0.15 mm.[1]

---

[1] Two discontinuous places in the observations are found at slit widths about 0.09 mm. and 0.13 mm., respectively. It was thought that these might have been due to diffraction bands within the useful space on the first mirror. If, however, we supply in the formula for the distance (x) of the diffraction band from the center of the beam

$$x = \lambda \frac{a}{s}$$

(3) From (1) it follows that if a given instrumental equipment produce a curve similar to fig. 13 (a) then a galvanometer n times as sensitive will record ordinates always n times as great as in fig. 13 (a), whence the depression observed in the curve will then be n times as deep. If meanwhile the battery record remains equally as good as at first, the limit of resolution will, according to (2b), have been increased.

(4) Conversely, if without change of sensitiveness the battery record should be improved so that the average depth of the accidental deflections becomes $\frac{1}{n}$ as great as before, the limit of resolution would be increased.

(5) Suppose, now, that the curve of intensity of energy in the bright image were a, b, c, d, e, fig. 13 (b), if the image were produced under optically perfect conditions, but that in practice, owing to imperfect optical arrangements, the real distribution of energy in the image is represented by the curve a, b, f, d, e, in which the minimum is less pronounced. Suppose, further, that owing to the width of the bolometer strip the actual bolograph is represented by a, b, g, d, e, in which the minimum is still less pronounced. It is clear that whatever instrumental changes tend to make the two latter curves more like the first will increase the effective bolographic "resolution."

(6) In view of the preceding, it is theoretically possible that bolographic "resolution" may exceed the visual.[1] For while a difference of brightness of one one-hundredth between the positions of maximum and minimum brightness would not suffice, according to Lord Rayleigh's definition, for visual resolution, yet, by the expedients mentioned in (3), (4), and (5), this small difference might be made sufficient for bolographic resolution as defined in (2b). Nevertheless, the inference by no means follows that the resolving power would then be $\frac{0.2}{0.01}$ (i.e. 20) times as great; for it can be shown, by superposing two theoretical intensity curves corresponding to two bright lines, that when the lines are so close as to subtend no more than eight-tenths the angle subtended by one wave-length at a distance equal to the horizontal aperture of the optical apparatus considered there is absolutely no minimum, and resolution is impossible. Thus the bolographic resolution can not exceed the visual by more than about 25 per cent.

It appears, then, that the limit of resolution fixed in the investigation by Lord Rayleigh depends upon a physiological property of the eye, and requires modification in its application to investigation by the bolometer. It appears further that the limit possible to be reached by the bolometric method of spectrum analysis is a question of the useful working sensitiveness of the recording apparatus which it may be possible to attain; but under ideally perfect conditions the bolographic resolution can not exceed

---

[1] It is interesting to note that in the table giving the results of this investigation there are instances where lines are found somewhat nearer than Lord Rayleigh's visible resolution limit would allow. These lines can in some instances be compared with those in Higgs's photographs with results strikingly in confirmation of their existence with the relative intensities found in the bolographs. In other cases bolographs with the glass prism substantiate the close lines discovered with the salt prism.

the visual by more than 25 per cent. The results obtained by a comparison of bolographs lead to the belief that we have in practice slightly exceeded the maximum visual resolution.

SOURCES OF ERROR.

From this discussion of limitations let us turn to a quantitive consideration of the actual sources of error and confusion now existing.

It will readily be seen that in order to obtain satisfactory results by the method of comparison already discussed it is necessary to produce curves which are exactly similar under like conditions, so that corresponding deflections shall lie opposite or over one another in the curves compared. It is necessary, in other words, that there shall be the least possible variation in the positions of genuine absorption deflections. The capacity of accurate determination of the position of the bolometer may be considered apart from its sensitiveness, and has led to very great attention being paid to the determination of the errors of position still likely to occur.

About the year 1895 it was decided to attempt to obtain improved instrumental conditions which should insure that curves having a length of 60 cm. should be free from any probable error of position of a magnitude greater than one-tenth of a millimeter, a quantity corresponding (under the usual speed ratio) to about six-tenths of a second of arc in the spectrum, as measured on the spectrometer circle. It may be said in anticipation that after a long struggle with instrumental difficulties these conditions have been so nearly reached and so high a degree of accuracy has been attained that the relative deviation of many of the infra-red absorption lines measured from A is known to at least the same order of accuracy that the deviation of any line in the visible spectrum is known.

In the remarks just made the method here followed is considered as a method of precision depending on instrumental conditions, but these instrumental conditions are themselves intimately connected with those arising from the site in which the instruments are installed, and it has already been more than once remarked that the greatest difficulty—a difficulty which at one time appeared wholly insurmountable—has arisen from the tremors set up by the traffic of the streets of the city in which the observatory building, designed at first for a temporary use, is still unfortunately situated.

It has been, then, an object of the utmost solicitude to reduce the accidental deflections, i. e., such as are shown by the "battery record," and the subsequent discussion will show how nearly they have been actually eliminated.

For this purpose I group the sources of error into six principal headings, designating each by a large roman numeral, while classifying the minor sources of error under each of the respective heads by arabic subscripts. The general divisions are as follows:

  I. Sources of error in the optical apparatus.
  II. Sources of error which introduce accidental deflections in the record curves.
  III. Sources of error from imperfect mechanism.
  IV. Sources of error in preparing linear photographs.
  V. Sources of error in comparator measurements.
  VI. Sources of error inherent in the bolographic method.

I.—SOURCES OF ERROR IN THE OPTICAL APPARATUS.

1. APPARENT DISPLACEMENTS OF THE SLIT.

(a) From refraction by air strata of varying density.
(b) From jarring or shifting of the apparatus.

In order to measure the disturbances from these sources a micrometer eyepiece was placed on the bolometer support instead of the bolometer, and visual observations of the slit image were conducted on several occasions. While it can not always be decided with certainty whether momentary displacements belong to the types (a) or (b), yet the importance of those of type (a) can be estimated by observing whether the slit image appears frequently to be made crooked, which could hardly be caused by shaking of the apparatus. In 1896–97, when the slit was outside the observatory, there was in this respect much difference in favor of the warmer months, when the temperature within and without the observatory is nearly the same, and in favor of still days, when gusts of wind do not force theirway through the joints of the double-walled tube through which the light entered the observatory. On the most favorable days it.could not be observed that such distortion of the image occurred. Observations of December, 1896, and January, 1897, showed, however, that on cold, windy days displacements from air currents of three-tenths to four-tenths of a millimeter at the bolometer, corresponding to 8″ to 12″ arc of spectrum, sometimes occurred. On still days, even the coldest, these displacements were much less and not to be distinguished from those due to shaking of the apparatus. Since the slit has been within the observatory—that is, since December, 1897—no displacements of type (a) have been observed.

Displacements of type (b) are much augmented by wind, but on still days are mainly caused by the passage of vehicles over the asphalt pavement near the observatory. Such passages caused, in 1896–97, rapid vibrations of the slit image, having an amplitude in aggravated cases of ten seconds of arc, and usually as great as two or three seconds. As the travel in question is very frequent, thirty passages an hour being not unusual, and as the disturbance lasts from a quarter to a full minute, according to the speed of the vehicle, the obstacle which was thus presented to the work may be understood. Fortunately, the displacements of type (b) have latterly been much reduced on account of some improvements in the mirror mountings. They now seldom exceed 1", and average about 0.4".

The effect on the slit image of air tremors and ground vibrations may be stated as follows: In the year 1896-97, on still days during the colder months (which were the only days then available for obtaining bolographs of high grade, as will be shown later), the spectrum by reason largely of passing vehicles was in a state of vibration averaging an amplitude of 1.5" of arc during intermittent intervals amounting to half the whole time, while during the remainder of the time the average amplitude was about 0.5". In 1897-98 these displacements seldom exceeded 1", and averaged 0.4" This state of things causes frequent obliteration of small deflections, but merely broadens without in general changing the position of those deflections still remaining.

Periodical shifting of the position of the slit image has been observed. Several series of measurements have been made with a micrometer eyepiece in place of the bolometer, which have shown that shifting is now chiefly caused by temperature changes in supports of the optical apparatus. In the following table is given the general results of some measurements of the shifting:

TABLE 5.—*Slow movements of the slit image.*

Date of observation.	Remarks relative to apparatus.	Time observed.		Temperature of spectrometer chamber.		Total angular displacement of slit image.	Average angular displacement for 1° rise of temperature.
		Beginning.	End.	Beginning.	End.		
		h. m.	h. m.	°	°	"	"
May 6, 1896	Eyepiece micrometer on bolometer stand and prism and plane mirror removed......	9 57	12 22	20.0	22.1	¹—22	—10
	Mirror, slit, and bolometer stand mounted as described on page 30....................	12 37	4 30	22.1	23.1	+ 6	+ 6
May 26, 1896	As on May 6..................................	9 39	1 47	21.0	23.8	²—50	—18
July 2, 1896	As on May 6, except that concave mirror and micrometer eyepiece were now mounted on metal entirely instead of part wood........	10 8	3 15	24.0	26.8	—0.5	— 0.2
Aug. 26, 1896	Collimated beam by introduction of two spherical mirrors. Otherwise as on July 2	9 45	3 30	23.1	27.0	+8.5	+ 2.2
Dec. 23, 1896	Arrangement described on page 98, differs from that of Aug. 26, in that the prism mirror and another plane mirror parallel to it on the same metal mounting are introduced........	10 30	12 0	19.2	19.1	— 1.2	+12.0
Dec. 29, 1896	As on Dec 23..................................	9 35	3 10	20.0	22.4	+15.3	+ 6.4
Dec. 30, 1896	As on Dec. 23.................................	1 0	3 20	22.4	22.5	+ 1.2	+12.0
Jan. 6, 1897	As on Dec 23, 1896............................	10 30	3 10	17.0	21.0	+24.6	+ 6.1
Mar. 10, 1897	As on Aug. 26, 1896...........................	9 55	12 6	20.9	21.4	+ 8.5	+17.0
Mar. 28, 1898	Arrangement described on page 44. Slit moved inside observatory. Two cylindric collimating mirrors........................	10 48	3 36	21.4	21.2	0.0	0.0

¹ A positive sign before an angular displacement indicates that the motion of the beam with rising temperature is in the direction to increase the apparent deviation of the prism.
² Readings taken during the time intervals, but not included in the table, show that in all cases the temperature change and the position change were both continuous in one direction during the entire interval as given by the table.

By inspection of the table it will be seen that the substitution of metal for wood in the mirror support and bolometer stand practically eliminated the shifting of the beam with the uncollimated arrangement in use prior to August, 1896.

The introduction of the two collimating mirrors in August, 1896, and the movement of the image-forming mirror in the clock pier introduced the shifting anew, but always in the contrary direction to that which, as we shall later see, is produced by a change of temperature in the prism. The mirror shifting was not constant with this arrangement of apparatus, or else the temperature conditions at the mirror supports were not known accurately enough to determine a constant, but the motion was roughly 10″ of arc at the bolometer for 1° rise of temperature in the spectrometer chamber.

With the apparatus in use in 1897-98 no shifting was observed.

2. IMPERFECT ADJUSTMENTS FOR MINIMUM DEVIATION.

(a) As affecting the lengths of curves.
(b) As affecting deviation measurements made as described on page 67.

(a) Having given a parallel beam, it is required to know how great a departure from minimum deviation is allowable in the adjustments, in order that the difference in deviation between two points 60′ apart in the infra-red spectrum shall not be varied from this cause by more than 0.6″ of arc.

From the data appropriate to the beginning and end of a 60 cm. bolograph starting near A and proceeding 60′ of deviation into the infra-red, the following table has been calculated by the general formula for the angle of deviation θ as expressed in terms of the angle of incidence i, refracting angle of prism α, and refractive index n.

$$\theta = i + \alpha + \sin^{-1}\left[n \sin\left(\alpha - \sin^{-1}\left\{ \frac{\sin i}{n} \right\} \right) \right]$$

The first values of i_1 and i_2 in the table are those for minimum deviation, and other values are as follows:

$$\alpha = 60° \ 0' \ 0.000''$$

$$\log n_1 = 0.865865969 \qquad \log n_2 = 0.1834014166$$

TABLE 6.—*Error from imperfect adjustment for minimum deviation.*

i_1	θ_1	Δi_1	$\Delta \theta_1$	i_2	θ_2	Δi_2	$\Delta \theta_2$	$\Delta \theta_1 - \Delta \theta_2$
° ′ ″	° ′ ″	′	″	° ′ ″	° ′ ″	′	″	″
50 12 20	40 24 40.000	0.0	0.0	49 42 20	39 24 40.000	0.0	0.0	0.0
17	40.402	5	.402	47	40.390	5	.390	.012
22	41.606	10	1.606	52	41.560	10	1.560	.046
27	43.609	15	3.609	57	43.507	15	3.507	.102
32	46.408	20	6.408	50 2 20	46.228	20	6.228	.180
37	50.000	25	10.000	7	49.720	25	9.720	.280
42	54.382	30	14.382	12	53.980	30	13.980	.402
47	59.551	35	19.551	17	59.005	35	19.005	.546
52	40 25 5.504	40	25.504	22	39 25 4.792	40	24.792	.712
57	12.238	45	32.238	27	11.338	45	31.338	.900
51 2 20	19.750	50	39.750	32	18.640	50	38.640	1.110
7	28.037	55	48.037	37	26.695	55	46.695	1.342
12	37.096	60	57.096	42	35.500	60	55.500	1.596

From an inspection of the table it appears that for a position of the prism in which the angle of incidence is 36' greater than that required for minimum deviation, the angle of deviation for the index of refraction n_1 has increased by 0.6" more than the angle of deviation for the index n_2 and that a 60 cm. bolograph would therefore be increased in length 0.1 mm. It also appears that an inaccuracy of about 7' in the adjustment for minimum deviation would produce 1" actual increase in the deviation. While no similar calculation has been made for decreasing angles of incidence, it may be safely assumed that the increase of deviation with decreasing incidence follows for some distance a sufficiently similar curve, so that if the minumum deviation adjustment of the prism be made with less than 5' inaccuracy, the deviation will be sufficiently near a minimum for purposes both of direct measurement and of bolographic extrapolation.

On several occasions a test of the accuracy of the theodolite method of adjusting the prism has been made as follows: Having first adjusted the prism mirror vertical and leveled the beam as described in the footnote on page 66, the prism mirror was thrown out of adjustment and again made vertical with the theodolite. The mirror was now caused to reflect back as described in the footnote on page 66 an image of the horizontal slit: In no case was the image found more than 2 mm. from the slit, showing a maximum error of a single theodolite setting of $\frac{1}{20000}=20''$. Thus it will be seen that the prism face may be easily be set to within 1' of the proper angle with the theodolite as described on page 67, so that errors in the length of bolographs thus far considered are absolutely negligible.

In what has been said we have supposed the beam incident at the prism to be perfectly parallel and the prism faces perfectly flat, so that all the light of a given wave-length continued to be parallel after passing the prism. Neither of these conditions was exactly fulfilled in practice, for the beam was slightly divergent before it reached the prism, and was still more so after it emerged from it, owing to the concavity of the prism faces. The effect of these imperfections is discussed in the small type which follows, but the general result found is that no appreciable error is introduced in the length of the curves, provided they are considered as corresponding to the prism angle measured at the middle of the prism faces; and it is shown that the prism angle actually used in the reduction of the bolographs was sufficiently close to this, so that the maximum error was less than 0.5" of arc.

The effect of concave prism faces and divergent beam on the length of bolographs.

After all the bolographs of 1897-98 had been compared, measured, and nearly reduced, attention was drawn to this matter by the difficulty experienced in measuring angles and absolute deviations. As will be shown a little later, very great care is essential to prevent error in these measurements on account of imperfect collimation and the invariable concavity of rock-salt prism faces intended to be flat. It appeared worth investigating whether these same causes did not produce appreciable inaccuracy in the reduction of the bolographic observations. It was known that the beam incident on the prism had been divergent when the bolographs were taken, for the focal length of the concave cylindric mirror used for collimation is such that there was not quite distance enough available in the spectrometer chamber to secure exact collimation and it had not been supposed necessary to correct this. But the most serious question was this: If the prism is so concave that its angle at the edge may be 30″ or more less than its angle near the back, what are we to assume as the proper angle to use in reducing the measurements?

As the analysis and numerical solution of the problem would occupy considerable space, and as the results would not be of general usefulness, it will be sufficient merely to give here an outline of the method of investigation and its general result.

It was assumed (1) that the beam incident upon the prism came from a virtual slit at a distance $u = 20,000$ cm. (somewhat exaggerating the actual state of things in 1898); (2) that the prism faces were of circular concave curvature, so that in a central section the angle at the middle of the faces was a mean between those at the edge and at the back, and (3) that the curvature was such that after passing the prism the rays came from a virtual slit at a distance $u' = 10,000$ cm.

This was a very large estimate for the curvature, and corresponds, indeed, to a difference of angle between edge and back of more than a whole minute of arc. It was further assumed that the bolometer track did not point directly at the image-forming mirror, but made an angle with this direction $\gamma = 100'$. This condition of affairs would be such that if a given spectral line, as A, were in the field in focus upon the bolometer strip, a movement of the bolometer either backward or forward would not only destroy the focus, but would cause the line to go out of the field, instead of remaining on the bolometer strip as long as the imperfect focus admitted of seeing the line. The assumption $\gamma = 100'$ is clearly far too great.

The salt prism, image-forming mirror, and bolometer assumed were those used in 1898. It was shown that the ray central in the incident beam would be within 0.02″ of central in the slit image of a given wave-length at the bolometer. As the bolometer is more than 1″ wide, evidently no appreciable error is introduced in assuming that when this ray falls upon its center the bolometer is central in the slit image, and we need consider only the ray incident at the center of the prism face in the analysis of the problem.

If in the absence of the prism the image of the slit falls in focus upon the bolometer strip, then, in order to cause the A line to fall in focus upon the bolometer strip after the insertion of the prism, the bolometer must be moved backward on its track and the central ray of the incident beam must be at an angle of incidence $i_A + di_A$ slightly greater than that, i_A, corresponding to minimum deviation. The small angle of dp_A through which the prism must be turned from a position where the incidence is i_A to that where the incidence is $i_A + di_A$, and the A line falls upon the bolometer, was determined, as was also the corresponding angle $dp_{(A-100')}$, corresponding to the same conditions for radiations of 100′ less deviation than A, which are at the extreme distance where absorption lines are found on the bolographs.

These small angles dp_A and $dp_{(A-100')}$ were found to be so small that the angles of deviation of the rays in these positions of the prism were not different from those of minimum deviation by as much as 0.01″, so that the deviations of the central ray might still be considered minimum, though its angle of incidence was slightly greater than its angle of emergence.

It now remains to determine how far the prism must be turned to pass from one of these positions to the other.

Imagine for the moment a perfect prism of the same size and mean angle, and a perfectly collimated beam to be used in connection with it. Then let this prism be set so that the beam makes the angle i_A. The A line will then fall in focus upon a bolometer strip suitably placed. Now turn the prism counter clockwise through an angle $p = 50'$, and rays of deviation 100′ less than A will fall upon this bolometer strip. Substitute for the ideal prism and beam in the first position the concave prism and divergent beam and turn the prism clockwise through the angle dp_A and the A line falls upon the real bolometer. Next turn the real prism counter clockwise $dp_A + \Delta p$ and a beam having a slightly different direction from the beam which was formerly central will now be central owing to the motion of the center of the prism face in space. Call this difference of direction dp, and turn the prism clockwise dp. The central ray now makes the angle of incidence $di_A - \Delta p = di_{(A-100')}$. Now turn the prism clockwise $dp_{(A-100')}$ and the ray of minimum deviation $\theta_{(A-100)}$ will fall upon the real bolometer.

Therefore the amount through which the clock turns the circle in the case of the real prism is greater than that in the case of the ideal prism by $dp_A - dp_{(A-100')} - dp$, and the error in the difference in minimum deviation is $2\,(dp_A - dp_{(A-100A)} - dp)$.

Computation showed that this increase in difference in deviation would correspond to a prism angle about 5″ greater than that at the center of the prism. The prism angle actually used in the reductions was, however, measured about halfway between the center of the prism faces and the back of the prism. Here the angle may have been as much as 10″ greater than at the center of the prism, so that the error was in all probability slightly overcorrected. However, the error of overcorrection would not amount to more than 0.6″ in difference in deviation in the whole 100′ of arc covered by the bolographs, and is negligible.

The general result of the inquiry then is that moderate inaccuracy of collimation and curvature of prism faces does not affect the accuracy of bolographs appreciably, and that the angle to be used in reduction is practically that at the center of the prism.

(b) Error in measurements of absolute deviation by the method of page 67.

Let us now consider what degree of acuracy may be expected of a single measurement of the angle of deviation of a visible spectrum line by the method described on page 67. This measurement requires, it will be remembered, first to adjust the circle and collimating mirrors so that the image of the slit is reflected upon the slit by the plane mirror on the back of the prism; second, to adjust the image forming concave mirror so that the slit image is made to fall on the bolometer strip in the absence of the prism; and third, to set the circle so that the spectrum line falls on the bolometer strip. Two circle readings corresponding to the first and third of these adjustments, respectively, give the data for determining one-half the angle of deviation.

Adjustment (1) accomplishes two things—first, to collimate the beam, and second, to set the prism mirror normal to the incident beam. The first of these is the more difficult and is highly important. For the adjustment (2) presupposes the optical distance of the slit infinite. It will be recalled that in passing the prism and its plane mirror the beam is shifted laterally about 14 cm. and thus strikes the image-forming mirror as though coming from a source at 14 cm. to one side of the real source. If the real source be optically at an infinite distance, i. e., if the beam be accurately collimated, this lateral shifting of the apparent source is immaterial; but if not, then the adjustment (2) does not insure that the beam which reaches the bolometer strip shall have passed the prism in minimum deviation. It has in practice been found difficult to insure collimation so exact as to avoid an error as great as 15″ from this cause. This would correspond to an optical distance of the apparent slit of 2,000 meters, or about 1¼ miles, and to an inaccuracy in the longitudinal position of the slit image at the bolometer of about 1 cm. This degree of accuracy is ample to insure correct abscissæ in bolographs (see page 88), but introduces 30″ error in measurements of absolute deviations.

The lateral error in adjustment (1) need not exceed 0.5 mm., corresponding to about 1″ of arc in the setting of the prism mirror.

Adjustments (2) and (3) may probably be in error by 1″ each. In order to obtain an idea of the probable acuracy of a circle reading, the following table is given, which consists of readings of the four verniers at various parts of the circle:

TABLE 7.—*Error of circle measurements.*
[Date of observations, May, 1896. Observer, F. E. F.]

	1	2	3	4	5	6	7	8	9
Number of observation.	Vernier readings.				Column 1, corrected by column 3.	Column 2, corrected by column 4.	Differences in column 5.	Differences in column 6.	Differences between columns 7 and 8.
	Vernier No. 1.	Vernier No. 2.	Vernier No. 3.	Vernier No. 4.					
	° ′ ″	° ′ ″	° ′ ″	° ′ ″	° ′ ″	° ′ ″	° ′ ″	° ′ ″	″
1.........	341 18 30	71 18 27.5	161 18 16	251 18 25	341 18 23	71 18 26.2	....	....	
2.........	20 39 20	110 39 5	200 39 7.5	290 39 15	20 39 17.7	110 39 10	39 20 50.7	39 20 43.8	6.9
3.........	71 58 47.5	161 58 36	251 58 45	341 58 45	71 53 46.2	161 53 40.5	40 14 32.5	40 14 30.5	2.0
4.........	109 47 40	199 47 35	289 47 35	19 47 47	109 47 37.5	199 47 41	37 53 51.3	37 54 00.5	9.2
5.........	159 9 15	249 9 25	339 9 25	69 9 29	159 9 20	249 9 27	49 21 42.5	49 21 46.0	3.5
6.........	201 1 5	291 1 7.5	21 1 17.5	111 1 10	201 1 11.2	291 1 8.7	41 51 51.2	41 51 41.7	9.5
7.........	253 43 22.5	343 43 20	73 43 20	163 43 15	253 43 17.5	343 43 17.5	52 42 10.1	52 42 8.8	1.3
8.........	289 41 15	19 41 25	169 41 20	199 41 15	289 41 17.5	19 41 20	36 57 56.2	36 58 2.5	6.3
Mean ..									5.5

Columns 1, 2, 3, and 4 are the vernier readings. In columns 5 and 6 the readings of the opposite verniers are combined to eliminate eccentricity of circle. Subtracting the separate readings in column 5 from each other in succession we obtain column 7, and column 8 is similarly obtained from column 6. Thus we have two independent measurements of the same angle as the result in each case of readings on two verniers. The differences between these determinations are given in column 9, and their mean is $5.5''$. If now two such angles had been determined with four verniers instead of two, it would be expected that they would differ on the average by $\frac{1}{\sqrt{2}} \times 5.5'' = 3.9''$, and it thus appears that $3.9''$ is the average deviation of a single determination of an angle by readings from four verniers.

Summing up and including the several sources of error enumerated, we may assign $16''$ as a fair value for the probable error of a single observation of half the deviation of a spectrum line, so that for the whole deviation the probable error of a single observation would be $32''$. That somewhat greater accuracy might be secured by more elaborate provision for collimating the beam and for setting at the slit and at the bolometer is not to be doubted, but as the apparatus is evidently not well suited for the accurate determinations of absolute deviations (which are not immediately in question in taking bolographs), no special trouble has been taken to secure extreme accuracy by this method, and recource has been had to the ordinary method of measuring twice the deviation by a telescope mounted on a movable arm of the spectrometer when accurate measurements of this quantity were required.

3. Error in Determining the Prism Angle and Angles of Minimum Deviation.

It may seem unnecessary to here discuss the errors in the ordinary method of determining indices of refraction, but several precautions not always observed are necessary in order to reach such accurate results as are here aimed at, and some of these are here given. It may be said in anticipation of them that the errors to be avoided arise from imperfect collimation of the beam and from imperfectly flat faces upon the prism. The latter difficulty is greater with rock salt than with prisms of substances having less coefficients of thermal expansion, but collimation so perfect that two parts of the same beam 5 cm. apart shall have the same direction within five seconds of arc is very difficult to secure except from uninterrupted starlight.

In the ordinary method of measuring refractive indices the prism is placed level upon a fixed prism table and the observing telescope is rotated with the circle. The refracting angle is determined by observing the reflected image of the slit from each face of the prism and taking the half difference of the corresponding circle readings. In these observations we do not use the same portion of the supposed collimated beam, and the resulting value of the angle is greater or less according as the beam is convergent or divergent. Furthermore, unless two symmetrically placed diaphragms are employed we introduce the edge error of the prism faces in our measurement of the angle.

Occasionally, however, another method is employed, in which the telescope remains fixed while the prism rotates with the circle, and the angle is given by 180° minus the angle through which the prism moves. Unless care is used in placing the prism with reference to the axis of rotation this method is liable to both the errors of the other and to a third, namely, that the reflection may take place near the edge of one face and near the back of the other, which again requires that the faces shall be flat.

In measuring deviations, the prism is set at minimum deviation first on the one side, then on the other, and the half difference of the circle readings between the two corresponding positions of the observing telescope gives the deviation. Here again we may readily use the different parts of the imperfectly collimated beam and different parts of the prism whose faces are not perfectly flat.

As an illustration of the magnitude of the errors into which we may thus be led, it is difficult, using Schuster's well-known method of collimating a beam under the most favorable circumstances, to distinguish any difference between cases where the apparent slit is at $+2,000$ meters, infinity, or $-2,000$ meters. Let us suppose that to avoid edge error in the prism we use diaphragms with apertures 2.5 (curtate) centimeters from the edge of the prism; then the resulting value of the angle is 5 seconds too small in the case where the apparent slit is at $+2,000$ meters. Again, a glass prism by the best makers may readily be concave or convex by one sodium wave-length. If we suppose its faces to be 5 cm. wide, this would mean that the difference of angle between the edge and back might be $15''$, provided the curvature was regular and not compensatory. At the edge, errors far in excess of this are common. Rock-salt prisms, however, are more difficult to figure than glass on account of their great thermal expansion. As the result of this, rock-salt prisms are usually concave to a greater amount than that just indicated.

Hence we may conclude that the ordinary methods of measurement do not suffice to obtain results on deviation and angle of rock-salt prisms closer than perhaps $30''$ of arc unless special precautions are adopted to measure the angle and the deviation at the same place upon the prism, and to prevent error from imperfect collimation. Let us now consider how these errors may be avoided.

First method.—The angle is measured by moving the telescope. A pair of diaphragms are provided with equal apertures, as narrow and small as are consistent with the definition of the line whose deviation is to be measured. These are placed symmetrically about the refracting angle and near the edge of the prism, but still far enough from the edge to escape the edge irregularity of the prism.

The prism is first set so that the axis of rotation is in the plane bisecting the refracting angle, and this is now measured as usual, giving a result too large or too small according as the beam converges or diverges.

The prism is next set for a minimum deviation measurement, so that the axis of rotation is in the plane bisecting the refracting angle, and at such a distance from the edge that when in minimum deviation the same ray which was central in the aperture of the diaphragm on one face of the prism for the angle measurement is now central in the aperture of the diaphragm on the other face of the prism for the deviation measurement. By this procedure a value of the deviation is obtained greater or less than the true deviation by the same amount that the angle is greater or less than the true angle. Now reset the prism so that the ray which was central in the aperture of the diaphragm of one face during the angle measurement shall be central in the aperture of the diaphragm of the same face when the prism is set for minimum deviation, and again measure the deviation. This gives a value in error from the true deviation by the same amount as the first measurement, but in an opposite direction. The half sum of the two results gives the true deviation, and their half difference the proper correction to the measured angle.

Second method.—The angle is measured by moving the prism. Here the apertures of the diaphragms should be placed at the center of the prism faces. In measuring the angle the prism is so placed upon the prism table that identically the same ray of the beam falls centrally in the aperture, whether the one face of the prism or the other reflects the beam to the observing telescope. Thus the result of the measurement gives the true angle of the prism at its center.

In measuring the deviation the same procedure is followed as in the first method, except that the prism is now placed so that identically the same beam is used upon the one side as the other for the two positions of minimum deviation. Thus the result of a single measurement is the true angle of minimum deviation at the center of the prism.

The second method is to be preferred, especially for large prisms, for the first method can not give the angle at the center of the prism faces unless the prism is very small. When measurements are carefully conducted by the second method the angle and deviation may readily be established with a probable error not greater than one or two seconds of arc. The results of both methods, as applied to the great salt prism R. B. I., will be found at a later page.

94   ANNALS OF ASTROPHYSICAL OBSERVATORY.

4. ERRORS FROM TEMPERATURE CHANGES OF THE PRISM.

Even when the long investigation which forms the subject of this volume was in its earlier years it grew evidently desirable to take account of the changing temperature of the glass prism, and when the use of rock salt became regular, the study of temperature changes in the prism could no longer possibly be neglected. My early work in this respect will be found at a later page.

In those early days the prism was subjected to extraordinary changes in temperature—that between summer and winter reaching at times in extreme cases 70 or 80 degrees Fahrenheit, for I was obliged often to work in the coldest of winter days without artificial warmth of any kind, observations being taken during hours together in a temperature of over 20 degrees Fahrenheit below freezing, when it was difficult for the observer to maintain the proper physiological and psychological conditions. (See table on page 96.)

Subsequently efforts were made to obtain a more uniform temperature, but the crude means at my disposal at Allegheny were wholly insufficient. It was early recognized that a uniform temperature was necessary, not only for the prism, but for other parts of the apparatus, and notably to cause a diminution of the drift in the galvanometer, yet the lowering of the temperature of a building like the observatory is an extremely difficult thing without the command of a regular cooling plant, which implies large expenditure. In this as in other respects there has been, however, a gradual improvement, culminating in the introduction in 1897 of the permanent cooling plant referred to, and what follows is to be understood of work conducted under almost incomparably more favorable circumstances than that of the early years.

It may be observed in further anticipation of what follows that where the changes of temperature in the apartment have been reduced, as here, to one or two tenths of a degree in an hour, or to quantities of that order, their effect on the accuracy of the record is inconsiderable. This somewhat full discussion would hardly be justified, then, had not all other errors of observation been reduced to a nearly similar degree, so that the temperature changes in the prism, though no longer absolutely important, might be considered relatively so.

We have first to observe that changes are of three kinds: (a) In the dispersion of the prism. These are here negligible, so that if, for instance, one bolograph were taken at a constant temperature of 20°, and another at a constant temperature of 20.1°, the effect would be inappreciable for the purposes of this research. That which we have next to consider (b) is the change in the refractive indices of the prism due to change in temperature during an observation, for if the temperature, for instance, were 20° at the commencement and 20.1° at the end of an hour of observation, the effect on a 60 cm. plate, though very minute, would be not entirely negligible. Finally, there remains (c) changes in the refracting angle of the prism due to rising or falling temperature.

It may here be observed that the temperature of the spectrometer chamber is, in general, very uniform, as the annexed thermographic records show; nevertheless it is not entirely uninfluenced by outdoor temperature, even with the automatic regulation of that of the observatory, so that the more recent bolographs have been taken at temperatures constant during the observation within one-tenth of a degree, chosen sometimes between 18° and 22°, but mostly between 19° and 21°.

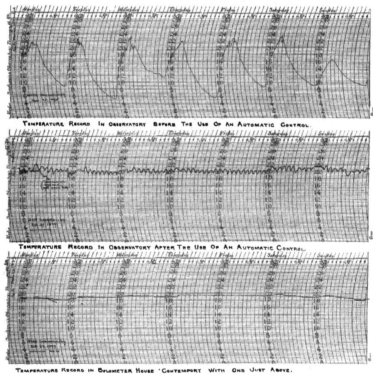

Fig. 14.—Thermographic records.

Owing to the presence of the observer at intervals in the spectrometer room, and to the opening of the door for his entrance and departure, the temperature of this room (which is customarily lower than that of the observatory) rises gradually on all days when bolographs are taken. A rise of temperature of from one-tenth to two-tenths of a degree in an hour is thus sometimes observed, and treating this minute change as having gone on at a constant rate, it becomes necessary to apply the small correction already spoken of to the observed lengths of some curves taken with the salt prism to reduce them to a constant temperature. No such correction is necessary with the glass prism, owing to its small temperature coefficient for refracting indices.

CHANGE OF DEVIATION OF ROCK-SALT PRISM WITH TEMPERATURE—ALLEGHENY OBSERVATIONS.[1]

Data for the reduction just described are to be found, as has already been observed, in the writer's publications from the Allegheny Observatory, and the table there given is here repeated, although it has never been considered by the writer to be as accurate as he could wish to make it:

TABLE 8.—*Allegheny observations on temperature and deviation of a 60° rock-salt prism.*

Temperature.	$\lambda = 0.4861\,\mu$ (F)	Temperature.	$\lambda = 0.5183\,\mu$ (b_1)	Temperature.	$\lambda = 0.5890\,\mu$ (D_1)
°	° ′ ″	°	° ′ ″	°	° ′ ″
−11.5	42 00 35	−11.5	41 42 42	−11.5	41 13 10
− 6.5	42 00 24			− 6.5	41 10 11
− 6.7	41 59 59	+21.2	41 36 23	− 6.7	41 11 06
+24.0	41 54 09	24.0	41 35 13	21.2	41 05 57
				24.0	41 05 01

Temperature.	$\lambda = 0.7594\,\mu$ (A)	Temperature.	$\lambda = 1.36\,\mu$	Temperature.	$\lambda = 1.82\,\mu$ (Ω)
°	° ′ ″	°	° ′ ″	°	° ′ ″
−11.5	40 32 38	− 6.7	39 43 02	− 6.5	39 39 12
− 6.5	40 31 07	+21.2	39 43 23	+21.2	39 33 08
− 6.7	40 30 52	+22.2	39 42 25	+22.2	39 31 35
+21.2	40 25 20				
+22.2	40 23 56				
+24.0	40 24 41				

As a simple method of combining these results, the mean of the deviations taken at high temperature was adopted as the deviation at the mean high temperature and a similar procedure was followed for those taken at low temperatures. This method gave for the several lines the following values of the temperature correction per degree centigrade between the given limits. The deviation decreases with 1° rise of temperature by the amount given in the column headed $d\,(d)$.

TABLE 9.—*Result of Allegheny observations on temperature and deviation.*

Line.	Between.	$d\,(d)$.	dn.
	° °	″	
F	− 8.2 and +24.0	11.5	−0.0000351
b	−11.5 and +22.6	12.1	371
D	− 8.2 and +22.6	13.0	401
A	− 8.2 and +22.5	13.3	413
ψ	− 6.7 and +21.7	13.1	409
Ω	− 6.5 and +21.7	14.5	454

[1] S. P. Langley, Mem. Nat. Acad. of Sci., Vol. IV, pt. 2, 1889, p. 211.

EXPERIMENTS AT THE ASTROPHYSICAL OBSERVATORY, WASHINGTON.

With the hope of obtaining results of greater accuracy some experiments have been made at this observatory. The simplest method of determining the changes is doubtless that pursued at Allegheny, where measurements of the absolute deviation are made at two temperatures as nearly constant and as widely different as possible. In the Allegheny observations the temperature of the salt was assumed to be that of its surroundings, and these were only approximately constant, but the error in temperature determinations was eliminated as far as possible by employing a temperature interval of nearly 30°. The prudence of subjecting the great prism here to such a temperature change was doubtful, and other methods were sought.

In attempting to obtain sufficient accuracy with a moderate temperature interval, three methods were tried, but while none proved wholly satisfactory, none seemed to be wholly without interest, and they are given here in the smaller type which follows. In all these it was intended to measure the difference in deviation of some particular line in the spectrum caused by a temperature change in the prism without measuring the deviation itself directly. In making these experiments two essential improvements over the Allegheny arrangements were possible. First, we have in our possession here a second block of salt of nearly the same size as the prism used, cut in the prismatic form and rudely polished, which enables the temperatures of the actual working prism to be closely inferred by noting those in the thermometer sunk in a vertical hole filled with mercury in the center of its substitute. In the second place the temperature of the apartment can here be kept constant for hours:

First method.—The arrangement of apparatus is indicated in the diagram.

The usual collimated beam falls in part on the prism and in part on a mirror m_1, which is at right angles to the beam. The upper half of the beam, after passing through the prism, falls upon the mirror m_2, set at such an angle with the prism that the ray under investigation after passing the prism in minimum deviation, falls perpendicularly on the mirror m_2, and thus is returned through the prism in minimum deviation. m_1 and m_2 are adjusted about their horizontal axes so as to cause the slit image and the spectrum line both to fall on a micrometer eyepiece r placed on the top of the slit s.

Fig. 16.—Determination of temperature correction to dispersion of rock salt.

Under these conditions the slit image serves as an origin from which to measure the change of deviation of the spectrum line with the temperature, and all errors from changes in the position of the collimating mirrors are eliminated. It is only requisite that m_1, p, and m_2 maintain the same relative position. As these were on excellent metallic mountings on the same stand this was thought not doubtful. It is apparent that the change of deviation is doubled by a second passage through the prism, and that the angular magnitude can be very accurately determined with the micrometer eyepiece, provided the definition is good enough. The temperature was intended to be kept (automatically) constant for at least ten hours prior to each reading.

On trial the device proved unsatisfactory as in the unfavorable conditions of observation at the slit (which was then outside the observatory), and with the weak spectrum resulting from two passages through the prism and magnification by the micrometer eyepiece the definition was quite too bad.

Second method.—Measurements of August 26, 1896, had shown that the slit image remained in a very constant position, so that it might be hoped that it would remain sufficiently so to make micrometer measurements at the bolometer stand of value, using the ordinary arrangement of apparatus. As a precaution a plane mirror was arranged parallel to the prism mirror to give an undispersed image of the slit at the micrometer as well as at the spectrum line. The arrangement is shown in the diagram (fig. 17).

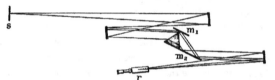

Fig. 17.—Determination of temperature correction to dispersion of rock salt.

A series of measurements taken with various temperatures at intervals of two weeks, about January 1, 1897, while they were quite useless to determine the effect of the temperature to change the deviation of the A line owing to movement of the mirror supports, yet furnished part of the data of page 87. These results were quite important, therefore, as they first drew attention to the effect of temperature on the position of the beam.

Third method.—In considering the results of the second method the idea presented itself that if in place of the mirror m_1, a second prism and mirror opposite to the first should be inserted, then variations of the beam would move both spectra alike in the same direction, but variations in the refractive index would produce motions in opposite sense, so that readings on the change of linear distance between corresponding lines in minimum deviation in both spectra would,

Fig. 18.—Determination of temperature correction to dispersion of rock salt.

when reduced to angular measure, give double the change of deviation. In this it is assumed that temperature affects the refractive indices of different samples of rock salt equally, a supposition which is plausible.[1] Accordingly a second 60° prism mirror combination was arranged on an adjustable brass mounting rigidly secured to that of the first. The prism employed, in addition to the great "R. B. I," described on page 45, was one of 8 cm. face made of Baden salt and marked "S. P. L. T." A diagram of the apparatus would be exactly like fig. 17, except at the prism table, which is as shown in fig. 18.

---

[1] See later observations, which show that three salt prisms at 20° C. have equal refractive indices. There is no assignable reason why this should be true at 20° and not true at all other temperatures.

A series of observations upon the A line of both spectra is given below. These observations, it will be seen, extended over an interval of seventeen days, during which time the adjustment of apparatus was not altered:

TABLE 10.—*Observations at the Smithsonian Astrophysical Observatory on changes of deviation of a 60° rock-salt prism with the temperature.*

Date.	Time of observation.	Temperature in salt prism.		Difference in position of A lines by micrometer.		Remarks.
		Mean of two readings before and after micrometer setting.	Difference in readings.	Mean.	Number of settings.	
	h. m. h. m.	°	°	mm.		
Jan. 26, 1897..	3 20 to 3 37	16.55	0.10	0.576	6	The apparatus was set up on this date and the temperature was very slowly rising.
Jan. 29.......	9 25 to 9 50	15.88	.07	.240	3	From thermograph record less than 0.2° temperature change had occurred in 20 hours next preceding.
	11 00 to 11 07	16.00	.05	.413	3	
	2 50 to 2 56	18.36	.07	1.642	3	The temperature of the spectrometer chamber was rapidly raised over 3° C. between 11ʰ and 12ʰ; then kept nearly constant, but slightly falling.
	3 58 to 4 06	18.44	.02	1.621	3	
Jan. 30.......	9 30 to 9 45	16.60	.00	.695	3	Thermograph record shows falling temperature through the night, but only 0.2° fall during the 4 hours next preceding the first observation.
	2 47 to 3 03	20.70	.00	2.926	3	
	3 58 to 4 11	20.89	.08	2.962	3	
						Between 10ʰ and 11ʰ the room was rapidly heated 6° C.; then allowed to cool slowly.
Feb. 1........	1 36 to 1 51	18.56	.12	1.614	3	Temperature constant from 8ʰ to 12ʰ, but clouds prevented observations. Temperature rising in afternoon.
Feb. 3........	9 17 to 9 32	21.10	.00	2.866	3	Temperature constant for 10 hours next preceding within 0.1°.
Feb. 4........	9 39 to 9 46	19.40	.00	2.331	3	Temperature has fallen steadily in the room, 0.5° in 10 hours next preceding.
Feb. 12.......	11 52 to 12 07	20.96	.08	3.364	....	Thermograph record constant within 0.2° for several days next preceding.

These twelve observations were plotted with temperatures as ordinates and differences in position as abscissæ. A representative straight line was found to pass through the points (temp. = 16.00°, position dif. = 0.330 mm.) and (temp. = 21.00°, position dif. = 3.040 mm.), so that the line might be represented by the equation

$$y = 1.845\, x + 15.39°,$$

where y corresponds to temperatures and x to position differences. Using this equation, the temperatures were calculated corresponding to the observed differences of position, and the residuals of temperature between the observed and calculated values obtained are given in the table following.

TABLE 11.—*Reduced Washington observations on temperature and deviation.*

Number of observation.	1 Observed distance between the A lines in the two spectra.	2 Calculated temperature corresponding.	3 Observed temperature.	4 Residuals, column 3 - column 2.
	mm.	°	°	°
1	0.576	16.45	16.55	+0.10
2	.240	15.83	15.88	+ .05
3	.413	16.14	16.06	− .08
4	1.642	18.42	18.36	− .06
5	1.621	18.38	18.44	+ .06
6	.695	16.67	16.60	− .07
7	2.926	20.67	20.70	+ .03
8	2.962	20.86	20.89	+ .03
9	1.614	18.37	18.56	+ .19
10	2.866	20.68	21.10	+ .48
11	2.331	19.69	19.40	− .19
12	3.364	21.59	20.96	− .63
Mean				.16

REDUCTION OF OBSERVATIONS.

If $f=$ the focal length of the image-forming mirror, $(D_2 − D_1)=$ the interval of position differences, $(T_2 − T_1)=$ the interval of temperature corresponding, and $m=$ the angular value of $1''$ of arc, then $d\theta=$ the change of deviation per degree centigrade is given by the equation

$$2\,d\theta = \frac{(D_2 − D_1)}{f\,m\,(T_2 − T_1)}$$

Substituting the values

$$f = 467.0^{cm}$$
$$(D_2 − D_1) = 0.271^{cm}$$
$$(T_2 − T_1) = 5.00°$$
$$m = 0.00000485$$

we have

$$2\,d\theta = 23.9''$$
$$d\theta = 12.0''$$

If we regard the temperatures at the beginning and end of the temperature interval to be subject to average deviations of $0.16°$, as determined in the table, then the temperature interval is subject to a probable error of $0.18°$. From this we obtain the probable error of the determination of $2\,d\theta$ as $0.85''$, so that we have—

For the A line $d\theta = 12.0'' \pm 0.4''$.

This result does not vary greatly from that in the table of page 96, but it is more nearly like that given in the earlier article of the writer on "Hitherto unrecognized wave-lengths,"[1] namely, $11''$.

RELATIVE CHANGE OF DEVIATION WITH CHANGE OF TEMPERATURE AT DIFFERENT REGIONS OF THE SPECTRUM.

The determination just given of the temperature change at the A line obviously fails to show whether bolographs taken at constant temperatures of $19°$ and $21°$ ought to vary appreciably in length. For this purpose we need to determine whether there is a sensible difference in the temperature coefficient at different portions of the infra-red spectrum.

---

[1] S. P. Langley, American Journal of Science, vol. 32, p. 97, 1886.

To settle this question fast-speed bolographs have been taken at constant temperatures of 13.4° and 21.6°, respectively, within a few days of each other in point of time, and under conditions as nearly identical, except for temperature, as could be attained. These bolographs, four in number, of which two were taken at each temperature, have been measured upon the comparator with the following results:

TABLE 12.—*Temperature and deviation for salt prism R. B. I.—Observed change of deviation with temperature in infra-red relative to change at A.*

Distances from A line.				Mean distance at 21.6°—mean at 13.4°.	Difference for 1°.
Mean from two curves at 13.4°.	Average deviation of mean.	Mean from two curves at 21.6°.	Average deviation of mean.		
° ′ ″	″	° ′ ″	″	″	″
00 00 00	0.00	00 00 00	0.00	0.0	-,-0.00
19 51.8	.48	19 53.	.00	1.2	.15
32 02.9	3.36	32 05.	1.68	2.1	.26
32 33.1	1.44	32 36.	2.88	2.9	.36
33 12.0	1.92	33 15.6	1.36	3.6	.45
35 35.5	.00	35 39.6	1.20	4.1	.51
38 08.2	.48	38 10.3	.72	2.1	.26
44 13.7	.24	44 17.3	1.20	3.6	.45
54 06.5	.24	54 08.2	.24	1.7	.21
1 08 58.8	.72	1 08 58.8	.72	— .2	— .02
1 34 55.2	.48	1 34 56.4	1.44	1.2	+ .15

A minute effect of temperature on the length of these bolographs appears certain. In order to eliminate errors of determination, so far as possible, a plot was made with column 1 of the table as abscissæ and column 6 as ordinates, a smooth curve was drawn fairly representing the observations, and the ordinates of this curve were read off at several points.

The results which have been obtained from this study of the temperature of the rock-salt prism will be found summarized in the following table, whence it appears that the change of deviation for 1° C. rise of temperature is very slightly over 12 seconds, and that it is nearly constant for all portions of the infra-red spectrum, so that bolographs taken at constant temperatures between 19° and 21° will be of practically equal length.

TABLE 13.—*Temperature and deviation for salt prism R. B. I.—Summary of results.*

Designation of line	Approximate deviation for 60° prism.	Approximate refractive index.	Deviation change for 1° C. rise of temperature.	Refractive index change for 1° C. rise of temperature.
	° ″		″	
A	40 25	1.5368	—12.0	—0.0000373
ρσr	40 05	.5350	12.2	382
φ	39 53	.5308	12.3	388
Ψ	42	.5287	12.3	388
Ω	32	.5268	12.3	385
X	22	.5250	12.2	384
Y	39 00	.5209	12.1	382
	38 45	1.5181	12.0	379

EXPERIMENTS ON THE THERMAL CONDUCTIVITY OF ROCK SALT.

To obtain a rough idea of the thermal conductivity of salt and how far the Allegheny observations just cited might be in error from imperfect temperature determinations the following experiment was tried:

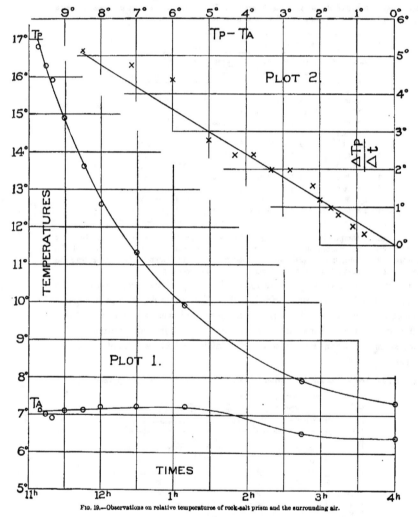

FIG. 19.—Observations on relative temperatures of rock-salt prism and the surrounding air.

The temperature prism, which is a 60° prism 18 cm. on each edge and 16 cm. high, was carried from a warm room to the cold basement and readings made at intervals until its temperature had nearly reached that of the surrounding air. These readings are given in the accompanying table, and the results appear also in plot 1, fig. 19, while plot 2 shows the derived function $\frac{\Delta Tp}{\Delta t}$ as ordinates with the difference of temperatures ($Tp-Ta$) as abscissæ, where Tp is the temperature within the salt prism, Ta that of the air, and t the time in hours.

TABLE 14.—*Observations on heat conductivity of rock-salt.*

Time of observation t.	Observations.		Data for plot 2. From interpolation in plot 1.			
	Temperature of prism Tp.	Temperature of air, Ta.	Δt.	ΔTp.	$\frac{\Delta Tp}{\Delta t}$.	$Tp-Ta$.
1895.	°	°	*Hour.*	°	°	°
Dec. 11, 11ʰ 10ᵐ	16.8	7.1	¼	1.3	5.2	8.5
15ᵐ	16.3	7.0	¼	1.2	4.8	7.1
20ᵐ	15.7	6.9	¼	1.1	4.4	6.0
30ᵐ	14.9	7.1	¼	.7	2.8	5.0
46ᵐ	13.6	7.1	¼	.6	2.4	4.4
12ʰ 1ᵐ	12.6	7.2	¼	.6	2.4	3.8
30ᵐ	11.3	7.2	¼	.5	2.0	3.3
1ʰ 10ᵐ	9.9	7.2	¼	.5	2.0	2.8
3ʰ 0ᵐ	7.9	6.5	¼	.4	1.6	2.2
4ʰ 15ᵐ	7.3	6.4	½	.6	1.2	2.0
Dec. 12, 10ʰ 40ᵐ	5.9	5.3	½	.5	1.0	1.7
			½	.4	.8	1.5
			1	.5	.5	1.1
			1	.3	.3	.8

From plot 2 one may obtain the change of temperature at the center of the prism which might be expected in one hour corresponding to a given difference of temperature maintained in the surrounding air. It will be seen that the change of temperature of the salt in an hour is about three-fifths of the (constantly maintained) temperature difference causing it.

CHANGES IN THE REFRACTING ANGLE OF THE PRISM DUE TO RISING OR FALLING TEMPERATURE OF THE SURROUNDINGS.

The following successive observations of the angle of the great salt prism, extending over about five hours' time, were made on a day when the temperature was rapidly rising:

[Date, July 18, 1898.]

Number of observation.	Temperature at center of temperature prism.	Observed value of 2a by 4 verniers.	Value of a.
		° ′ ″	° ′ ″
1	Not observed	119 49 05.5	59 54 33
2	do	119 49 09	59 54 34.5
3	do	119 49 12	59 54 36
4	do	119 49 15	59 54 40
5	25.4°	119 49 20	59 54 40
6	26.04°	119 49 26	59 54 43

The rise of temperature in the air near the prism, as recorded by the thermograph, was almost 4° during the five hours occupied by the series. Thus it appears that a rise of temperature of 0.8° per hour in the air causes an increase of prism angle amounting to 10" of arc.

That the change of angle could not be attributed to error in measurement, and that measurements at constant but differing temperatures would agree was shown by seven observations of July 27 and 28. On the latter date, after several observations at constant temperature agreeing within an average deviation of 2", the temperature was raised quickly, with the result that the angle increased 7", thus confirming the observations of July 18.

The reason for this change of the prism angle lies, doubtless, in the somewhat large temperature coefficient of rock salt taken in connection with its slow conductivity to heat.[1]

This small change of angle would increase the deviation by about the same amount. It has been shown that the increase of angle is in the neighborhood of 10" of arc for a rise of temperature of 0.8" per hour. Hence when, as in taking bolographs, the rise of temperature is, at most, only 0.1° or 0.2°, the corresponding change of deviation would be 1" to 2", a change slightly smaller in amount and of contrary sign to that due to the temperature change of refractive index.

The coefficient of thermal expansion of glass is so much smaller than that for salt that no error of this kind would be expected. And, indeed, no change of prism angle like that recorded for salt has ever been observed with the great flint prism.

SUMMARY OF EFFECTS OF TEMPERATURE CHANGES.

(a) *Different constant temperatures.*—The differences between different parts of the infra-red spectrum in temperature coefficient of refractive index are so small that it is unnecessary to apply a correction for the reduction of observations from a constant temperature of, for instance, 19° or 21° to the standard temperature of 20°. Such a correction would in any case not exceed seven-tenths of a second of arc.

(b) *Varying temperature during observations.*—As the temperature of the prism not infrequently rises from 0.1° to 0.2° during an observation an hour long, the desirability of a correction of bolographs taken with the salt prism, amounting to from 1" to 3" in deviation, corresponding to 0.2 to 0.5 mm. on the plate, is indicated. This correction is, however, counteracted, since the rise of temperature in the room causes an increase of the prism angle—i. e., by making the faces convex—and also a shifting

---

[1] As a result of the observations recorded, the effect of temperature on the flatness of the prism faces became a matter of attention. It was immediately found that though left by the polisher very flat, as shown by the Newton's rings test, yet upon cooling to the standard temperature of the observing room the faces were considerably concave. In polishing it is now the practice to leave the faces convex to a considerable degree, experience having shown that thus plane faces may be secured when the constant temperature is assumed.

of the beam, which are found to introduce errors each of slightly less magnitude and fortunately in an opposite sense to that from rise of temperature of the prism, so that both, taken together, rather more than neutralized the first-mentioned source of error in bolographs of 1896-97. No separate corrections were judged desirable for these causes. In the bolographs taken in 1897-98, and measured for the purpose of determining the positions of absorption lines, no correction has been found necessary on account of the very constant temperature conditions under which most of the bolographs were taken, and on account of the fact that the opposing sources of error were even more nicely balanced than in 1896-97, owing to a decrease in the amount of error from shifting of mirrors.

II.—SOURCES OF ERROR WHICH INTRODUCE ACCIDENTAL DEFLECTIONS IN THE RECORD CURVES.

We now pass from a comparatively unimportant discussion (that of the changes of temperature in the prism) to a most fundamentally important one, namely, the disturbances affecting the galvanometer needle. Formerly the most conspicuous effect of these was "drift" of the galvanometer needle, but this being chiefly due to variations in the temperature of the observing room has of recent years largely subsided.

The attempts which the writer had commenced to make years before with most insufficient means of obtaining a constant temperature were finally carried out in 1897. The effects of the introduction of a perfect system of temperature control were immediate, and that "drift" of the galvanometer which had interfered with the accuracy of every observation previously and vitiated them in measurable proportions, and like an inherited malady affected everything done for twelve years preceding, was finally brought to a practical cessation, for the drift which in 1882 had amounted frequently to an average of 100 millimeters in a minute was now reduced to 1 millimeter in ten minutes, or a little more than one one-thousandth of what it had been.

Thus "drift," formerly the greatest difficulty in bolometric research, has now become comparatively insignificant. With this preliminary we pass to a detailed consideration of present conditions.

1. Disturbances Primarily Affecting the Galvanometer Needle Independent of the Bolometer Circuit.

These include, first, earth tremors; second, electrical disturbances. These two classes of disturbances are not readily separable, and their effects will not be distinguished in what follows.

In Pl. XVII are exhibited a series of photographic records, full size, of the galvanometer trace when the instrument is not connected with the bolometer circuit. In these records, which are generally selected to represent the best results obtained at different periods, the conditions of the experiment are quite varied.

(a) Taken October 2, 1895, shows the best such curve taken in the daytime with the earlier system of galvanometer support when it rested directly on a pile of four square flat stones, separated at the corners by rubber, and themselves resting on the pier. The speed of the plate is 1 cm. in one minute, and the record is very free from small deflections. So excellent a day record was, with this earlier system of galvanometer support, very unusual.

(b) Taken November 25, 1895, shows a record taken just after midnight at the same speed and with the same support for the galvanometer. While the light used (a calcium light) was so much less intense than sunlight that the difficulty of reproducing this record acts somewhat in its favor, yet the irregularities are, in the original plate, as in the reproduction, much less than those in (a).

(c) Taken January 28, 1896, shows a day record similar to (a), but with the galvanometer supported on the Julius three-wire suspension now used. While not better than (b), it is distinctly better than (a).

(d) Taken April 16, 1896, is like (c), except that the inner chamber had been extended to include the galvanometer. The resulting absence of "drift" is noteworthy. All these four records were taken with the galvanometer needle described on page 31, and a time of single swing of three to four seconds. The weight of that needle is 56 mgs.

(e) Taken November 14, 1896, was made with the needle in use from July, 1896, to August, 1897, and described on page 61. The weight of this needle is 31 mgs., and the time of single swing was five seconds. This record is in all respects taken under the same circumstances as (d).

(k) Taken December 7, 1897, is with the new galvanometer used in 1897–98, as at first constructed with double (but not air-tight) case, and with a needle of 6.5 mgs. weight. The support is the Julius suspension, as in (e).

(l) Taken December 27, 1897, is like (k), except that the mercury floating of the galvanometer had been introduced in addition to the Julius suspension.

(m) Taken March 9, 1898, is like (k), except that the present air-tight galvanometer case had replaced the double walled one.

A comparison of these three records shows clearly that advantage was gained by the introduction of the Julius suspension, and that the needle of 31 mgs. weight was not too light for an almost perfect record, but that a reduction of the weight to 6.5 mgs. was attended with considerable increase in disturbance. Some improvement resulted from the introduction of mercury floating, and also from making the galvanometer case air-tight. I recur to considerations about a constant temperature because this is a specific against "drift," which is associated with these tremors so far as both disturb the needle though in very different ways.

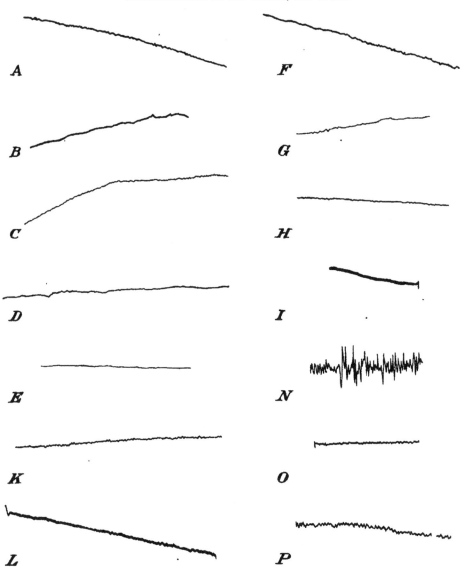

GALVANOMETER AND BATTERY RECORDS, 1895 TO 1898.

2. Disturbances Secondarily Affecting the Galvanometer Needle through a Primary Disturbance of the Bolometer Circuit.

(a) Those causing "drift."[1]
(b) Those causing accidental deflections in the instrumental records.
(c) Accidental deflections from clouds.

(a) As has already been said on page 63, a change in the battery current strength is, with all bolometers, productive of more or less "drift" of the galvanometer needle, owing probably to the different capacities for the radiation and convection of heat of the strips. By care a bolometer may be prepared relatively undisturbed by such slow current changes as occur from a storage battery. Slow change of temperature of the surrounding water jacket produces little drift with a bolometer which is unaffected by slow current changes.

The bolometer No. 20, which was used for all the bolographs taken in 1896-97, as well as all later ones prior to March 31, 1898, is very free from disturbances by either of these causes, and with it the "drift" which remains has been shown to be caused by the rise of temperature in the galvanometer and spectrometer rooms, always attending the taking of bolographs, or to a slow fall of electromotive force of the battery when the battery has but recently been charged.

Bolometer No. 21, which has been used since March 31, 1898, while not badly affected by drift, is still not so free from it as No. 20, so that alterations of the electromotive force of the battery have produced considerable drift in some bolographs. The presence of the observer near the bolometer for the purpose of setting the circle frequently creates a temporary "drift," lasting perhaps five minutes, and this is shown on the records at the beginning and end of bolographs when insufficient time is given for the apparatus to regain its status.

The heating of the observatory, while automatically controlled, is necessarily irregularly distributed, no steam radiator being allowable near the galvanometer. Thus the quite large galvanometer chamber is usually several degrees colder than the surrounding air, and the frequent entrance of the observer gradually warms the air throughout the day, so that a "drift" from the change of temperature of the magnets is often observed when the bolographs are taken. However, even this exceptional "drift" rarely exceeds 2 cm. an hour with the very sensitive apparatus now employed. Oftentimes no "drift" at all will be observed in an hour's run. In the earlier work with the bolometer (as at Allegheny), the "drift," even with a comparatively insensitive galvanometer, often exceeded 10 cm. a minute, and even in more recent practice the light-spot had to be brought back on the scale by "rebalancing" at the resistance box twice or thrice daily.

---
[1] It is significant of the improvements introduced in the last few years that this disturbance caused by "drift," once the most important of all, and which would have needed a chapter for its discussion, has chiefly through the obtainment of uniform temperatures fallen to such insignificance as to be here designated by this subtitle.

Under the conditions of use prevailing in 1896–97, when a millimeter deflection corresponded to 6×10^{-10} amperes in the galvanometer circuit, the spot might be expected to appear on the 20 cm. long scale every day without any change of balance in the bolometer circuit; and it was very seldom that the apparatus got so far out of balance in a week as to give a deflection of 10 cm. on closing the galvanometer switch.

With the galvanometer as employed in 1897–98 at a sensitiveness six or seven times as great, it has only occasionally been necessary to rebalance during the day, though nearly always to balance before beginning the day's work. Thus it will be seen that the "drift," so long the great enemy of the research, is effectually subdued, as far as present conditions are concerned.

(b) Accidental deflections of the galvanometer when the whole circuit is connected appear in Pl. XVII.

(f) Taken September 9, 1895, shows the best record of 1895 with a galvanometer employed at such sensitiveness that a current of 14×10^{-10} amperes gave a deflection of 1 mm. The battery current was 0.075 ampere.

(g) Taken April 16, 1896, shows the performance of the same galvanometer needle with the same current after the introduction of the great storage battery, special rheostat, and inner chamber about the galvanometer.

(h) Taken November 16, 1898, shows a good record taken under the same circumstances as the curves used in preparing the results of 1896–97 here communicated. The battery current was 0.075 ampere, and 1 mm. deflection corresponds to 6×10^{-10} amperes in the galvanometer circuit.

(i) Is a record taken January 9, 1897, with a battery current of 0.098 ampere and the same sensitiveness of galvanometer. The increase of current over that employed in (h) is not without some increase in the magnitude of the accidental deflections.

(o) Is a record taken December 7, 1897, with a battery current from the great storage battery of 0.040 ampere, bolometer No. 20 ($\frac{1}{100}$ mm.), and with the new galvanometer No. 3. The needle used weighs 6.5 mgs. and the time of swing was $4\frac{1}{4}$ seconds. A deflection of 1 mm. corresponds to a current of 1×10^{-10} amperes.

(p) Taken May 18, 1898, is similar to (o), except that bolometer No. 21 ($\frac{1}{100}$ mm.) was used. The current is 0.030 ampere and the time of swing $3\frac{3}{4}$ seconds.

(q) Taken April 1, 1898, is similar to (p), except that ten cells of "Cupron" battery were used. The battery current was 0.028 ampere and time of swing $3\frac{3}{4}$ seconds.

EFFECT OF THE WIND.

(n) Taken January 24, 1898, under the same conditions as (o), but on a day when there was a very moderate breeze, illustrates an effect of the wind. This was noticed in former years, but not in so great a degree owing to the inferior degree of sensitiveness then employed. The galvanometer needle is not affected primarily, as might be supposed, but only through the bolometer circuit. Nor is the mischief done by drafts blowing over lead wires or contacts, or upon the battery, as at one time surmised. The true method of action and the proper remedy was not discovered here until February, 1898, after it had been feared that it would be impossible to use the sensitive galvanometer to advantage on this account, for no day was so entirely calm as to obviate disturbance from this cause. That the original place of the disturbance, the bolometer strips, should have been so long overlooked is due to the fact that they had already been so well protected as to seem impregnable to the assaults of moving masses of air. As I arranged them at first they were inclosed by a brass tube which had a series of diaphragms in front of the bolometer, and the cylinder containing the leads and eyepiece closed the rear end of the tube. The cylindric salt lens was connected by a tight box with the front end of the tube, thus entirely closing that end. The leads and eyepiece were wrapped with thick felt. The case, thus protected, was inclosed in a double-walled chamber, itself within the main walls of the observatory.

It was not until February 23, 1898, when the space occupied by the bolometer was rendered so thoroughly air-tight that it would preserve for many hours a difference of air pressure of one-third of an atmosphere, that the effect of the wind ceased to be noticeable. There is now no disturbance from this cause, and the strongest winds are now no obstacle to successful bolographic work. Atmospheric pressure is employed about the bolometer strips, though it seems probable that a vacuum would render the bolometer more sensitive, but probably would cause some delay of the galvanometer needle in returning to the condition of rest after a deflection.

Measurements on Instrumental Records.

Measurements have been made on "battery records" accompanying bolographs used in the preparation of final results here published. The following table, prepared from such measurements, gives an idea of the magnitude of accidental inflections encountered. Of the records embraced in the table, the first and second are the average and best of type (h) (see page 108), the third and fourth the worst, and an average one of type (i); the fifth an average one of type (o), and the sixth and seventh ones of type (q):

TABLE 15.—*Measurements on accidental inflections of battery records.*

No.	Date of record.	Length of record.	Number of visible inflections of various amplitudes.	Amplitude from crest to trough.	Total length occupied by visible inflections.	Average Length.	Average depth.
				mm.	mm.	mm.	mm.
1	Oct. 26, 1896	45 mm. or 4.5 min.	6	0.1–0.3	8	1.0	0.1
			22	.0–0.1	19		
			Total..28		Total..27		
2	Oct. 31, 1896	60 mm. or 6 min.	22	.0–0.1	11	.5	.05
3	Dec. 10, 1896	54 mm. or 5.4 min.	6	.3–0.5	11	1.5	.14
			4	.1–0.3	7		
			22	.0–0.1	30		
			Total..32		Total..48		
4	Dec. 17, 1896	50 mm. or 5.4 min.	5	.3–0.5	6	1.0	.14
			13	.1–0.3	15		
			22	.0–0.1	19		
			Total..40		Total..40		
5	Mar. 14, 1898. III.	55 mm. or 5.5 min.	6	.8–1.2	....	1.0	.40
			20	.4–0.8	....		
			33	.0–0.4	....		
			Total..59		Total..55		
6	April 1, 1898. IV.	34 mm. or 3.4 m'n.	7	.8–1.4	....	1.1	.46
			6	.4–0.8	....		
			18		....		
			Total..31		Total..34		
7	April 1, 1898. V.	35 mm. or 3.5 min.	6	.8–1.2	....	1.0	.37
			6	.4–0.8	....		
			26	.0–0.4	....		
			Total..38				

If there were no such accidental deflections as those indicated in the preceding table the battery record would be a smooth line and any displacement in bolographs would be due to solar or telluric absorption, and would always have a fixed place.

Such accidental deflections, however, always exist, and being superposed upon the "real" ones corresponding to absorption, they must usually change both the place and amplitude of the latter. These "real" ones are many of them distinguished by their much greater amplitude and all by their fixity of place, but their magnitudes may have the same or even a less value than the average accidental ones, though in the last-mentioned case it can not be shown that they are "real" with certainty. In the illustration which follows we have for clearness supposed the accidental ones to be half as great in amplitude as the "real." We treat, for convenience, all deflections as having the same algebraic sign. The superposition is quite fortuitous and an unlimited number of combinations may result. We consider in illustration only one usual and typical case, where the accidental minimum, without exactly coinciding with the "real one," falls so near it as to displace it.

Fig. 20 is the bolographic curve as it would appear were there no accidental deflections, but showing one real one.

Fig. 21 is the "battery record" curve with its small deflections nearly coinciding in one instance with the real one in the ideal record (fig. 20).

Fig. 22 shows the result of the superposition of the accidental deflections in fig. 21 on the curve (fig. 20), thus producing the actual effect we are discussing—of shifting the position of the "real" minimum in the bolograph. It will be seen that from a single curve, like fig. 22, we can never recover the exact place of the "real" deflection.

The maximum possible shifting of the "real" minimum will be half the period of the accidental deflection superposed on it. This amount of shifting can only occur with "real" deflections of much smaller amplitude than the accidental ones; and as very few of the deflections are considered to be probably "real" that are as small as the larger of the accidental deflections this case will rarely occur. For deflections of equal amplitude the shifting can not exceed one-fourth the period of the accidental deflection superposed; but for the comparatively very deep "real" deflections of which this class is largely composed the shifting will become negligible in all cases.

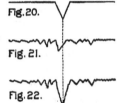

Effect of accidental disturbances.

(c) *Accidental deflections from clouds.*—I had occasion to recognize even in the earlier work that passing clouds which are distinctly visible were less to be apprehended in their effects than the rapid changes which were due to the invisible clouds (if I may call them so); that is, changes of heat transmissibility in what is to the naked eye an almost uniformly clear sky, and in this connection only such small invisible clouds are here referred to, as the effect of visible ones is quite too palpable to require comment, usually reducing the deflections almost or quite to zero. The effect of these invisible disturbers, to which I have long since directed attention, appears in the bolographs beginning with the A line very clearly. Thus in Pl. XX, it will be seen that very considerable deflections exist in the first five or six minutes of arc in the curve below A, but are never found twice alike, and are clearly greater than the battery record would lead us to expect so frequently. Another region, similarly barren of "real" deflections, yet much indented in the separate curves, is the portion between $\rho\ \sigma\ \tau$ and φ. All that can be said with regard to these irregularities is that they exist, but are less uniform in their occurrence than the accidental instrumental deflections, and their spurious indications are therefore more easily recognized in a comparison of curves.

112   ANNALS OF ASTROPHYSICAL OBSERVATORY.

Summing up this discussion of accidental deflections, it may be said (giving values referring to bolographs of 1896–97 first in brackets and those for 1897–98 after) that real deflections less than [0.2] 0.4 mm. amplitude are not identifiable among them; that "real" deflections between [0.2] 0.4 and [0.5] 1.0 mm. are rendered more or less doubtful, and that certainty as to the "reality" of deflections between [0.5] 1.0 and [1.0] 2.0 mm. is attainable only with more pains owing to their presence. "Real" deflections of less than 1.0 mm. in amplitude are in the best plates liable to shifting of from 0.0 to 0.3 mm. by the superposition of accidental deflections, but their effect on the greater "real" deflections is in general negligible.

III.—SOURCES OF ERROR FROM IMPERFECT MECHANISM.
1. Irregularities in the Motion of the Plate.

Several determinations of the error from this source have been made by the method to be immediately described, which, it will be seen, is more rigorous than actually required, since it shows whether or not the plate moves at a uniform rate with respect to a second's pendulum, not with respect to the circle.

Method of determination.—An electric current is arranged from a seconds pendulum clock, so that at every other second, or at every minute, as may be desired, an electro-magnet removes a shutter from before the slit in the front of the plate carrier box. (See Pl. XIV.) A suitable source of illumination is provided by means of a gas lamp and reflecting mirror, so that at the instant the shutter is thus removed a flash of light falls on the photographic plate, moved as usual by the driving clock. After the plate has received several of these records it is developed and measured on the comparator and the deviation of its intervals from equality thus obtained.

By repeated tests of this kind it has been abundantly proved that no deviations from regularity exceeding 0.1 mm. occur in the whole 60 cm. motion of the plate running at the usual speed. The accompanying table shows the results obtained from one such determination:

TABLE 16.—*Test of Plate motion.*
[Observations of May 16, 1896.]

No.	Distance of interval from start.	Length of interval, mean of two measurements.	Deviation of intervals from their mean.	Algebraic sum of residuals.	No.	Distance of interval from start.	Length of interval, mean of two measurements.	Deviation of intervals from their mean.	Algebraic sum of residuals.
	cm.	cm.	cm.	cm.		cm.	cm.	cm.	cm.
1	0.00	0.998	−0.0028	−0.0028	31	29.86	0.997	−0.0018	+0.0000
2	1.00	.9975	− 23	− 51	32	30.85	−.996	− 08	− 08
3	1.99	.997	− 18	− 69	33	31.85	.995	+ 02	− 06
4	2.99	.9965	− 13	− 82	34	32.84	.9935	+ 17	+ 11
5	3.98	.995	+ 02	− 80	35	33.84	.9925	+ 27	+ 38
6	4.98	.996	− 08	− 88	36	34.83	−.9975	− 23	+ 15
7	5.97	.9965	− 13	−.0101	37	35.83	.994	+ 12	+ 27
8	6.97	.9955	− 03	− 104	38	36.82	.9965	− 13	+ 14
9	7.96	.996	− 08	− 112	39	37.82	.9945	+ 07	+ 21
10	8.96	.993	+ 22	−.0090	40	38.81	.9955	− 03	+ 18
11	9.95	.995	+ 02	− 88	41	39.81	.9925	+ 27	+ 45
12	10.95	.991	+ 42	− 46	42	40.80	.9945	+ 07	+ 52
13	11.94	.999	− 38	− 84	43	41.80	.995	+ 02	+ 54
14	12.94	.9935	+ 17	− 67	44	42.79	.994	+ 12	+ 66
15	13.93	.9925	+ 27	− 40	45	43.79	.9975	− 23	+ 43
16	14.93	.996	− 08	− 48	46	44.78	.9935	+ 17	+ 60
17	15.92	.9955	− 03	− 51	47	45.78	.9975	− 23	+ 37
18	16.92	.9945	+ 07	− 44	48	46.77	.994	+ 12	+ 49
19	17.91	.995	+ 02	− 42	49	45.77	.996	− 08	+ 41
20	18.91	.9955	− 03	− 45	50	48.76	.9955	− 03	+ 38
21	19.90	.9925	+ 27	− 18	51	49.76	.9945	+ 07	+ 45
22	20.90	.993	+ 22	+ 04	52	50.76	.998	− 28	+ 17
23	21.89	.995	+ 02	+ 06	53	51.75	.9975	− 23	− 06
24	22.89	.996	− 08	− 02	54	52.75	.9945	+ 07	+ 01
25	23.88	.9935	+ 17	+ 15	55	53.74	.998	− 28	− 27
26	24.88	.9965	− 13	+ 02	56	54.74	.9935	+ 17	− 10
27	25.88	.9915	+ 37	+ 39	57	55.73	.995	+ 02	− 08
28	26.87	.9945	+ 07	+ 46	58	56.73	.995	+ 02	− 06
29	27.87	.998	− 23	+ 16					
30	28.86	.995	+ 02	+ 18		¹0.9952			²−0.0112

¹ Mean.   ² Maximum numerical value of irregularity.

2. IRREGULARITIES IN THE MOTION OF THE CIRCLE WITH RESPECT TO THE PLATE.

Several methods of conducting this test have been proposed and tried, but none have proved wholly satisfactory, owing to the extreme degree of precision required to determine irregularities *of one-half second of arc* in the rate of a continuously moving circle whose total travel in an hour is only thirty minutes.

Various methods of multiplying the circle motion by increasing the effective radius, or even doubling the motion, were tried, but proved to be attended with too much error to be useful. Recourse has therefore been had to a less elegant but more trustworthy expedient. In the absence of a suitable light grating a small glass plate was covered with a very weak solution of hydroquinone and dried, so that its surface was thus coated with very minute crystals. This glass plate was affixed in a horizontal position to a brass arm which extended radially slightly beyond the edge of the circle and was firmly clamped so as to turn with it. The brass arm was open beneath the center of the glass plate so as to allow of good lighting of the latter from beneath. A compound microscope was then solidly clamped to the base casting of the circle and focused upon the crystals covering the glass plate. In the eyepiece of the microscope was a micrometer.

The system of time flashes upon the photographic plate described on the preceding page 112 was made to operate, and a second shutter similar to that there described was adjusted to partially cut off the beam of light, except when opened by a second electro-magnet in a separate circuit operated at pleasure by the observer at the circle with a telegraph key. The part played by the observer was as follows: Having started the driving-clock and taken his station at the microscope, he selected a crystal in the field and observed and telegraphed upon the moving plate its transit across a certain line of the eyepiece micrometer, and again its transit across a second line after an interval of about a minute. A new crystal was then selected and its passage across the same pair of lines similarly recorded, and the process was repeated till the end of the test.

So good was the definition, but withal so high the magnification, that it was found that an accuracy reaching to one part in a hundred in the intervals thus registered was in general attainable, corresponding to something less than one-tenth millimeter on the plate.

By the application of this method it was shown that the motion of the circle was so far irregular (in June, 1896) as to give rise to sensible errors in the positions on the 60 cm. plate, and further experiment showing that this irregularity was not due to the clock work, a new worm and wheel segment by Warner & Swasey was procured and installed in September, 1896. This new mechanism moves the circle within the limits of accuracy of the means of testing its motion, as is shown by the following table. Two tests are presented in the table. The second column gives the distances in centimeters from the beginning of the plate to the space intervals determined from observed transits as above described and tabulated in column 3. Column 4 gives the residuals between the observed space intervals and their mean, and in column 5 is presented the algebraic sum of these residuals which would, if the observed intervals covered the plates continuously, show the total departure from regularity in the motion as in the table of page 113. Under the circumstances, however, as the end of one interval is not the beginning of the next, a single faulty determination makes this column appear very unsatisfactory for a considerable distance. Though the maximum value of the variation from uniform motion as given by the table (0.36 mm.) is considerably above the 0.1 mm. limit aimed for, yet there seems to be great probability that the error may be due to inaccuracy of the method of testing rather than to inaccuracy in the circle motion, and the maximum error in the latter, certainly below 0.4 mm., is probably not greater than 0.1 mm:

TABLE 17.—*Test of circle motion.*

	I. Date, September 18, 1896, F. E. F. and C. G. A.					II. Date, September 19, 1896, F. E. F.			
1	2	3	4	5	6	7	8	9	10
No.	Mean plate distance of interval.	Interval on plate mean of two comparator measurements.	Deviation of interval from mean.	Algebraic sum of residuals in column 4.	No.	Mean plate distance of interval.	Interval on plate; mean of two comparator measurements.	Deviation of interval from mean.	Algebraic sum of residuals in column 9.
	cm.	*cm.*	*cm.*	*cm.*		*cm.*	*cm.*	*cm.*	*cm.*
1	0.00	0.523	−0.012	−0.012	1	0.00	0.625	−0.002	0.002
2	0.86	.558	+ .015	+ .003	2	0.90	.618	− .005	− .003
3	1.56	.550	+ .007	+ .010	3	2.13	.630	+ .007	+ .004
4	2.79	.540	− .003	+ .007	4	3.04	.622	− .001	+ .003
5	4.04	.548	+ .005	+ .012	5	4.40	.632	+ .009	+ .013
6	4.94	.560	+ .013	+ .025	6	5.37	.625	+ .002	+ .015
7	5.80	.538	− .005	+ .020	7	7.05	.612	− .010	+ .005
8	7.61	.538	− .005	+ .015	8	7.69	.618	− .005	+ .000
9	9.10	.532	− .011	+ .004	9	8.37	.615	− .008	− .008
10	10.99	.535	− .008	− .004	10	8.72	.600	− .023	− .031
11	10.73	.553	+ .010	+ .006	11	9.32	.632	+ .009	− .022
12	11.68	.540	− .003	+ .003	12	10.42	.630	+ .007	− .015
13	12.90	.555	+ .012	+ .015	13	11.58	.632	+ .009	− .006
14	14.33	.548	+ .005	+ .020	14	12.32	.640	+ .017	+ .011
15	15.97	.510	− .033	− .013	15	12.44	.622	− .001	+ .010
16	16.67	.548	+ .005	− .008	16	14.53	.622	− .001	+ .009
17	17.33	.568	+ .025	+ .017	17	14.83	.618	− .005	+ .004
18	18.07	.550	+ .007	+ .024	18	15.62	.635	+ .012	+ .016
19	20.35	.538	− .005	+ .019	19	15.93	.628	+ .005	+ .021
20	21.17	.552	+ .009	+ .028	20	16.63	.610	− .013	+ .008
21	22.13	.535	− .008	+ .020	21	17.48	.610	− .013	− .007
22	23.66	.530	− .013	+ .007	22	18.04	.625	+ .002	− .005
23	24.33	.535	− .008	− .001	23	18.78	.622	− .001	− .006
24	25.55	.542	− .001	− .002	24	19.06	.635	+ .012	+ .006
25	27.22	.545	+ .002	+ .000	25	20.34	.615	− .008	− .002
26	27.97	.535	− .008	− .008	26	20.85	.612	− .011	− .013
27	29.21	.560	+ .017	+ .009	27	22.10	.615	− .008	− .021
28	30.26	.535	− .008	+ .001	28	21.46	.620	− .003	− .024
29	31.56	.545	+ .002	+ .003	29	22.90	.618	− .005	− .029
		¹ 0.543			30	24.11	.628	+ .005	− .024
	F. E. F ends; C. G. A. begins.				31	24.90	.638	+ .015	− .009
					32	27.03	.612	− .011	− .020
30	36.15	0.740	+ .004	+ .007	33	27.94	.625	+ .002	− .018
31	37.95	.758	+ .022	+ .029	34	29.89	.610	− .013	− .031
32	39.92	.722	− .014	+ .015	35	30.53	.620	− .003	− .034
33	40.96	.738	+ .002	+ .017	36	31.95	.622	− .001	− .035
34	42.11	.760	+ .024	+ .041	37	35.75	.622	− .001	− .036
35	44.27	.745	+ .009	+ .050	38	36.11	.620	− .003	− .039
36	45.14	.730	− .006	+ .044	39	37.18	.620	− .003	− .042
37	47.32	.725	− .011	+ .033	40	38.15	.628	− .005	+ .037
38	48.94	.738	+ .002	+ .035	41	39.37	.635	+ .012	− .025
39	50.14	.722	− .016	+ .019	42	39.98	.645	+ .022	− .003
40	51.05	.718	− .018	+ .001	43	41.40	.620	− .008	− .006
41	53.90	.732	− .004	− .003			¹ 0.623		² −0.036
42	56.44	.740	+ .004	+ .003					
		¹ 0.736		² +0.047					

¹ Mean. ² Maximum value of the irregularity.

3. A DETERMINATION OF THE EXACT RATIO OF PLATE TO CIRCLE SPEED.

The method and result of the determination of this constant are here given, since the method comes in a similar category with those just cited.

The plate is 600 mm. long, and as it is wished to secure an accuracy of 0.1 mm., it is necessary that the ratio of circle to plate motions be known to one part in 6,000. While this degree of accuracy of position on the plates was sought primarily for the purpose of their comparison, yet it will be desirable to retain it, if possible, in passing from linear measurements to angular ones.

The method employed in measuring the ratio of linear distances, on the plate, to corresponding differences in prismatic deviation is as follows: The clock gears are changed so as to give the most rapid rate of circle motion compared with that of plate. A photographic plate is then inserted and the driving part of the clock disconnected, so that there is simply a train of gears moving circle and plate simultaneously. With these arrangements the observer marks the plate at one end (photographically) through the slit in front of the plate carrier, reads the circle with care, turns the gearing till the other end of the plate is reached, and again marks it and reads the circle. Thus for the usual interval of plate, sixteen times the usual interval of circle, is read, and the errors of reading the circle become unimportant. In this way the following values have been obtained as the mean of eight determinations on December 22, 1896, and of six determinations on January 6, 1898, respectively:

December 22, 1896 : 1 cm. of plate $=30.001'' \pm 0.001$ of circle.

January 6, 1898: 1 cm. of plate $=29.999'' \pm 0.0016$ of circle.

Whence, remembering that $1°$ of circle motion corresponds to $2°$ of spectrum motion, we have:

December 22, 1896, 1 cm. of plate $=01' \, 00.002'' \pm 0.002''$ of spectrum.

January 6, 1898, 1 cm. of plate $=00' \, 59.998'' \pm 0.003''$ of spectrum.

IV.—SOURCES OF ERROR IN PREPARING LINEAR SPECTRA FROM BOLOGRAPHS.

While the errors introduced in this process are of less importance since the final data for the position of lines are made to rest on comparator measurements on the curves themselves, yet the investigations made in the determinations of these errors are perhaps worth a word of allusion.

1. ERRORS IN POSITION FORMERLY INTRODUCED BY THE USE OF A CYLINDRIC LENS.

As has been remarked, it was found that in "cylindrics" or linear spectra prepared by the aid of a cylindric lens, lines corresponding to deflections near the edge of the plate were somewhat curved and at a very considerable angle with those corresponding to deflections at the center of the plate. To determine the full amount of distortion a plate 60 cm. long and 20 cm. wide was made opaque except at certain dots at measured positions, so situated as to fairly represent all parts of the surface. A "cylindric" was then made with the distance from the original plate to the cylindric lens and from this lens to the copy, each 100 cm. Upon the film of the copy after development were scratched lines parallel to the axis of ordinates at the measured abscissæ of the dots on the original plate.

Comparing the scratched lines with the photographed ones, it was found that if a strip running lengthwise through the central portion of the "cylindric" should be cut out, as would be done in making a line spectrum, lines corresponding to dots on the extreme edges of the original plate would be out of place from 2 to 3 mm. relative to those representing dots at the center of the plate.

Besides this, a progressive crowding together of the lines of the copy plate was observed in passing from the center of the plate toward the ends, owing to the different thickness of glass in the path of the rays of light producing the lines.

This experiment was conclusive against the use of the cylindric lens even in the production of composites for illustration. The remaining discussion will assume the use of the slit as described on page 73.

2. BROADENING OF THE LINES FROM THE USE OF THE SLIT.

(a) *From geometrical considerations.*—If a be the width of a spot of light on the original plate, b the width of the slit, c and d the distances from the original plate to the slit and to the copy, respectively, then the total width of the line corresponding to the original spot will be:

$$\frac{b\ d + a\ (d-c)}{c}$$

or, if $d = 2\ c$ as is usual, simply:

$$a + 2\ b.$$

The illumination of this width will not, however, be uniform, but will fall off at the edges.

(b) *Broadening of the lines from diffraction.*—The broadening from this source will be less the wider the slit we are using in place of the lens. Hence it is clear that the best width of slit to employ will be a question of balance between the effects (a) and (b), the one increasing, the other decreasing, with increasing slit width. Accordingly a series of exposures with various slit widths was made, and it was found that the most satisfactory definition was obtained at a width of 0.4 mm. The excess of the width of the lines over the width of the original blocked-in deflections was scarcely noticeable, since the illumination at the edges was too faint to affect the plate during the time of exposure.

3. IMPERFECT ADJUSTMENTS OF THE CAMERA.

Let the figure represent at A the original plate with its blocked-in curve, at A' the copy plate, and midway between them the slit, all represented in side elevation. Let $\alpha = \alpha'$, and let α be the maximum value allowable from considerations of length of exposure and rigidity of apparatus, which is about 120 cm. Let us first see what effect is produced by an angular displacement of A' B'.

Assuming $x = 30$ cm., or one-half the length of a 60 cm. plate, and $\alpha = \alpha' = 120$ cm., then if A' B' = A B + 0.1 mm., we have

$$\Delta \theta = \frac{\alpha\ (A'\ B')}{x^2} = 0.0013 = 4.5' \text{ of arc.}$$

The part below A' will be somewhat shortened. To determine what accuracy will be required in the relative lengths α and α', we will assume that the plate copied and the copy plate are accurately perpendicular to the line AA', then for an accuracy of 0.1 mm. in the length of the copy plate,

$$\Delta \alpha' = \frac{\alpha}{x} \Delta x' = \frac{120}{30} \times 0.05 \text{ mm.} = 0.2 \text{ mm.}$$

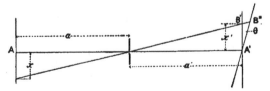

FIG. 23. Adjustments of camera for production of linear spectra.

It is clear that if the planes of the two plates and of the slit are not parallel the lines appearing upon the copy plate will not be parallel. No analytical investigation of the degree of perfection required in this adjustment has been made, since the camera may be so adjusted by trial that errors from this source may be minimized.

From the above considerations it is thought improbable that a greater photographic accuracy in linear translation of full-length plates than 0.2 mm. can be obtained. But as the amount of error from each of the sources enumerated, excepting the want of parallelism between slit and plates, decreases with the length of plate employed, it is probable that for short plates of 30 or 20 cm. the error will not amount to more than 0.1 mm.

4. Personal Error in Blocking in.

The composite process is practically automatic and independent of individual judgment, except for the need of "blocking in" by hand and the introduction of the "personal equation" of the person who conducts this part.

To determine the magnitude of the personal errors arising from this source a curve about 20 cm. in length, in which deflections occurred typical of all forms which are met with, was blocked in successively and independently by three persons who were ignorant of each other's work until the respective linear potographs were compared. Each observer measured on the comparator the positions of the minima on the curve and the corresponding lines on his linear translation, making two series of measurements in each case. The average difference between the means of these two series of measurements was considered to be the measure of the personal error introduced by the observer in blocking in.

The results obtained are as follows:

Observer.	Mean personal error of blocking in.
	mm.
F. E. F.	0.16
R. C. C.	.27
C. G. A.	.32

so that the mean of these personal errors is 0.25 mm

5. Errors in Drawing Line Spectra by Hand.

The composite process involves so much labor that the linear spectra illustrating the results of 1897–98 were drawn to scale from the tabulated values, and appear in Pls. XX, XXIV and XXV. Pls. XXIV and XXV are generally accurate to within 0.1 mm.; that is to say, wave lengths of lines read from them by the aid of the annexed scale are usually correct to 1 Ångstrom unit. Pl. XX is accurate to about 0.4 mm., and is to be looked upon rather as an illustration than as a standard of reference.

V.—ERRORS IN COMPARATOR MEASUREMENTS ON THE CURVES.

1. Errors Due to Inaccuracy in the Graduation of Scales and in Reading them.

The scales of the comparator were graduated by Messrs. Brown & Sharpe with their very accurate dividing engine. This firm reports that no errors greater than one two-hundredth millimeter were found in the scales upon comparison. Since all measurements are made with the same instrument, it is evident that it is sufficient for present purposes that the scale should be accurately divided, which it appears by the above report is the case, and its absolute length in terms of the standard meter need not be here discussed.

The scales are graduated to half millimeters, and are read by verniers to fiftieths of a division; that is, to hundredths of a millimeter. The graduations are very fine and the vernier surfaces continuous with the scales, so that there is not the slightest difficulty in reading with certainty to hundredths of a millimeter.

2. Errors from Changes of Temperature.

The coefficient of linear expansion for silver is nearly 0.00002. Hence a change of temperature of 1° would produce a change in length of a 600 mm. scale of nearly 0.012 mm. The coefficient for glass is, however, nearly half as great, and as the room temperature is very uniform and not more than 12° different from that of the observer's body, it is not probable that errors as great as 0.05 mm. are ever introduced in the total length of the plate, so that no correction is applied.

3. Personal Error in Determining the Positions of Minima on Bolographic Curves.

The inflections in the bolographic curves are of all characters from acute angles to slight depressions, with no very definite minimum point. Hence the magnitude of the error from differences in personal judgment as to the situation of a minimum point is very different for different inflections. However, to gain a general idea of the divergence between the opinions of several observers, thirty-four inflections were selected upon a certain curve which well represented all cases likely to occur, and three observers—F. E. F., R. C. C., and C. G. A.—each measured the positions of these inflections, making two separate series of observations in each case. The present very accurate comparator was not then available, so that it is presumable that the same test if now carried out will show even closer agreement. However, the results obtained were but little divergent, and are as follows:

Average deviation of the observations of each observer from his mean: F. E. F.'s observations, 0.035 mm.; R. C. C.'s, 0.156 mm.; C. G. A.'s, 0.022 mm.

Average deviation of each observer's mean from the general mean: F. E. F.'s observations 0.056 mm.; R. C. C.'s, 0.076 mm.; C. G. A's, 0.067 mm.

Average personal error: 0.066 mm.

Average probable error of general mean of three observers: 0.031 mm.

VI.—SOURCES OF ERROR INHERENT IN THE BOLOGRAPHIC METHOD.

It is evident that the deflection of the galvanometer needle must be subsequent to the passage across the bolometer of a point of the spectrum in which there has been an increase or dimunition of its energy. Among the causes of such tardiness may be mentioned the very slight delay of the bolometer strip in assuming the new temperature, the electrostatic capacity of the system, and the inertia of the galvanometer needle, and these and still other causes may conceivably act with unequal results as the change of energy is greater or less.

In fig. 24 the ordinate ending at B is $F(x)$, that at C, $F(x+a)$ and that at D, $F_{(x-a)}$. Then $AB = F_{1(x)}$ where A is the intersection of $F_{(x)}$ with the chord AC. In a similar way we may obtain $F_2(x)$ from $F_1(x)$ when the latter is reduced to the form of a curve. If necessary one may extend this series to additional terms in which $F_3(x)$, etc., would be obtained in a manner analagous to $F_2(x)$.

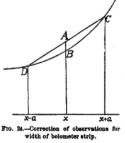

Fig. 24.—Correction of observations for width of bolometer strip.

In the writer's earliest work both slit and bolometer strip were relatively so wide that such considerations were important, but in the recent work of this observatory the slit and bolometer have been so very narrow in angular measure that but slight corrections were to be expected when the procedure indicated by Paschen is followed out. However, to make sure of the matter, several typical deflections in curves of 1897–98 taken with the widest bolometer used (3.0″) were thus treated. It appeared that while the ordinates of the curves were sometimes affected to the extent of a half millimeter or less, the positions of minima were shifted in deviation almost imperceptibly, and never by as much as 0.1 mm., corresponding to 0.6″ of arc. Hence it was concluded that the correction for bolometer and slit widths is negligible.

SUMMARY OF THE RESULTS OF THE PRECEDING DISCUSSION OF THE LIMITATIONS OF THE METHOD, AND OF THE ERRORS IN ITS APPLICATION.

A.—LIMITATIONS.

1. The visual resolution of the great salt prism actually employed is such that about 1,300 absorption lines could be discriminated between $\lambda = 0.76\mu$ and $\lambda = 5.3\mu$, if they exist, if the region was visible to the eye, and if the spectrum was sufficiently pure. Of the lines, 300 could be found below 1.8μ. The resolution of the glass prism employed between $\lambda = 0.76\mu$ and $\lambda = 1.8\mu$ is such that about 2,000 lines would be found in this region under the conditions above cited, or twice as many as with the salt prism. Thus nearly 2,500 lines are theoretically possible to visual observation were the eye sensitive to these radiations.

2. There is reason to believe that bolographic resolution may under conceivable circumstances exceed the visual resolution, and indeed that it has actually slightly exceeded it.

3. The present sensitiveness of the actual bolographic apparatus is so far limited that a slit image of 1.2″ width and a bolometer of 1.2″ width are employed in order that a suitable galvanometer deflection may be obtained when using the salt prism between $\lambda = 0.76\mu$ and $\lambda = 2.2\mu$. For other regions of the spectrum, and for the glass prism in this region, greater widths are necessary.

4. Owing to the refraction of air currents, and to vibrations of the optical apparatus (the latter mostly caused by the passage of vehicles between the observatory and the United States National Museum building), the slit image is nearly always in vibration to the extent of from 0.4" to 1", though at times the confusion is much less and may even sensibly disappear.

5. Electrical and mechanical disturbances of the galvanometer produce a nearly continuous series of accidental inflections in the bolographic curves. The average magnitude of these accidental deflections was in 1896–97 about 0.15 mm. in amplitude and 1.0 mm. in length, while with the more sensitive arrangements of 1897–98 the average amplitude was 0.4 mm. and the average length 1.0 mm.

6. Since the spectrum is thus rendered to some extent impure by tremor, and the smaller deflections corresponding to absorption lines are obscured by accidental inflections of the bolographic curve, the bolometer does not attain the results it could reach under other local conditions.

B.—ERRORS.

1. Temperature changes in the mirror supports produced in 1896–97 slight shifting of the slit image, but under usually good temperature conditions of the observatory it appears that such shifting did not exceed in an hour's run 1" to 2". The shifting observed in 1898 is negligible.

2. Rise of temperature in the salt prism produces a change of deviation of $-12"$ of arc per degree due to decrease in refractive indices, and, under ordinary temperature conditions (a rise of $0.0°$ to $0.2°$ per hour), this would give a change in length of bolographs corresponding to $-0.0"$ to $-3"$.

Such a rising temperature as just mentioned produces the paradoxical effect of an increase in *all three* angles of the prism, causing a change of deviation of 0.0" to 2".

3. These temperature errors (1 and 2) counteract each other so exactly as to make it undesirable to apply separate corrections for them. Residual errors between the limits 0.0" to 1.0" may be tolerated.

4. Instrumental records taken at various times show that advantage in steadiness of the galvanometer record has come, (*a*) from the use of a Julius suspension with the further improvements which Mr. Abbot has introduced in that method; (*b*) from the system of mercury flotation; (*c*) from using an air-tight galvanometer case. The usual day record is nevertheless still inferior to the night record by reason of tremors set up by neighboring traffic, but how slight the result of these tremors has become with the Julius suspension mercury flotation the illustration in Pl. XVII will show.

5. The present conditions are such that drift, once a most formidable and apparently most hopeless difficulty, is no longer a serious drawback. Few curves of an hour's duration are taken in which the drift is less than 1 cm., but where a drift of more than 3 cm. occurs it is unusual, and due either to rise of temperature in the gal-

vanometer room, or to a gradual fall in the electromotive force, such as may occur for several weeks after the battery has been freshly charged. This latter cause may be avoided by a preliminary partial discharge through a suitable resistance.

6. Among the total instrumental accidental deflections must be included those due to the singular effect of the wind upon the battery record (traced to its obscure cause and entirely overcome only in February, 1898) as well as to fluctuation of the current employed in the circuit and other conditions. In the best bolographs of 1896–97 the accidental deflections in the battery record averaged in amplitude 0.15 mm.; in period 1.0 mm. In those of 1897–98, taken with a galvanometer six times more sensitive, the average magnitude was in amplitude 0.40 mm.; in period 1.0 mm.

7. In bolographs of 1896–97 all "real"[1] deflections of less than 0.15 mm. in amplitude were lost among these false ones, and those of less than 1 mm. amplitude were liable, by reason of the superposition of the accidental ones, to a displacement of from 0.0 mm. to 0.25 mm.

The corresponding limits for bolographs of 1897–98 are: To be obscured 0.4 mm.; displacement for deflections below 2.0 mm. amplitude, from 0.0 mm. to 0.25 mm.

8. Irregularities in the motion of the plate may be neglected, as experiments show they never exceed 0.1 mm.

9. Irregularities in the motion of the circle are so slight as to be not susceptible to certain measurement by either of several methods used, but are shown to be at their maximum within 2.0″, and are probably much less. The probable error does not exceed 0.5″.

10. As the so-called "cylindric" or automatic linear translation is only used here in illustration, it will be enough to observe that each line of such a translation is subject to displacement from "blocking in," the average displacement varying with different persons, but having a mean value of 0.25 mm. Photographic processes involved in the production of linear bolographs may produce displacements at the extremes of a 20 cm. plate of 0.1 mm. and proportionally greater displacements for longer plates.

11. In determining the position of minima on bolographs the average deviations of the mean of two comparator observations by one observer from the mean of six observations on the same curve by three observers was found to vary with different persons from 0.022 to 0.16 mm., but to give a mean of about 0.07 mm. With the comparator at present used the agreement of such observations would be without doubt still more exact.

12. Curves have been taken in opposite directions through the spectrum to eliminate errors from tardiness of the galvanometer indications, and the error not thus eliminated is confined to a comparatively small number of unsymmetrical deflections where it may reach 0.1 mm. to 0.2 mm.

---

[1] By "real" deflections are to be understood those due to actual telluric or solar absorption.

13. Errors from the width of slit and bolometer strip are, as has been shown here, negligible, on account of the narrowness of the slit and bolometer strip here actually employed.

From the preceding discussion it may be expected that the relative angular deviation of any well-marked absorption line in the infra-red is capable of determination from six bolographs with a probable error of less than one second of arc, and an examination of the tabular results given later will show that this expectation is more than realized.

Chapter VI.

THE ABSORPTION LINES IN THE INFRA-RED SOLAR SPECTRUM—RESULTS OF BOLOGRAPHIC SPECTRUM ANALYSIS IN 1897-98.

In presenting the following results of a bolometric investigation to determine the position of absorption bands in the infra-red solar spectrum, it should be remarked at the outset that they are not to be regarded as the most complete which can be obtained; for throughout the investigation experience has continually suggested improvements in apparatus or methods of procedure which have yielded on trial valuable gains in result. It is a sufficient illustration of this to say that in the last year of the investigation, by the construction of a more sensitive galvanometer and by other improvements, more lines were discovered than were previously found in all the preceding years which have elapsed since bolometric investigation of the infra-red solar spectrum was inaugurated. It is safe to say that there is reason to suppose that by a patient continuance of the research, though preferably in a more undisturbed locality, the number of lines discovered in this region beyond the reach of photography might be increased several fold. It has, however, been felt desirable to bring this long investigation to at least some provisional halt and to proceed to other lines of work in which the experience and equipment obtained in the present research may be made immediately available. The results already obtained are therefore published in the hope that they contain a useful contribution to astrophysical knowledge.

It has already been said that it had been intended to publish in 1897, but owing to reasons connected with the relations of the Astrophysical Observatory to the Government the necessary funds for the publication were not available. The delay, though regretable, has not proved wholly unfortunate, because the results obtained have in the meantime been greatly enriched. Nevertheless, inasmuch as the spectrum map prepared in 1897 was ready for the press, and as the results of that year were more fairly applicable than those since obtained for a comparison of the dispersion of fluorite and rock salt included in Part II, such of these results of 1896–97 as are of interest are retained and will be found at various points where they remain applicable, especially in Part II. We will next examine the work of the years 1897 and 1898.

EARLY BOLOGRAPHS OF THE INFRA-RED SOLAR SPECTRUM.

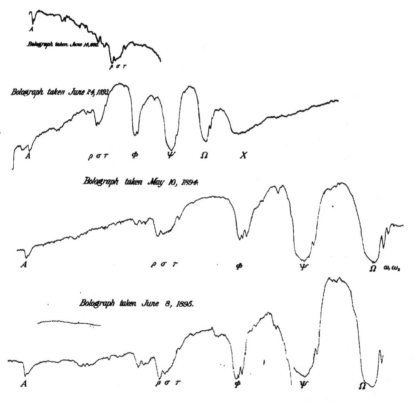

PLATE XIX.

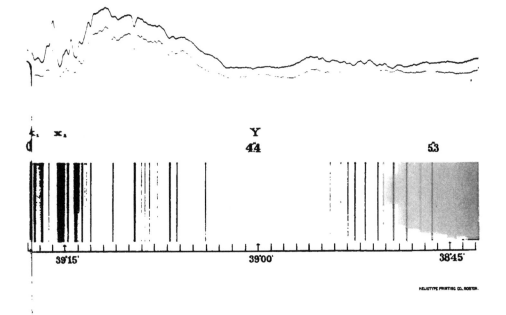

ILLUSTRATIONS.

Plate XIX: To obtain a general idea of these bolographs and of their results, Pl. XIX should be first consulted, which is confined to observations made not later than 1896 and exhibits for the convenience of the reader in the same view both the linear spectrum and the superposed energy spectra from which it is derived. By a comparison, therefore, of Pl. XIX with Pl. XX, the great gain in detail obtained with the latest and most sensitive apparatus will be realized.

Plate XX: This plate exhibits the final result of a long research originally due to a discovery made by the writer upon Mount Whitney at an altitude of 12,000 feet, in September, 1881,[1] a discovery consisting of a great and hitherto unsuspected region since known by the general name of the lower infra-red spectrum, for, I repeat, it was at this time (1881) not merely unknown in its details, but not known as having any existence. The discovery, once made in the clear air of this great elevation, could be and was followed at lower altitudes by a prolonged research which was intended to map out what may be called this newly discovered country, and the writer commenced this work at Allegheny immediately upon his return. One of the earliest graphic reports of his results was communicated to the British Association for the Advancement of Science, at its Southampton meeting in 1882, and will be found reproduced in Nature of September 11 of the same year.

The present map consists of two parts on the same plate. In the upper part of the plate we have a transcript of the energy spectrum formed by the great salt prism as it appeared on two days in 1898 to the bolometer. The more refrangible portion extends from the limit of the visible spectrum (A at 0.76μ) to the limit of the formerly known spectrum[2] (Ω at 1.8μ).

The less refrangible portion is that which includes the lower infra-red spectrum, first discovered by the writer in 1881 on Mount Whitney. This lower portion beginning at Ω is extended on this map (to the right) by subsequent observations made

---

[1] The writer is not now certain of the earliest printed reference to this discovery, but one will be found in a letter of September 11, 1882, to l'Academie des Sciences, and published in the Comptes Rendus des Séances de l'Académie des Sciences (vol. 95, p. 482).

It had been previously inferred by the writer that the spectrum would be found continued to a length of as much as 3μ, but at this time no certain observations appear to have been made below a wave length of 12,000, although it seems that M. Mouton had observed a cessation of heat at what is now known as about 1.8μ, being, in fact, the approach to Ω. The essence of the author's discovery on Mount Whitney was that beyond this a great till then quite unknown region existed where the spectral heat, whether in that of of a prism or a grating, if less than in the upper portion of the infra-red, was of indefinite extent.

See also Report of the Southampton meeting of the British Association for the Advancement of Science, 1882, p. 459; Nature, October 12, 1882; Annalen der Physik und Chemie, Neue folge 22, p. 598, 1884; American Journal of Science, Vol. XXV, March, 1883; Professional Papers of the Signal Service, No. 15, p. 132; American Journal of Science, XXVII, March, 1884.

[2] It will be remembered that in the view of eminent authorities at this time (1881) the utmost theoretical limit of the infra-red spectrum would not be greater than 1.2μ. So far as is known to the writer, it was his own observation with the bolometer in the prismatic spectrum, used subsequently in connection with the grating, which first demonstrated that the point to which the solar heat spectrum extended at least included this wave length of 1.8μ, and that therefore rays passed through the prism which the formula of Cauchy declared could not do so.

with the bolometer to the vicinity of 5.3μ. It is well known by the results of recent years that some solar heat, though in very limited quantity, is found below this point. The writer has indeed described this in previous writings, but it is insignificant in amount and is not included in the present map.

In the lower part of the same plate and immediately beneath the energy spectrum is the linear spectrum. To prevent misapprehension, let it be repeated that it is not derived solely from the two just described curves given above it, which are direct reproductions from observation, given merely in illustration, but is obtained from the collective means of very many such curves in the way which is about to be described and of which the general features have already been given. Of the nearly 600 lines given in the linear spectrum, every one, without exception, has been subject to the comparison already stated, establishing the coincidence of each of the corresponding 600 deflections on a great number of independent bolographs. It is the result of the entirely independent comparison of at least two observers, each using at least six equally independent original records. The linear spectrum on Pl. XX may be taken, then, as the graphic representation of the results found in the table shortly to be given. It will be understood from what has just been said that the energy curves in the upper part of the plate have not the same authority as the linear spectrum, as they are only transcripts of single observations, given for the convenience of the student, who may wish to get an approximate idea of the energy involved in the production of each line of the linear spectrum in the lower portion of the plate. They were taken with the great salt prism R. B. I., at the speed ratio 1 cm. of plate$=1'$ of spectrum$=1$ minute of time.

Pls. XXI A and XXI B are a similar couple of superposed pairs of bolographs, taken with the great 60° glass prism at the same speeds, but owing to the greater dispersion of the glass these cover only the region between A and Ω. Pl. XXII, also from bolographs taken with the glass prism, shows the bands ω_1 and ω_2, which were supposed until recently to be single lines, but are now found to be similar to the A line, containing many fine lines, and, like A, much influenced in character by the altitude of the sun.

It may be seen by comparison of Pls. XIX, XX, and XXI both what gain in detail was made between the years 1896 and 1897, and what advantage is gained by using the glass prism. In all these curves the reader will see that in certain portions minute casual deflections appear in which the irregularities of the so-called battery record are mingled with minute solar ones. He will, in seeing and comparing these latter with the larger undoubtedly solar ones, be better able to judge how large a number of lines may be recognized, even at the first inspection, as undoubtedly real; and if he will consider that the discrimination does not depend upon this first visual impression from two curves, but, on the contrary, on a systematic study and comparison of at least six such curves, by the aid of the micrometric comparator, as described

BOLOGRAPHS OF THE SOLAR SPECTRUM ABSORPTION BANDS ω_1 AND ω_2 SHOWING THE INCREASE OF ABSORPTION IN ω_1 WITH DECLINING SUN.

OBSERVATIONS OF S. P. LANGLEY WITH THE ASSISTANCE OF C. G. ABBOT.

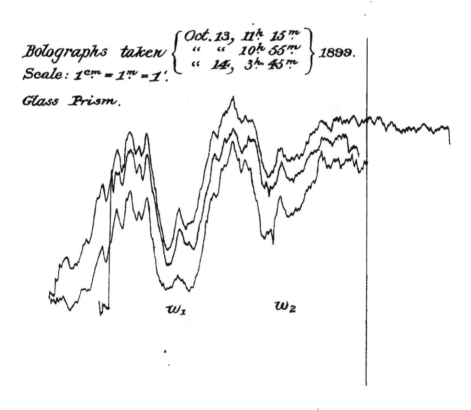

in Chapter IV, he will be in a position to form some personal judgment as to the confidence with which he may accept the lines given as "real." I regret that the comparison will still be unfavorable to the curves, in that these are lithographically printed and are unavoidably greatly inferior to the original glass plates in detail, so that they by no means give the impressions that the originals themselves do in regard to the reality of the very small deflections.

Before proceeding to further discuss the results, attention is called to Pl. XXIII, which gives an idea of the energy curve of the rock-salt prismatic solar spectrum from the F line in the green to near Y in the distant infra-red. The two bolographs here shown superposed were taken immediately in succession at the speed 1 cm. $= 4'$ arc $= \frac{1}{2}'$ time. These speeds are quite unsuitable to show detail, but nevertheless the originals show above one hundred "real" deflections between F and A, and about the same number below A in the infra-red. The D lines are of course shown as one deflection here owing to the rapid speeds. The slit width was altered three times during each run in order to vary the height of the curve so as to keep it on the photographic plate. Thus the scale of ordinates is 100 from F to near C; 50 from thence to below A; 20 from thence to X, and 100 from thence to the end. This has produced several marked breaks in the curve which will be immediately recognized.

Pls. XXIV and XXV give upon the normal or wave-length scale the positions of the lines discovered. No attempt has been made to shade this map, but the positions of great absorption bands are indicated by brackets, and the relative intensity of lines is approximated.

COMPARISON OF BOLOGRAPHS.

We here enter on a matter of great importance to the reader who desires to satisfy himself of the degree of credibility of the entire record.

When we approach the limits of vision or audition, or of perception by any other of the human senses, no matter how these may be fortified by instrumental aid, we finally perceive, and always must perceive, a condition still beyond, where certitude becomes incertitude, although we may not be able to designate precisely where one ceases and the other begins.

This is always the case, it would seem, on the boundaries of our knowledge in every department, and it is so here.

It is impossible, for instance, to look at the great and notable deflection of a line such as A, or to the deflections corresponding to yet larger bands below it, and to see these in exactly the same place on scores of plates taken for years together without feeling an absolute certitude of their real existence as regions of special absorption in the solar or terrestrial atmosphere. After longer study it is found that as absolute a certainty exists as to many hundred smaller lines seen in the same conditions, and yet as we improve our apparatus and recognize still minuter solar deflections, we finally come to a condition where these are reduced to the same order of magnitude as those which may be due to earth tremors and to similar accidental distubances, which are here represented by the irregular line which is called the "battery record."

But, it may be asked, are we not entitled to demand that these last should somehow be eliminated altogether and the "battery record" become a perfectly smooth line? The answer is, that this can never be.

As seismography improves, it becomes more clear that there is no part of the earth's surface free from constant tremor; as the refinements of electrical science advance we constantly discover earth currents where they were not perceived before; as we multiply the sensitiveness of our measuring apparatus, till it comes to what seems almost indefinite delicacy, we find that the most massive apparatus and the most refined precautions which we may take, do not prevent the existence of all but infinitesimally small accidental disturbances, nor of the notation of their sensible effects if the record itself be only minute enough, for this record is a testimony, in fact, to the sensitiveness of the apparatus itself, and minute disturbances are always to be found if the observation itself which deals with them provides in itself the means of detecting them.

It fell to the writer once to establish a permanent meridian instrument whose supports he desired to build up with every condition of stability which experience and caution could suggest. He personally looked to the obtaining of the required blocks of granite at the quarry and to laying them in the same way in the foundation of the observatory on its bed rock as they lay in the original bed, and he superintended the placing of those, one upon the other, until the foundation was laid for the piers which finally supported the instrument, and which were chosen with the same care. He believed that this instrument was as solidly mounted as anything on the earth could be. He used it for many years in his observations with a confidence justified by the results; but these observations required a powerful telescope, and there was no time at which a tap of the fingers on the side of the monolithic piers which carried the telescope would not be accompanied by an apparent leap in the heavens of the star on which it was directed—a statement which will not surprise any professional astronomer. It is made here to emphasize the like statement that there is, then, no limit to our power of perception of tremors. These are, it will be remembered, instances which may be paralleled in illustrations drawn from the use of other senses, and not peculiar to the present observation.

Clearly, we may never distinguish the entire number of solar lines which exist here more than we could in visible spectra by the use of the eye or by photography. In every case there must finally come a time when we must stop our investigations because we have reached a degree of minuteness in the solar lines corresponding to the intervening disturbances due to terrestrial causes, which we can never eliminate.

These considerations, it is hoped, will be present with the reader in connection with the method which is here given of determining which of the smaller lines are due to solar or telluric disturbances and which to the tremors and disturbances shown in the battery record.

Twenty-one bolographs, all taken at the usual speed, 1 cm. $= 1' = 1'$, have been measured upon the comparator to determine the ordinates and abscissæ of the deflections judged "real" in preparing the results that follow. Each curve was thus twice measured by each of two observers—F. E. F. and C. G. A.—making in all some 44,000 observations. Besides the curves actually measured in both ordinates and abscissæ, the measures of F. E. F. on the ordinates of two other bolographs are included in the tables, and some twenty additional bolographs were examined in fixing the opinions of the observers as to the "reality" of the deflections whose positions were measured upon the twenty-one bolographs above mentioned. By this means they assured themselves of the reality of minor lines otherwise in some doubt. These facts, and the fact that the bolographs inevitably lose in detail in reproduction, must be kept in mind by the reader who finds this or that line included in the tables, but not, in his opinion, clearly indicated by the bolographs here given as illustrations.

The method of comparison and measurement of bolographs has been already given at page 74, but it will be well perhaps to give a brief review of it here. Two observers—F. E. F. and C. G. A.—each examined the bolographs and together selected some which seemed the most suitable for measurement to the number of 21. The grounds of selection were the excellence of sky conditions and the freedom of the "battery records" accompanying the bolographs from accidental deflections.[1] Curves taken when the battery record had an average accidental deflection (measured from crest to trough) greater than 0.5 mm. would be rejected. Of these 21 excellent bolographs, 13 were taken with the great salt prism "R. B. I." and 8 with the great glass prism. Of the 13 of the first lot 6 covered the region $A - \Omega$, and 7 that from Ω to the region beyond 5.3 μ, where the solar radiation reaching the earth is too feeble for bolographic spectrum analysis with the present apparatus. The 8 bolographs with the glass prism were equally divided between the upper and lower half of the region $A - \Omega$. Each observer independently superposed and carefully examined these bolographs, together with others as good and perhaps in some parts even better than some measured, such as bolographs which have been locally spoiled for measurement by a cloud or an accident which would leave the most of the curve uninjured. About twenty of these auxiliaries were employed. When satisfied by the evidence of magnitude, similar form, and corresponding position and universal or frequent occurrence that any deflection is certainly or very probably "real," i. e., caused by an absorption band in the spectrum, the observer placed a dot of ink opposite to it on each of the bolographs to be measured, together with its number. No deflections were included as probably

---

[1] These "battery records" accompany all the bolographs, and are taken under the same conditions as the curves themselves, except for the absence of solar radiations.

"real" which were smaller than the average accidental deflection in the record curves or "battery records." This lower limit for the amplitude of "real" deflections is 0.4 mm.

Each of the 21 bolographs first mentioned was twice independently measured upon the comparator (verifying in the second measurement all discordant measures), and the positions of the selected "real" deflections were determined both in abscissæ (corresponding to prismatic deviations) and ordinates (corresponding to intensity of radiation). The position read in both ordinates and abscissæ is at the minimum ordinate or bottom of each deflection. In recording these measures there was annexed to the characteristic number a subscript, either a, b, c, or d, whose significance is as follows: Class a includes about a dozen great depressions like Ω, places where for whole minutes of arc in deviation the energy is largely absorbed. Class b includes deflections less marked than these, but still so prominent as to be above 5 mm. in depth. Class c includes deflections between 2 and 5 mm. in depth. Classes a, b, and c include all those where the depth is more than five times the average depth of the accidental deflections of the record curves, and thus, by the ordinary convention of probability, all those assumed to be certainly real on the evidence of their amplitude alone. Class d includes deflections between 0.4 and 2.0 mm. in depth, which are judged to be real because of their universal or nearly universal occurrence with a similar form, and in the same position in the spectrum.

When the first observer has completed his measurements on the four series of plates already described, the duplicate measures on each bolograph are first averaged and the distances in abscissæ all expressed as so many centimeters from some well-marked symmetrical deflection of class b chosen as a reference for the series of similar bolographs. The observer now compares his results for all the bolographs, rejecting or questioning such deflections as seem from divergence in measured position to be spurious or doubtful. The second observer, having independently gone through the same steps (all traces of personal marks on the bolographs used by his predecessor being first removed), the results of the two are compared and all deflections which they both agree to be questionable are eliminated. Some few small deflections are estimated to be real by one observer, but not by the other, and such are designated in the final tables, for the readers notice, as in some degree doubtful. Deflections upon which both observers unite after independently completing all the long procedure of comparison and measurement as described above have, it will be conceded, very strong claims to "reality." The reader will note that in the procedure just described there are furnished four checks on the "reality" of deflections. These depend upon, first, magnitude; second, similarity of form; third, coincidence of position on many bolographs, and fourth, upon the independent judgment of two observers. The reader will clearly bear in mind that the question of "reality" thus decided only arises in case of a comparatively few of the lines here given.

NOTES ACCOMPANYING BOLOGRAPHS.

Passing now to the results, we have first the notes recorded when the measured bolographs were taken. These are given in somewhat abbreviated form in the small type which follows. They are given thus fully so that the conditions may be known under which every bolograph used in these results was taken, if desired.

Bolographs of the region A to Ω taken with rock-salt prism.

[Date, April 1, 1898. Observers, C. G. A. and F. E. F.]

Apparatus.—Slit aperture, 100 mm. by 0.50 mm. Angular width at bolometer, 1.2″. Prism, "R. B. I." System of collimation, cylindric. Galvanometer No. 3. Time of single vibration, $3\frac{3}{4}$ seconds. Bolometer No. 21, 12 mm. by 0.3 mm. Angular width, 1.3″. Current of battery, 0.028 ampere. Spectrum speed 1′ in 1 minute. Plate speed, 1 cm in 1 minute. 1′ = 1 cm.

Bolograph II. A—ω_2.
Temperatures, $\begin{Bmatrix} \text{In prism} = 19.90° \\ \text{In air} = 20.5° \end{Bmatrix}$ at $9^h\ 30^m$.
Start $10^h\ 2^m\ 15^s$. High wind.
Close $11^h\ 2^m\ 30^s$.
Temperatures, $\begin{Bmatrix} \text{In prism} = 20.10° \\ \text{In air} = 20.15° \end{Bmatrix}$
Grade of bolograph, 2_b.[1]

Bolograph XI. A—ω_2.
Temperatures, $\begin{Bmatrix} \text{In prism} = 20.30° \\ \text{In air} = 20.45° \end{Bmatrix}$ at $2^h\ 14^m$.
Start $2^h\ 41^m\ 0^s$.
Close $3^h\ 41^m\ 40^s$.
Temperatures, $\begin{Bmatrix} \text{In prism} = 20.30° \\ \text{In air} = 20.30° \end{Bmatrix}$
Grade of bolograph, 1_b.

[Date, April 2, 1898. Observers and apparatus as on April 1, except that time of single swing of galvanometer is four seconds.]

Bolograph II. A—ω_2.
Temperatures, $\begin{Bmatrix} \text{In prism} = 19.9° \\ \text{In air} = 20.5° \end{Bmatrix}$ at $9^h\ 40^m$.
Start $10^h\ 10^m\ 0^s$.
Close $11^h\ 12^m\ 0^s$.
Temperatures, $\begin{Bmatrix} \text{In prism} = 20.3° \\ \text{In air} = 20.8° \end{Bmatrix}$ at $11^h\ 18^m$.
Grade, 1_b.

---

[1] It may be recalled that the figures 1, 2, etc., in "Grade of bolograph" refer to the state of the sky, the subfigures *b*, etc., to instrumental conditions such as the battery records, and that "Temperatures in prism" are obtained by a thermometer sunk in a neighboring prism of salt similar in size to that used for forming the spectrum.

[Date, April 6, 1898. Observers and apparatus as on April 1, except that spectrum now passes over the bolometer from long wave radiations to short.]

Bolograph II. ω_2—A.

Temperatures, $\begin{cases} \text{In prism} = 18.9° \\ \text{In air} = 19.3° \end{cases}$ at $9^h 45^m$.

Start $9^h 56^m 30^s$.
Close $10^h 56^m 40^s$.

Temperatures, $\begin{cases} \text{In prism} = 19.1° \\ \text{In air} = 19.8° \end{cases}$ at $11^h 0^m$.

Grade, 1_b.

[Date, May 18, 1898. Observers and apparatus as on April 6.]

Bolograph II. ω_2—A.

Temperatures, $\begin{cases} \text{In prism} = 20.4° \\ \text{In air} = 21.4° \end{cases}$ at $9^h 45^m$.

Start $10^h 0^m 0^s$.
Close $11^h 0^m$.

Temperatures, $\begin{cases} \text{In prism} = 20.7° \\ \text{In air} = 21.5° \end{cases}$ at $11^h 5^m$.

Grade, 1_b.

Bolograph IV. ω_2—A.
Slit width, 0.45 mm. Subtends 1.1″.
Temperatures as above.
Start $11^h 26^m 45^s$.
Close $12^h 27^m$.

Temperatures, $\begin{cases} \text{In prism} = 20.9° \\ \text{In air} = 21.1° \end{cases}$ at $0^h 55^m$.

Grade, 1_b.

Bolographs of region Ω to 7 μ taken with salt prism.

[Date, December 7, 1897. Observers, C. G. A. and F. E. F.]

Apparatus.—Slit aperture, 100 mm. by high. Width and angular width at bolometer given below. Prism No. 3. Time of single vibration, $4\frac{1}{4}$ seconds. Bolometer No. 20, 10 mm. by 0.08 mm. Angular width, 3.4″. Current of battery, 0.040 ampere. Spectrum speed, 1′ in 1 minute. Plate speed, 1 cm. in 1 minute. 1′ = 1 cm.

Bolograph V. Ω-7 μ.

Temperatures, $\begin{cases} \text{In prism} = 20.2° \\ \text{In air} = \text{———} \end{cases}$ at $0^h 55^m$.

Start $1^h 21^m 0^s$.
Slit width, Ω-X 0.10 mm. Subtends 2.5″.
Slit width, X-7 μ 0.40 mm. Subtends 10.0″.
Close $2^h 21^m 35^s$.
Temperature in prism, 20.4°.
Grade of bolograph, 1-2_b.

[Date, December 9, 1897. Observers and apparatus as on December 7, except that battery current is now 0.075 ampere, and time of single vibration of galvanometer needle 5 seconds.]

Bolograph II. Ω-7 μ.
Temperature in prism, 20.4° at $9^h 40^m$.
Start $10^h 0^m 0^s$.
Slit width, 0.10 mm. Ω-X. Subtends 2.5″
0.20 mm. X-Y. 5.0″
0.40 mm. Y-end. 10.0″
Close $11^h 0^m 15^s$.
Temperature in prism, 20.5° at $11^h 2^m$.
Grade, 1-2_c.

Bolograph IX. Ω-7 μ.
    Temperature in prism, 20.55° at 11ʰ 2ᵐ.
    Start 0ʰ 57ᵐ 50ˢ.
    Slit width, 0.10 mm. Subtends 2.5″.
                0.30 mm.   7.5″
                0.40 mm.   10.0″
    Close 1ʰ 58ᵐ 5ˢ.
    Temperature in prism, 20.55° at 2ʰ 0ᵐ.
    Grade, 1-2ᵦ.

[Date, March 14, 1898. High wind. Observers and apparatus as on December 7, except as follows: Collimating system cylindric. Time of single swing of galvanometer needle, 3¼ seconds. Current of battery, 0.080 ampere.]

Bolograph II. Ω-7 μ.
    Temperatures, $\left\{ \begin{array}{l} \text{In prism}=20.6° \\ \text{In air }=20.7° \end{array} \right\}$ at 9ʰ 20ᵐ.
    Start 9ʰ 29ᵐ 45ˢ.
    Slit width, Ω-X 0.40 mm. Subtends 1.0″.
                X-Y 1.50 mm.   3.7″.
                Y-end 3.00 mm.   7.5″.
    Close 10ʰ 30ᵐ 30ˢ.
    Temperatures, $\left\{ \begin{array}{l} \text{In prism}=20.6° \\ \text{In air }=20.7° \end{array} \right\}$ at 10ʰ 33ᵐ.
    Grade, 1-2ᵦ.

Bolograph VIII. Ω-7 μ.
    Temperatures, as above, at 12ʰ 5ᵐ.
    Start, 12ʰ 24ᵐ 0ˢ.
    Slits, as above.
    Close 1ʰ 24ᵐ 25ˢ.
    Temperatures, $\left\{ \begin{array}{l} \text{In prism}=20.7° \\ \text{In air }=21.0° \end{array} \right\}$ at 1ʰ 26ᵐ.
    Grade, 1-2ᵦ.

[Date, April 6, 1898. High wind. Observers and apparatus as on March 14, except as follows: Time of single vibration of galvanometer needle, 3½ seconds. Bolometer No. 21. 12 mm. by 0.03 mm. Subtends 1.3″. Current of battery, 0.028 ampere. Direction of spectrum motion from long wave radiations to short.]

Bolograph V. 6μ-Ψ.
    Temperatures, $\left\{ \begin{array}{l} \text{In prism}=19.1° \\ \text{In air }=19.8° \end{array} \right\}$ at 11ʰ 0ᵐ.
    Start 11ʰ 30ᵐ 0ˢ.
    Slit width, 6 μ-Y 4.00 mm. Subtends 10.0″.
                Y-X 2.00 mm.   5.0″.
                X-end 0.60 mm.   1.5″.
    Close 0ʰ 41ᵐ 0ˢ.
    Temperatures, $\left\{ \begin{array}{l} \text{In prism}=19.5° \\ \text{In air }=19.7° \end{array} \right\}$
    Grade, 1-2ᵦ.

Bolograph XI. 6μ-Ψ.
    Temperatures, $\left\{ \begin{array}{l} \text{In prism}=19.7° \\ \text{In air }=20.7° \end{array} \right\}$ at 2ʰ 7ᵐ.
    Start 2ʰ 25ᵐ 30ˢ.
    Slit as above, except X-end 0.50 mm. Subtends 1.2″.
    Close 3ʰ 25ᵐ 40ˢ.
    Temperatures, $\left\{ \begin{array}{l} \text{In prism}=19.7° \\ \text{In air }=19.8° \end{array} \right\}$ at 3ʰ 29ᵐ.
    Grade, 1-2ₐ₋ᵦ.

Bolographs of region A to φ with glass prism.

[Date, May 9, 1898. Observers and apparatus as on April 6, except: Prism, great glass. Time of single vibration of galvanometer needle, 3¼ seconds.]

Bolograph X. φ-A.

Temperatures, $\begin{cases} \text{In prism} = 20.7° \\ \text{In air} = 20.7° \end{cases}$ at $12^h\ 50^m$.

Start $1^h\ 11^m\ 15^s$.
Slit width 1.0 mm. Subtends 2.5″.
Close $2^h\ 11^m\ 15^s$.

Temperatures, $\begin{cases} \text{In prism} = 20.6° \\ \text{In air} = 20.6° \end{cases}$

Grade, 1_{b-c}.

[Date, May 13, 1898. Observers and apparatus as on May 9.]

Bolograph I. φ-A.

Temperatures, $\begin{cases} \text{In prism} = 21.6° \\ \text{In air} = 22.0° \end{cases}$ at $3^h\ 30^m$.

Start $3^h\ 34^m\ 45^s$.
Close $4^h\ 34^m\ 45^s$.

Temperatures, $\begin{cases} \text{In prism} = 21.7° \\ \text{In air} = 21.8° \end{cases}$

Grade, $1-2_{a-b}$.

[Date, May 16, 1898. Observers and apparatus as on May 9, except spectrum moves from short wave radiations to long.]

Bolograph II. A-φ.

Temperatures, $\begin{cases} \text{In prism} = 21.3° \\ \text{In air} = 21.7° \end{cases}$ at $10^h\ 29^m$.

Start $10^h\ 32^m\ 0^s$.
Close $11^h\ 32^m$.

Temperatures, $\begin{cases} \text{In prism} = 21.4° \\ \text{In air} = 21.0° \end{cases}$ at $11^h\ 39^m$.

Grade, 1_{a-b}.

[Date, May 17, 1898. Observers and apparatus as on May 15.]

Bolograph VIII. A-φ.

Temperatures, $\begin{cases} \text{In prism} = 21.3° \\ \text{In air} = 21.35° \end{cases}$ at $11^h\ 59^m$.

Start $0^h\ 25^m\ 30^s$.
Close $1^h\ 26^m\ 30^s$.

Temperatures, $\begin{cases} \text{In prism} = 21.20° \\ \text{In air} = 21.10° \end{cases}$ at $1^h\ 30^m$.

Grade, 1_{a-b}.

Bolographs of the region φ to Ω with glass prism.

[Date, April 13, 1898. Observers and apparatus as on May 9, except as follows: Slit width, 0.70 mm. Subtends at the bolometer 2.1″. Current of battery, 0.025 ampere.]

Bolograph VIII. Ω-φ.

Temperatures, $\begin{cases} \text{In prism} = 21.4° \\ \text{In air} = 21.8° \end{cases}$ at $1^h\ 0^m$.

Start $1^h\ 5^m\ 20^s$.
Close $2^h\ 6^m\ 20^s$.

Temperatures, $\begin{cases} \text{In prism} = 21.5° \\ \text{In air} = 21.8° \end{cases}$ at $2^h\ 20^m$.

Grade, $1-2_{b-c}$.

[Date, May 14, 1898. Observers and apparatus as on May 9, except: Time of single vibration, 3¼ seconds. Current of battery, 0.030 ampere.]

Bolograph VIII. Ω-φ.

Temperatures, $\begin{Bmatrix} \text{In prism} = 21.8° \\ \text{In air} = 22.1° \end{Bmatrix}$ at $0^h\,30^m$.

Start $0^h\,50^m\,30^s$.
Close $1^h\,50^m\,30^s$.

Temperatures, $\begin{Bmatrix} \text{In prism} = 21.8° \\ \text{In air} = 21.8° \end{Bmatrix}$ at $2^h\,15^m$.

Grade, 1-2$_{a.b.}$

[Date, May 17, 1898. Observers and apparatus as on May 14, except spectrum moves from short wave radiations to long.]

Bolograph V. φ-Ω.

Temperatures, $\begin{Bmatrix} \text{In prism} = 21.2° \\ \text{In air} = 21.5° \end{Bmatrix}$ at $11^h\,35^m$.

Start $11^h\,58^m\,0^s$.
Close $0^h\,58^m\,55^s$.

Temperatures, $\begin{Bmatrix} \text{In prism} = 21.3° \\ \text{In air} = 21.55° \end{Bmatrix}$

Grade, 1$_{a.b.}$

Bolograph XI. φ-Ω.

Temperatures, $\begin{Bmatrix} \text{In prism} = 21.20° \\ \text{In air} = 21.10° \end{Bmatrix}$ at $1^h\,30^m$.

Start $1^h\,55^m\,15^s$.
Close $2^h\,50^m\,0^s$.

Temperatures, $\begin{Bmatrix} \text{In prism} = 21.2° \\ \text{In air} = 21.4° \end{Bmatrix}$ at $3^h\,10^m$.

Grade, 1$_{a.b.}$

It will be observed from these notes that two bolometers having different angular widths have been employed; that the angular width of the slit has been varied to suit the intensity of energy, and that about half the bolographs were taken with the spectrum passing from long wave-lengths to short and the remainder in the contrary direction. The speed ratio has been constant throughout, and is such that 1 cm. in abscissæ $= 1'$ arc of spectrum $= 1$ minute time.

MEASUREMENTS ON BOLOGRAPHS.

The three following tables contain the results of measurements upon the bolographs. Tables 18 and 19 give the abscissæ, corresponding to differences of deviation of the lines discovered. Table 18 contains the results obtained by the use of the salt prism, and table 19 those from the glass prism. No reduction has been made except that indicated on page 75; that is, the pair of check observations made by each observer (rarely differing more than 0.03 mm.) have been averaged and the resulting mean measurements on all the bolographs reduced to a common origin of abscissæ. The system of numbering to designate the lines in the glass prism is wholly independent of that for the salt prism, but in the final tables of reduced results the lines in the glass spectrum may be connected through their wave-lengths with corresponding ones in the salt spectrum:

TABLE 18.—*Bolographs of rock-salt prismatic solar spectrum—Observed positions of minima.*

[Initials refer to observers.]

Designation.	Distance.	April 1, 1898. Bolograph II.		April 1, 1898. Bolograph XI.		April 2, 1898. Bolograph II.		April 6, 1898. Bolograph II.		May 16, 1898. Bolograph II.		May 16, 1898. Bolograph IV.	
		F.E.F.	C.G.A.	F.E.F.	C.G.A.	F.E.F.	C.G.A.	F.E.F.	C.G.A.	F.E.F.	C.G.A.	F.E.F.	C.G.A.
	cm.												
1 d ?	32	......	762	......	738	......	734	......	722	718	728	717	712
2 d		706	649	663	658	690	......	641	......	648	......	660	658
3 d } b		644	......	613	......	621	622	593	596	606	604	616	606
4 d		560	569	......	546	545	546	......	......	548	550	556	554
5 d ?		......	......	......	......	......	......	509	510	504	506	......	......
6 d		......	......	469	466	451	442	......	......	......	461	471	468
7 d		......	......	355	332	......	......	369	362	......	......	347	336
8 d		356	297	.292	......	......	292	263	260	292	289	296	288
9 d	A	295	205	250	249	266	......	218	......	230	......	237	234
10 d	a	......	......	......	......	......	196	......	150	192	196	189	183
11 d		206	......	156	......	......	......	150	......	132	......	149	......
12 d		126	......	104	......	160	......	085	......	102	......	065	......
13 d } b		052	057	029	030	081	078	017	017	046	052	011	060
14 d	31	918	923	940	862	931	980	892	......	942	992	983	975
15 d		834	823	884	......	873	......	857	864	832	839	859	856
16 d		......	732	776	777	763	765	751	752	742	750	794	794
17 d		648	......	692	......	657	......	639	......	680	......	635	......
18 d		602	635	636	625	616	610	......	646	......	......	616	616
19 d		545	......	526	......	501	......	519	......	......	......	......	......
20 d		450	......	430	......	452	......	......	......	422	......	......	......
21 d ?	29	......	......	......	999	......	080	......	982	......	966	......	990
22 d	29	245	244	216	218	253	254	......	254	......	264	......	......
23 d ?	27	......	010	......	954	......	988	......	......	......	030	......	968
24 d ?	26	......	......	......	322	......	306	......	808	......	......	......	......
25 d ?	25	......	......	......	291	......	292	......	......	......	309	......	335
26 d		......	190	......	178	......	192	......	205	208	207	204	
27 d		102	} 999	056	} 062	113	} 025	075	} 074	......	054	103	102
28 d		003		988		027		980		036			
29 d	24	900	895	895	......	906	908	925	......	952	952	933	930
30 c		830	839	826	824	819	825	839	833	843	844	828	822
31 d		757	751	744	......	724	748	769	768	743	748	751	749
32 d ? } b		658	} 657	658	} 656	621	} 622	653	} 656	677	} 629	661	} 622
33 c		608				583				630		625	
34 d ?		534	535	......	514	544	......	579	} 568	594	......	587	......
35 d		504	......	512	......	......	......	565		532	......	533	......
36 d		408	} 419	420	} 426	415	} 441	479	} 404	440	} 376	451	} 416
37 c		361		388		373		403		376		395	
38 d		......	......	189	......	177	......	165	......	220	......	......	......
39 d		074	......	062	......	065	......	073	......	128	......	105	......
40 c		015	019	008	010	013	014	023	032	998	999	015	010
41 d	23	844	......	886	......	851	851	829	......	828	843	838	......
42 d		......	666	......	675	......	680	......	664	666	662	659	660
43 d		499	} 454	443	} 444	435	} 435	......	......	465	} 460	......	} 406
44 d		414		380		371						405	
45 d		288	291	280	280	267	271	......	......	......	......	289	......
46 d		020	061	082	079	001	060	070	076	......	......	099	......
47 d	22	878	877	902	911	880	......	924	......	926	918	949	948
48 d ?		......	......	......	......	......	......	......	......	806	802	800	813
49 d ?		......	......	......	......	......	......	710	......	......	691	......	726
50 d ?	22	......	......	......	407	......	404	......	......	......	398	......	280
51 d	22	......	111	......	087	......	102	......	118	......	188	......	168
52 d ?	21	732	......	791	......	769	......	......	......	......	......	......	......
53 d ?		......	191	......	184	......	170	......	......	......	212	......	211
54 d ?		......	015	......	022	......	060	......	004	974	976	986	980
55 c	20	846	845	852	850	880	870	891	892	877	878	879	875
56 d		726	665	700	685	728	738	683	688	690	696	701	730
57 d		554	542	577	576	565	566	567	541	530	526	......	558
58 d		475	......	484	480	......	......	507	......	......	......	469	460
59 b		388	388	385	383	395	396	399	401	388	395	390	390

ANNALS OF ASTROPHYSICAL OBSERVATORY.

TABLE 18.—*Bolographs of rock-salt prismatic solar spectrum—Observed positions of minima*—Continued.

[Initials refer to observers.]

Designation.	Distance.	April 1, 1898. Bolograph II.		April 1, 1898. Bolograph XI.		April 2, 1898. Bolograph II.		April 6, 1898. Bolograph II.		May 18, 1898. Bolograph II.		May 18, 1898. Bolograph IV.	
		F.E.F.	C.G.A.	F.E.F.	C.G.A.	F.E.F.	C.G.A.	F.E.F.	C.G.A.	F.E.F.	C.G.A.	F.E.F.	C.G.A.
	cm.												
60 d ?										328		342	
61 d		260	205	250	250	283	288	297	218	280	290	251	265
62 d ?		196		136		151							
63 d ?	19				928		960		942				908
64 d ?					804		777				846		806
65 d		672	668	650	650	679	678						
66 d ?							358		380				378
67 d ?		284	271	302	203		244		236	278	271		256
68 d		199		211		209		240		203		257	
69 b		098	099	094	094	093	096	103	110	112	112	115	114
70 d ?					961	002	004	024	921	021	024	015	974
71 d	18	963		956		957				926		965	
72 d		870	852	828	826	857	860		822	834	835	839	834
73 d ?			736		735		698				732		735
74 d			521		512		494		487		588		596
75 d		311	313	352	354		344		312		316		328
*76 d			189		153		186		216				
77 d			045				071		078				
78 d	17		964		946		944		934		978		940
*79 d					774		798		786		770		777
80 c		612	611	626	624	617	620		636	626	628	639	640
81 d ?		442		430									
82 d ?		332		309	306		364				316		294
83 d ?					115		129						158
84 d ?					006		085				049		063
*85 d	16						912			950	950	928	926
86 d ?		860		878									
87 d		778		781	784		744			765	764		771
88 d ?		602	554	573	574		605			574	575		
89 d }			411	464	459		420	453	457	432	426	431	424
90 d ? }										261		293	
91 c			231	178	180			191	193	187	184	213	212
92 d		077	075	090	082			121	122	080	072	105	102
93 c } b	15	978	974	967	954			939	956	018	017	008	010
94 c }		858	857	854	851			861	864	864	864	881	876
95 d		722	729	712		696		738	738	739	726	738	732
96 d		678	669	652	670	649		657	666			606	
97 d		622			602	600	589	609		616	618	629	626
98 d		538	503	539			533	521	527	550	555	542	544
99 d		346	343	326	327	339	339	349	351	381	380	388	380
100 d		234 ?		230	228	237	226	253		258	248	296	274
101 d }	15	122	127	141	137	155	152	160	156	174	162	162	158
102 d ? }		070		064		090		083		106	101	103	104
103 d					060		088		636		036		014
104 d	14	996	995	979	976	990	985	935	934	958	950	964	
105 d		909	915	892	894	921	923	866	870	862	875	869	870
106 d		832	759	823	824	835	828	799	798	792	796	789	790
107 d ? } b		757		727		763		735					
108 c		666	667	640	730	660	662	693	682	676	677	681	678
109 c		524	523	532	531	528	530	569	571	534	535	551	552
110 d		392		422		407		417					
111 c		324	322	337	336	329	329	339	342	352	350	365	364
112 d		202				201		217		234 ?			
113 c /		156	154	167	165	151	154	172	188	182	182	184	183
114 c		122	057					103	105	098	} 102	113	118
115 d		061				072				037			
116 c	13	978	977	981	962	986	962	987	988	983 ?	044	006	008
117 c		849	814		703		796	828	833	820	819	837	836

* Note by F. E. F.: Band from 17.868 to 18.708. Band in II, April 1, 16.852-17.402.

S. Doc. 20——13

TABLE 18.—*Bolographs of rock-salt prismatic solar spectrum—Observed positions of minima*—Continued.

[Initials refer to observers.]

Designation.			Distance.	April 1, 1898. Bolograph II.		April 1, 1898. Bolograph XI.		April 2, 1898. Bolograph II.		April 6, 1898. Bolograph II.		May 18, 1896. Bolograph II.		May 18, 1898. Bolograph IV.	
				F.E.F.	C.G.A.	F.E.F.	C.G.A.	F.E.F.	C.G.A.	F.E.F.	C.G.A.	F.E.F.	C.G.A.	F.E.F.	C.G.A.
			cm.												
118 d				763	....	784	....	795	....	719	....	....	....	728	....
119 d				642	640	676	680	658	660	665	664	678	674	685	683
*120 d				562	....	592	586	557	560	594	....	618?	618	619	....
121 d				507	527	536	472	480	480	533	536	560	558	495	....
122 d	)			412	417	400	394	371	368	444	448	430	432	435	437
123 d				229	217	204	196	229	214	305	....	230	220	251	248
124 d				139	....	124	115	161	160	170	168	....	....	189	....
125 d				084	094	....	....	....	....	075	064	078	081	067	084
126 d				031	022	041	....	039	032	....	....	019	010	021	015
127 d	}b	ρ	12	913	909	926	918	907	916	921	924	930	931	931	934
128 d				852	848	848	844	831	826	848	843	871	865	885	882
129 d?				721	....	....	....	721	....	....	....	776	....	....	....
130 c				694	705	729	728	686	720	727	728	739	740	745	744
131 c				570	566	592	590	576	576	621	624	614	614	617	618
132 c	)			467	466	472	472	468	470	....	502	518	516	507	510
133 d				404	396	410	....	415	....	439	441	435	450	443	442
134 d				182	181	176	172	191	181	249	249	232	234	235	234
135 d?				128	098	....	054	....	090	133	114	192	154	181?	170
136 d				070	....	077	....	100	....	100	....	155	....	131	....
137 c			11	982	981	984	982	977	976	017	020	027	026	033	032
138 c				883	883	886	884	877	876	918	918	920	920	926	923
139 c				748	751	750	752	747	746	769	772	791	792	790	789
140 d				671	674	634	640	654	....	678	678	696	....	661?	....
141 c	}b	σ		569	569	580	560	572	574	585	590	616	616	611	610
142 b				432	431	431	428	428	427	456	452	466	467	469	467
143 b				231	231	250	251	251	252	249	251	289	290	296	294
144 c				046	049	070	067	062	061	071	074	104	105	099	096
145 c		a	10	960	957	972	975	967	968	002	002	016	019	015	014
146 c				900	....	906	900	916	922	907	911	936	938	927	929
147 d				....	....	....	....	772	774	793	....	792	804	806	802
*148 c				702	701	708	721	701	703	715	720	721	734	733	731
149 d				536	528	602	606	574	578	590	591	636	642	634	639
150 c				458	457	468	466	465	466	469	475	488	490	489	484
151 c			10	323	323	332	330	331	332	349	332	357	359	375	370
152 d?				246	....	244	....	275	....	291	....	300	....	314	....
153 c				152	155	172	170	149	150	191	190	178	181	197	196
154 d				032	035	042	048	015	016	035	036	044	050	054	056
155 c			9	852	859	859	858	875	878	901	904	893	894	906	906
*156 d?				....	....	....	664	....	664	714	680	714	712	705	698
157 d				....	....	668	....	676	....	659	....	....	....	....	....
158 c				598	597	570	572	589	590	579	581	588	601	613	616
159 c		r		450	445	454	452	484	464	441	443	486	482	462	461
160 d	}b			400	....	394	....	430	....	385	....	....	....	....	....
*161 d				264	259	251	224	270	273	241	229	....	248	301	211
162 d	}c			212	....	200	....	206	....	178	....	249	....	213	....
163 d				126	....	121	....	136	....	111	....	....	....	158	....
164 d			8	967	927	902?	....	957?	960	945?	949	....	972	979	972
165 d				830	829	822?	818	819?	824	825?	822	873	828	813	810
166 d				....	676	....	605	....	....	....	701	....	747	....	644
167 d				....	....	....	....	....	....	....	531	....	502	....	510
168 d			7	973	971	941	944	....	023	908	960	....	932	....	624
169 d				830	827	822	822	....	....	755	757	....	....	....	....
170 d?				....	....	....	611	....	....	....	....	600	598	640	638
171 d				544	534	547	539	....	....	517?	....	523	....	531	532
172 c				384	381	349	352	369	372	391	394	382	382	401	396

* Notes by F. E. F.: Lines 120-121, triple in II April 1 and II April 2; possibly line 120 double. Readings on II, April 1, 13.546, 13.578; mean=13.562 used. Readings on II, April 2, 13.544, 13.571; mean =13.557 used. Line 148 double in XI, April 1. Readings, 10.690, 10.725; mean =10.708 used. Line 148 double in IV, May 18. Readings, 10.706, 10.735; mean = 10.721 used. Line 156 and 157 possibly the same line. Lines 161-162 triple in II, April 1.

ANNALS OF ASTROPHYSICAL OBSERVATORY.

TABLE 18.—*Bolographs of rock-salt prismatic solar spectrum—Observed positions of minima*—Continued.

[Initials refer to observers.]

Designation.	Distance.	April 1, 1898. Bolograph II.		April 1, 1898. Bolograph XI.		April 2, 1898. Bolograph II.		April 6, 1898. Bolograph II.		May 18, 1898. Bolograph II.		May 18, 1898. Bolograph IV.	
		F.E.F.	C.G.A.	F.E.F.	C.G.A.	F.E.F.	C.G.A.	F.E.F.	C.G.A.	F.E.F.	C.G.A.	F.E.F.	C.G.A.
	cm.												
173 d		216	207	170	172	244				200			206
174 d	6				898		868		910		654		072
175 d		710†	721	706†	705	747†	746				696		712
176 d†					552		553				562		609
177 d					404		394				432		434
178 d†	5				950		976						045
179 d†					490		586		440		452		482
180 d			297		351		290		284		262		270
181 c					110		096		118		114		095
182 d	4				986		987		972		912		922
183 d					778		759		842		790		763
184 d			291		282		230		338		276		
*185 d					140		104				156		190
186 c	4	022†	019	001	000	994	998		024	027	028	031	029
187 c	3		891		878		850		914	924	920	906	904
188 d									782	797	791	777	773
189 d			723		684		704		679		689		
190 d		552	549	551	553	543	548		575	602	590	593	593
191 d†		470				443†							
192 c		416	412	406	405	393	401	405	414	420	424	441	438
193 c		167	166	168	162	181	189	203	207	212	209	237	230
194 c		050	050	052	050	051	067	068	090	086	088	104	112
195 c	2				91A		858	881	886	901	900	898	894
196 c		696	695	744	739	693	700	739	745	734	739	763	763
197 c		586	589	624	628	621	625	631	637	624	628	622	618
198 c		409	409	437	436	449	451	419	420	449	452	478	468
199 c			221	286	284	263	262	315	318	264	261	298	290
200 d									176		112		119
†201 r	2	024	021	042	040	015	016	058	063	063	028	077	072
202 d	1	948	945	976		955	957	950	963	960	954	968	968
203 d		848		875	870	820	820	831	835			839	837
204 c		783	781	757	754	712	709	716	718	776	780	775	770
205 c		608	601	612	610		588	606	604	630	638	637	641
206 d		501	501	503	502	491	478	495	494		536	535	531
207 c		360	359	394	392	391	404	407	406	432	432	439	437
208 d†										308		325	
209 d		245	239	275	272	263	252	291	289		308	275	285
†210 d		116	119	136	121	149	145	151	202	169	170	171	168
211 d		004	005	014	014	031	031	053	056	040	042	027	026
212 d	0	906				981							
213 c		845	843	880	876	874	872	882	884	892	892	884	882
214 c		717	713	737	734	729	732	747	746	742	745	745	744
215 b } b		567	567	594	592	597	597	605	605	602	603	601	600
216 c		424	423	431	432	431	431	439	440	436	443	445	444
217 c } φ		290	290	304	304	302	304	328	330	334	330	347	347
218 c }						267		256	262	258	260	268	269
219 b		000	000	000	000	000	000	000	000	000	000	000	000
220 d } a	0	194	196	182	188	169	172	150	148	145	140	139	138
221 c		322	325	324	328	318	320	314	313	319	318	305	308
222 c } b		496	499	524	513	515	516	507	512	476	480	481	490
223 c		602	603	586	586	577	574	581	577	568	568	567	566
224 c		726	729	727	728	727	727	713	712	706	707	701	703
‡225 c		858	857	852	852	838	831	839	837	810	847	812	850
226 c		914		910		913		♭17		880		884	
227 d		983	975	968	978		915	043	032	066		048	051	050

* Note by C. G. A.: Curves at line 185 d bad in April 1, II, and April 6, II; hence measures omitted.
† Notes by F. E. F.: Lines 201-202; note change in interval between these two lines, five first to last. Line 210 possibly double in II, April 1—1.101, 1.129.
‡ Note by C. G. A.: Line 225 double in II and IV of May 18. Readings on II, 0.883, 0.811; means = 0.847 used. Readings on IV, 0.884, 0.815; mean = 0.850 used.

ANNALS OF ASTROPHYSICAL OBSERVATORY.

TABLE 18.—*Bolographs of rock-salt prismatic solar spectrum—Observed positions of minima*—Continued.

[Initials refer to observers.]

Designation.	Distance	April 1, 1898. Bolograph II.		April 1, 1898. Bolograph XI.		April 2, 1898. Bolograph II.		April 6, 1898. Bolograph II.		May 18, 1898. Bolograph II.		May 18, 1898. Bolograph IV.	
		F.E.F.	C.G.A.	F.E.F.	C.G.A.	F.E.F.	C.G.A.	F.E.F.	C.G.A.	F.E.F.	C.G.A.	F.E.F.	C.G.A.
	cm.												
228 d	1	127	117	087	085	069	060	155	152	157	128	163	162
229 d		206	215	168	169	157	157	213	206	......	190	217	203
230 d		378	382	359	356	......	......	......	371	......	......	373	......
*231 c		460	463	432	436	} 416 {	414	445	438	426	426	423	427
232 d		560	......	593	......	560	......	593	......	562	......	521	......
233 d		608	619	......	592	......	577	......	590	610	614	......	608
234 d		652	......	708	......	637	......	......	......	650	......	615	......
235 c		822	825	813	818	817	811	829	......	815	812	799	800
236 c		994	999	970	970	985	981	975	985	974	972	961	964
237 c	2	180	185	178	180	179	172	149	146	142	142	......	140
238 c		292	292	269	272	261	257	254	250	246	243	241	240
239 c		421	425	430	432	421	418	409	404	412	410	402	412
240 d		522	545	......	537	......	520	......	538	506	496	497	489
241 d		571	......	545	......	539	......	549	......	557	......	547	......
242 d		671	673	......	640	680	683	......	644	621	620	663	598
243 c		748	751	736	767	737	742	729	726	716	712	708	708
244 c		880	879	886	886	872	870	877	878	869	864	861	865
245 d		962	......	001	......	017	......	953	......	918	......	901	......
246 c	3	089	093	076	078	067	066	063	061	050	050	053	054
247 c		254	257	250	251	264	263	233	231	246	236	247	248
*248 d		392	386	351	......	392	383	317	332	342	336	359	362
249 d		528	534	......	......	508	500	472	474	503	494	471	478
250 d		......	......	......	......	......	......	......	......	662	......	634	......
251 d		606	699	......	712	......	730	697	700	698	......	709	602
252 d		......	847	......	801	813	816	795	790	792	794	793	799
253 d		843	......	......	......	......	......	......	......	870	......	849	......
254 c	4	008	055	010	......	009	005	985	977	005	004	977	981
255 d ?		......	......	......	......	......	116	......	......	......	100	......	071
256 d		......	174	206	259	199	212	215	212	......	252	223	228
257 d		346	348	410	408	......	362	348	345	......	391	355	356
258 d		480	447	560	559	527	522	487	508	......	463	......	506
259 d		640	647	......	638	627	626	669	657	606	604	585	595
260 d ?		......	725	......	803	777	770	813	......	777	777	818	......
261 d ?		......	854	......	862	......	......	......	816	......	862	......	826
262 d		947	951	988	930	927	921	944	......	961	954	971	988
263 d	5	047	061	......	996	049	049	039	043	......	......	......	......
264 d		127	143	138	142	......	188	123	130	102	102	......	193
265 d		202	......	......	......	178	......	......	......	......	......	185	......
266 d		245	......	......	......	......	......	246	......	232	......	248	......
267 d		314	318	343	......	327	330	277	272	348	316	347	307
268 c		384	387	392	393	397	400	391	390	401	399	399	400
269 d		487	489	472	486	467	......	......	......	......	......	......	......
270 b	b	550	552	538	538	565	558	537	536	548	548	543	541
271 d		697	701	699	702	692	......	655	650	722	658	677	678
272 d ?		879	887	862	861	......	......	883	880	826	......	863	862
273 d		946	......	......	......	949	......	927	......	......	......	......	......
274 c	6	018	021	014	014	019	016	027	025	020	018	991	996
275 d		206	211	176	141	183	169	269	270	268	260	233	186
276 d		340	341	288	283	324	274	391	390	376	380	357	410
277 d ?		......	494	......	522	......	556	481	487	......	542	505	528
278 d		492	......	522	......	551	......	559	......	544	......	543	......
279 d		615	623	......	......	653	654	597	596	608	646	......	639
280 d		684	......	......	......	......	......	......	......	646	......	633	......
281 d		788	......	818	......	761	760	769	769	752	752	766	764
282 d		......	793	......	821	873	874	847	848	852	852	839	8.18
283 d		944	951	980	986	983	982	988	992	959	960	947	960
284 d ?	7	054	064	......	......	......	......	......	......	......	......	049	052
285 c		128	129	095	096	091	091	137	136	096	094	097	100

* Notes by F. E. F.: Line 248 double in II, April 6—3.335, 3.297.

ANNALS OF ASTROPHYSICAL OBSERVATORY.

TABLE 18.—*Bolographs of rock-salt prismatic solar spectrum—Observed positions of minima*—Continued.

[Initials refer to observers.]

Designation.	Distance.	April 1, 1898. Bolograph II.		April 1, 1898. Bolograph XI.		April 2, 1898. Bolograph II.		April 6, 1898. Bolograph II.		May 18, 1898. Bolograph II.		May 18, 1898. Bolograph IV.	
		F.E.F.	C.G.A.	F.E.F.	C.G.A.	F.E.F.	C.G.A.	F.E.F.	C.G.A.	F.E.F.	C.G.A.	F.E.F.	C.G.A.
	cm.												
286 d										130		130	
287 c		260	262	263	266	287	286	267	316	272	273	259	265
288 c		420	427	382	386	409	408	411	408	400	400	394	394
289 d		579	577	581	555	589	566	563	566	600	604	589	593
290 c		667	669	650	651	671	672	673	674	683	682	670	674
291 d		838	839	858	864	857	858	859	870	835	838	818	825
292 d		928	959	962	965	977	972	957	962	940	940	935	938
293 d	8	100	105	076	082	097	095	091	093	082	080	069	078
294 d		212	209	203	206	208	213	186	190	184	182	179	181
295 d		298	296	354	328	313				294	298	271	330
296 d		406	412	452	450	422	422	368		388	388		424
297 d		516	319	506	506	589	517	576	584	594	592	555	
298 d		618	623	582	583	687	588	659	660				
299 d		743	732	688	682	763	690	739	740	707	706		
300 d		795		803				844		810			
301 d		858	855	856	852	831	831		846			834	855
302 d		996	002	974	974	945	945	967	026			963	962
303 d	9	118	121	110	129	070	070					090	088
304 d		200	203	219	224	194	192	195	191	210	212	179	171
305 d		330	331	364	300	331	325	349	349	314	312	355	358
306 d				461	467	491	488	501	506			470	493
307 c		531	534	512	510	541	537	583	580	548	548	511	526
308 c		701	703	676	681	692	692	715	714	680	682	656	659
309 c	ψ	826	828	826	824	821	822	842	840	810	816		788
310 c		948	951	944	942	938	940	947	945	927	970		929
311 c	10	076	077	080	076	083	084	097	096	088	089	071	072
312 c	a	204	204	204	203	215	217	218	218	206	208	187	190
313 d		327	325	328	326	338	340	352	350	294	298	322	320
314 c		434	434	440	438	451	450	468	470	432	432	429	428
315 c		570	569	565	565	569	570	580	576	574	575	564	567
316 d		692	692	684	678	685	682	711	708	714	672	689	684
317 d		782	781	792	787	789	784	796	789	785	782	775	774
318 c		924	921	893	886	901	900	915	916	902	900	892	891
319 c	11	038	033	038	037	029	030	051	050	040	040	041	042
320 d		189	185	150	149	153	150	165	160	154	150		152
321 c		332	335	316	320	321	320	340	342	311	312	309	312
322 c*		468	469	446	444	461	464	471	466	454	454	445	446
323 c }b	ψ	567	568	548	545	556	554	566	564	560	560	553	554
*324 c		728	729	706	707	718	715	729	758	720	755	713	738
325 d		794		794		773		787		789		763	
326 c		934	933	932	932	927	924	961	958	948	946	931	933
327 d	12	045	037	068	066	073	074	054	042	092	081	058	045
328 d		118	120		150	164	152	135	131	145	138	126	122
329 b		352	352	344	344	355	352	359	356	352	350	345	345
330 d		504	501	459	465	483	482	475	471	444	432	452	442
331 d					566			577	575	566	561	549	554
332 d		660	662	656	653	654	648	653	660	671	662	663	664
333 d		730						728		716			
334 d ?		840	849	847	882	865		853	850			821	819
335 d		953	961	986	978	951	942	993	989				
336 d ?	13	104	103			088	085	165				078	053	042
337 a		212	213	182	186	177	180		156	184	184	156	147
338 d ?		310	313	278	288	307		398				299	
339 d ?		378	383	409	418	419	368		388	367	368	351	350
340 d		532	533	480	486	515	508	557	474	512	504	509	508
341 d		667	669			656	644	637	606	716	684	679	687
342 d ?		799				819				835		822	

*Note by C. G. A.: Line 324 double in II and IV of May 18. Readings on II, 11.789, 11.722; mean = 11 755 used. Reading on IV, 11.763, 11.713; mean = 11.738 used.

TABLE 18.—*Bolographs of rock-salt prismatic solar spectrum—Observed positions of minima*—Continued.

[Initials refer to observers.]

Designation.	Distance.	April 1, 1898. Bolograph II.		April 1, 1898. Bolograph XI.		April 2, 1898. Bolograph II.		April 6, 1898. Bolograph II.		May 18, 1898. Bolograph II.		May 18, 1898. Bolograph IV.	
		F.E.F.	C.G.A.	F.E.F.	C.G.A.	F.E.F.	C.G.A.	F.E.F.	C.G.A.	F.E.F.	C.G.A.	F.E.F.	C.G.A.
	cm.												
343 d		859	842	840	847	898	896	864	866	884	886	861	857
344 d ?		954		928		954				920	926	965	921
345 c		988	979	972	950	001	998	991	989	004	002	002	965
346 d ? b	14		113		122		150		150	122	119	101	094
347 d ?		115		126		159		158		168		149	
348 c		242	243	259	259	267	263	273	270	272	270	249	250
349 d ?		323		328						372		371	
350 d				420	432	411	447	435	433	421	396	426	414
351 c		524	523	525	526	529	526	533	547	538	588	513	517
352 c b		672	673	675	676	675	670	682	681	682	680	687	672
353 d		815	823	880	882	841	836	833	828	862	856	871	
354 d		919	919			909	910	912				916	920
355 d ?		973		978	982	963		955		964	956	965	
356 d	15	077		075		076		029				067	
357 d		166	171	194		155	152	159	166	141	152	171	181
358 d		278	280	266	271	273	285	242	244	305		290	290
359 d		374		324	334			347	345		306		
360 d ?		432		439	438	471	466	405	406	402	402	411	414
361 d ?		502		512	512	577	574	502	502	502	510	547	548
362 d ?		687		690	664	706	714	703	701	706	710	687	696
363 o		790	788	787	794	787	790	821	820	820	822	789	805
364 d		930						901		872		942	
365 o		988	994	987	968	977	970	027	024	996	992	016	018
366 o	16	100	101	104	106	131	132	140	143	134	133	140	148
367 d		237	233	284	286	249	242	297	349	280	278	275	279
368 d		382	387	410	418	405	424	456	458	425	423	416	402
369 d		468		448		515						482	
370 d		522		613	610	599	592	654	651	635	636	619	626
371 d		744	751	722	724	693	692	756	756	702	702	713	714
372 d ?										857		821	
373 o		884	885	846	846	868	866	905	896	899	904	879	882
374 d		994		967		995		014					
375 d	17	083	075	072	108	058	058	089	086	098	074	075	080
*376 d		135	138	111	199	119	135	169	202	147	168	158	132
377 d		210		192		199		222		184		229	
378 o		260	285	275 {	276	251	274	317	314	297	296	295	296
*379 d		320			285			361		327			
*380 d c			451	408	440		476		496	457	496	442	460
381 d		452		442		477		497		494		470	
382 d ?		564		546									
383 c		634	637	605	606	613	614	633	640	666	668	629	631
384 d		746	765	778	778	796	797	778				802	748
385 d		839	861	865	858	859	872	866	866	868	864	847	858
386 c		961	961	948	948	972	970	987	986	977	976	953	965
387 d	18	102	109	113	115	152	112	127	131	.30	134	111	115
388 d		250	245	237	239	249	249	323	326	277	282	272	277
389 d		406	413	426	398	425	428	447	446	462	463	445	453
390 c		519	521	546	548	522	522	536	590	553	550	581	582
391 d		709	711	711	711	709	731	737	736			726		708
392 d ?		792	797		814	837	838	775	870				808
393 d ?		873						869				823	
394 d		977	979		986	975	992	959	964			961	958
395 d	19	112	113		138	173	122		215			147	148
396 d		301	305	298	298	267	277	335	336			295	296
397 d		386	384	411	409	363	374	428	430	429	430	390	392
398 d		526	527	508	508	544	501	524	520			501	506	494
399 d		662	663	646	650	669	668	700	702	708	704		643
400 d a		837	833	864	864	839	882	910	850	848	861	885	

* Notes by F. E. F.: Lines 376–377, 378–379, 380–381, sets of doubles.

TABLE 18.—*Bolographs of rock-salt prismatic solar spectrum—Observed positions of minima*—Continued.

[Initials refer to observers.]

Designation.	Distance.	April 1, 1898. Bolograph II.		April 1, 1898. Bolograph XI.		April 2, 1898. Bolograph II.		April 6, 1898. Bolograph II.		May 18, 1898. Bolograph II.		May 18, 1898. Bolograph IV.	
		F.E.F.	C.G.A.	F.E.F.	C.G.A.	F.E.F.	C.G.A.	F.E.F.	C.G.A.	F.E.F.	C.G.A.	F.E.F.	C.G.A.
	cm.												
401 d } Ω	19	972	971	964	964	986	977	010	011	989	986	999	984
402 d	20	080	078	098	092	089	089	126	132	......	......	117	118
403 d ?		128	......	......	......	141	......	......	......	......	......	......	......
404 d		197	201	238	238	230	226	211	283	244	246	249	253
405 c		347	349	356	352	371	364	400	400	382	383	367	367
406 d		481	481	495	486	503	496	511	498	490	492	472	475
407 c		630	628	636	633	644	644	633	636	645	645	641	642
408 d		738	......	746	738	764	......	711	708	728	720	724	721
409 d		852	854	866	866	875	863	873	872	872	871	875	874
410 d		956	955	954	954	005	998	009	008	986	983	980	978
411 d	21	119	127	080	084	131	131	187	195	138	139	147	152
412 b } ω_1		348	337	311	314	335	333	375	383	371	370	351	356
413 c } b		444	443	446	444	459	465	539	536	510	509	484	496
414 d		710	707	725	716	735	717	804	806	736	730	703	703
*415 d		854	863	870	882	889	908	935	930	916	919	863	864
416 b } b		026	029	052	053	059	059	089	090	098	096	067	072
417 c } ω_2		191	191	196	196	207	204	263	264	241	242	228	235
418 d	22	280	275	310	310	325	320	369	366	......	352	372	368

Designation.	Distance.	December 7, 1897. Bolograph V.		December 9, 1897. Bolograph II.		December 9, 1897. Bolograph IX.		March 14, 1898. Bolograph II.		March 14, 1898. Bolograph VIII.		April 6, 1896. Bolograph V.		April 6, 1896. Bolograph XI.	
		F.E.F.	C.G.A.	F.E.F.	C.G.A.	F.E.F.	C.G.A.	F.E.F.	C.G.A.	F.E.F.	C.G.A.	F.E.F.	C.G.A.	F.E.F.	C.G.A.
	cm.														
412 b } ω_1	0	000	000	000	000	000	000	000	000	000	000	000	000	000	000
413 d		......	096	......	......	092	088	137	115	124	116	124	132	094	122
414 d		......	......	......	......	394	......	......	......	370	......	386	......	432	......
415 d		......	550	......	......	530	562	523	538	548	558	540	554	548	564
416 b } b ω_2		720	710	719	716	740	741	718	715	700	700	716	713	723	720
417 c		......	821	824	820	847	836	894	889	845	849	845	846	850	846
418 d		......	......	......	......	......	......	955	......	943	......	936	......	942	
419 d		......	074	170	......	132	133	107	096	070	060	115	110	142	145
420 d		......	222	279	266	......	304	185	184	192	182	233	223	250	249
421 d ?		......	371	} 294	} 392	......	......	} 389	{	386	} 466	334	} 447	} 348	{ 396
422 d ?		......	502		489	444	{ 450		470	465		496	402		
423 c		......	600	576	568	584	586	......	614	......	561	560	558	531	540
424 d		......	......	......	678	681	681	697	698	......	......	738	692	700	......	680
425 d	2	......	......	046	051	085	098	093	096	0e7	088	045	043	032	043
426 c		......	228	256	252	228	230	235	232	209	210	237	234	220	218
427 d		......	456	442	446	396	407	440	442	466	462	396	396	......	......
428 d		......	......	529	534	480	488	505	509	528	525	486	477	489	484
429 d		......	666	......	688	......	......	683	678	662	657	664	674	652	......
430 d		......	......	670	......	722	720	699	......	718	717	737	......	748	710
431 d ?		......	......	870	873	944	886	898	921	......	......	898	898	......	......
432 d ?	3	......	130	122	122	140	153	......	......	......	......	174	172	110	109
433 d		......	520	......	......	596	596	611	607	609	609	599	596	......	......
434 d		......	681	769	734	750	778	780	722	716	724	728	......	702	
435 d ?		......	......	......	......	816	831	859	......	866	865	884	882	......	......
436 d ?		......	920	930	929	920	924	921	......	......	......	......	......	......	......
437 d		......	998	......	......	982	986	......	978	994	992	984	980	990	990
438 c	4	......	216	235	234	253	260	247	241	196	198	220	216	213	219
439 d		......	......	......	......	......	354	......	......	351	336	296	302	369	365	330	330
440 d		......	482	502	508	541	552	467	485	473	486	494	489	465	468
441 d } o		......	624	602	......	......	......	632	630	626	622	586	586	610	605
442 d }		......	......	668	666	663	662	......	......	680	680	655	654	......	......

*Note by F. E F.: Line 415 triple in II, April 2; 21.889, 21.923, 21.855.

TABLE 18.—*Bolographs of rock-salt prismatic solar spectrum—Observed positions of minima*—Continued.

[Initials refer to observers.]

Designation	Distance	December 7, 1897. Bolograph V.		December 9, 1897. Bolograph II.		December 9, 1897. Bolograph IX.		March 14, 1898. Bolograph II.		March 14, 1898. Bolograph VIII.		April 6, 1898. Bolograph V.		April 6, 1898. Bolograph XI.	
		F.E.F.	C.G.A.	F.E.F	C.G.A.	F.E.F.	C.G.A.	F.E.F.	C.G.A.	F.E.F.	C.G.A.	F.E.F.	C.G.A.	F.E.F.	C.G.A.
	cm.														
443 d			902	890	892	920	924	899	894	851	872	878	863	836	835
444 d	5		023	020	012 078	012	036	038	035 105	032	032 098	028	994 046	016	923 085	008
445 d																
446 d				322	290	296	318		332	302		300	304	306	296	295
447 c				423	431	434	430	428	421	417	396	393	399	394	413	412
448 d				695			696	708	675	666	632	628	636	633	622	620
449 c		835	830	844	850	840	836	824	819	814	805	803	799	813	814
450 d ?	6								057		010		010			
451 d ?	6			108		068		059		015		066		049		
452 d ?			102		115				132		134		106		138
453 d ?						176		132		140		196		134		
454 d			260	247	264			247	240	256	260		198		252
455 c			340	338	350	311	308	306	303	314	312	314	312	314	316
456 d			509		522			519	526	516	508	482	490	502	510
457 d			646					619	644	638	642	624	626	612	638
458 d ?								683		754		708		684	
459 a x	6 10		646 952	484 878	577 845	795 695	522 721	824 684	782 592	800 600	756 578	875 566	785 602	800 550	741 601
460 d	10								592		578		602		601
461 d	10				845	866	857	845	824	840	822	829	816	804	786
462 d	10		952		968		958	983	984	969	983	999	987	020	018
463 c	11		224	250	256	262	252	239	224	206	203	233	231	242	242
464 d			388	380	398	402	396	425	404	416	365	436	433	431	424
465 d						480		499	492			498		530	526
466 c		734	726	744	753	748	746	719	714	698	694	726	722	744	744
467 d				860	869	856	845	817	810	810	812	813	809	840	845
468 d ?					966		042				028			004	012
469 b	12	076	070	090	100	114	111	044	041	080	074	086	082	080	(81
470 d				175	179			187	177	191	184				
471 d ?								349	351		322		368	336	338
472 c			454	393	401	449	440	439	438	394	392	423	486	444	448
473 c					490							506			
474 c			615	638	650	680	678	614	609	612	610	632	628	626	624
475 c			860	920	934	930	926	861	857	822	820	856	857	874	885
476 c		007	000	048	062	040	036	989	985	995	992	013	010	007	008
477 c x?	13	194	188	256	266	240	237	223	216	197	180	203	202	213	211
478 d a				477		468		488	422	414	413	438	435	434	437
479 c			479		488		468		482	480	476	502	512	524	526
480 d ?						564		561	558	592	590	598		606	
481 c			664	656	716	702	704	670	650	659	657	665	658	678	676
482 c			898	870	865	946	930	891	886	886	882	893	895	900	902
483 c	14	130	125	165	174	176	174	125	118	114	112	146	137	132	131
484 d				331	265	306		295	293	344	263	245	244	256	216
485 d			468	544	553	556		481	478	465 530	476	483 518	504	405 500	500
486 d															
487 d				624	642			565	549	620	616	632	624		
488 d						835	830	802	795	791	868	828	834	792	792
489 c		912	905				911	903	900	869		918	912	900	903
490 d				975		986		981		948					
491 d	15		045	125		104	094	081	068	058	054	090	087	077	075
492 d ?					133								138		136
493 c x?		350	342	378	386	384	385	362	359	329	328	367	362	350	350
494 b		636	630	642	650	602	661	605	600	590	586	616	614	589	593
495 d a				713	753			756	650	698	691	674	686	720	724
496 d ?	16		052		090	115	110	900 053	996	003 041	015		038	032	034
497 c															
498 b		234	236	266	276	294	289	231	227	230	225	252	244	251	254
499 c			500	672	682	645	640	603	596	600	590	610	578	632	582

? Note by F E. F.. Line 447 double in XI. April 6: 5 441, 5 385.

ANNALS OF ASTROPHYSICAL OBSERVATORY.

TABLE 18.—*Bolographs of rock-salt prismatic solar spectrum—Observed positions of minima*—Continued.

[Initials refer to observers.]

Designation	Distance	December 7, 1897. Bolograph V.		December 9, 1897 Bolograph 11.		December 9, 1897. Bolograph IX.		March 14, 1898. Bolograph II.		March 14, 1898. Bolograph VIII.		April 6, 1898. Bolograph V.		April 6, 1898. Bolograph XI.	
		F.E.F.	C.G.A.	F.E.F.	C.G.A.	F.E.F.	C.G.A.	F.E.F.	C.G.A.	F.E.F.	C.G.A.	F.E.F.	C.G.A.	F.E.F.	C.G.A.
	cm.														
500 c		......	721	811	815	788	785	747	740	759	751	800	769	765	742
501 d				929		934		961	958	962	930	943	930		
502 d	17	......	020	048	064	048	042	069	053	043	031	062	056	......	062
503 b		398	390	406	414	418	420	369	361	358	352	361	362	382	374
*504 c		} 775	{ 728	734	778	764	......	......	} 746	{					
505 c			962	943	632	985	968								
506 i	18	} 256	{ 195	} 380	{ 239	} 368	{ 232	......	......	} 320	......	402	{		
507 d					524		528								
508 i	19	022	026	......	008	028	037	......							
509 c		107	092	132	102	150	150	......	086	......	086	074	090	......	186
510 d†		270	......	......	212	290	288	......	......	......	220				
*511 c		524	516	} 616	{ 560	558	558	} 350	{ 589	} 508	{ 534	} 546	......	805	......
512 c		698	686		723	721	716								
513 c		994	984	......	002	009	006	......	001	......	932	961	......		
514 c	20	232	222	284	302	306	307					252			
515 d†		520				548						400			
516 d†		640				660									
517 b		842	834	865	863	874	874	793	799	784	782	794	804	816	817
518 d†	21	062	010	059	062	110	072	......	080	......	040	......	002		
*519 d		328	325	332	338	344	346	} 275	{ 358	} 289	......	271	{ 366	......	347
520 d } c		409	404	422	426	416	415								
521 c		653	656	670	678	684	680	622	636	586	586	588	644		
522 c		962	976	904	934	020	016	927	936	948	950	001	011		
523 d†	22	122	114	......	......	146	140								
524 c		410	364	387	372	404	398					348			
525 c		690	682	661	710	694	699	701	689	......	......	694	680		
526 d	23	012	078	004	062	064	054	008	040	......	026	932	948		
527 d†		159				146									
528 d		274	276	......	310	255	258	......	254	......	302				
529 d†		386				386									
530 d†		504				492									
531 c		592	590	606	608	576	641	550	565	482	......	564	539	516	618
532 d†		688				686									
533 d†		760				747									
534 c	24	166	200	258	270	174	176	083	108	172	192	180	310	072	220
535 d		776	898	......	892	884	984	25.132		25.048	825	......	......	782	......
536 d	25	439	562	......	642	570	706	......	......	584	......	......	......	632	......
537 a Y	27	......	......	730	296	28.148	868	834	745	625	555	824	390	875	688
	32	......	......	574	811	33.435	808	33.424	210	425	762	425	33.925	375	33.862
538 c	33	......	......	202	924	127	884	968	858	754	762	992	962	026	930
539 d	33	......	......	......	698	......	678	......	......	......	......	549	......	......	683
540 d	34	......	438	394	396	533	432	394	387	390	385	390	......	......	......
541 d	34	842	......	866	876	033	034	......	......	912	......	857	......		
542 c	35	286	300	306	316	359	338	237	312	238	237	268	286	289	272
†543 c		758	......	832	899	895	892	729	771	784	790				
544 d } c	36	} 019	{ 912	} 096	{ 004	} 124	{ 982	} 981	{ 970	} 951	{ 936	} 790	......	827	......
545 d					199				168		142				
546 d		......	......	......	......	256	......	287	......	282	......	......	......	298	
547 d		376	376	403	414	472	463	......	400	332	362	......	358		
548 d	37	......	703	......	628	......	741								
549 c		050	026	040	050	081	074	913	952	959	962	900	960	990	992
550 c		450	449	500	514	565	532	419	422	450	442	979	477	402	450
551 d	38	......	622	......	634	......	654	......	......	......	......	......	581	......	636

*Notes by F. E. F.: Lines 504 and 505 single in VIII, March 14, and V, December 7; mean 17.760. Lines 511 c and 512 c double at 19.710 and 19.541. Lines 519 and 520 single in last one; mean 21.278.

† Notes by F. E. F.: Lines 543, 544, and 545, single in V, April 6, XI, April 6 35.804

TABLE 18.—*Bolographs of rock-salt prismatic solar spectrum—Observed positions of minima—Continued.*

[Initials refer to observers.]

Designation.	Distance.	December 7, 1897. Bolograph V.		December 9, 1897. Bolograph II.		December 9, 1897. Bolograph IX.		March 14, 1898. Bolograph II.		March 14, 1898. Bolograph VIII.		April 6, 1898. Bolograph V.		April 6, 1898. Bolograph XI.	
		F.E.F.	C.G.A.	F.E.F.	C.G.A.	F.E.F.	C.G.A.	F.E.F.	C.G.A.	F.E.F.	C.G.A.	F.E.F.	C.G.A.	F.E.F.	C.G.A.
*552 d ⎱ c	cm.	....	....	} 038	{ 976	968	958	} 879	{ 888	876	905	} 936	{ 866	880	884
553 d ⎰		....	....		{ 126	213	....			982	....		{ 023	968	990
454 d ?		....	....	....	254	....	....	....	....	....	198	....	196	....	208
555 b	39	809	....	824	832	892	854	717	747	702	696	754	824	752	769
556 d		428	400	519	426	491	490	419	445	428	404	344	....	368	....
557 b		804	....	759	770	874	718	737	734	706	726	744	718	724	670
558 d ?		....	....	....	861	....	839	....	796	....	819	....	807	....	790
559 d ?		....	....	....	970	....	014	....	964	....	014	....	964	....	966
560 c	40	164	....	206	140	192	136	129	112	116	110	119	097	168	172
561 d		....	175	....	216	....	238	....	282	....	220	....	234	....	....
562 d ?		....	....	....	526	....	504	....	534	....	496	....	448	....	478
563 c		....	....	....	640	698	718	680	524	677	514	590	490	576	....
564 d ?					852		830		794		817		756		738
565 d ⎫					931		918		888		894		870		827
566 d ⎬ b	41	....	....	} 002	{ 026	} 031	{ 926	} 969	{ 022	} 852	{ 056	} 884	{ 929	} 840	{ 923
567 d ⎭					202		176		141		....		136		....
568 d ?		....	....	....	438	....	430	....	....	....	388	....	833	....	816
569 d		....	....	....	608	....	549	455	483	417	....	420	535	528	....
570 d ?		....	....	....	780	....	....	....	788	....	756	....	740	....	....
571 c		....	....	....	922	....	036	959	539	884	939	924	908	940	936	974
572 d	42	....	....	....	404	....	....	330	347	330	316	330	337	404	470
573 c		....	....	....	881	....	956	895	920	925	898	948	950	936	978
574 d	43	....	....	....	....	....	....	419	412	384	382	404	....	384	....
575 d	44	....	....	....	} 440	....	....	{ 949	238	{ 004	} 192	{ 999	} 244	{ 031	} 328
576 c	44	....	....	....		....	....	{ 603		{ 550		{ 548		{ 600	
577 c	45	....	....	....	....	....	....	332	411	384	462	494	542	478	556
578 d	46	....	....	....	....	....	....	331	339	482	412	754	....	500	470
579 d	47	....	....	....	118	....	086	....	066	....	065	....	006	....	006

* Notes by F. E. F.: Lines 552 and 553 single in II, March 14; V, April 6; II, December 9; mean = 37.951.

ANNALS OF ASTROPHYSICAL OBSERVATORY. 149

TABLE 19.—*Bolographs of glass prismatic solar spectrum—Observed positions of minima.*

[Initials refer to observers.]

Designation	Distance	May 16, 1898. Bolograph II.		May 17, 1898. Bolograph VIII.		May 9, 1898. Bolograph X.		May 13, 1898. Bolograph I.	
		F. E. F.	C. G. A.	F. E. F.	C. G. A.	F. R F.	C. G. A.	F. E. F.	C. G. A.
	cm.								
1 d	31	742	....	738	742	751	....	703	699
2 c		666	660	652	650	638	679	644	642
3 d		608	....	598	....	606	....	....	....
4 d		568	564	554	....	554	554	592	584
5 d		531	....	509	....	502	....	506	....
6 d		474	478	456	....	446	446	....	408
7 d		400	....	418	....	....	....	415	....
8 d		....	....	366	....	338	....	....	....
9 d		180	173	170	165	....	....	139	....
10 d		094	090	089	086	102	100	095	093
11 d		....	....	054	....	030	027	039	038
12 d A	30	966	926	977	973	956	954	943	940
13 d		903	901	884	882	876	874	890	892
14 c a		797	792	799	792	770	768	778	780
15 c		692	693	692	689	656	656	650	650
16 c b		579	576	596	584	566	562	544	541
17 c		458	456	466	422	446	440	432	430
18 c		353	351	351	351	343	336	328	321
19 c		241	238	223	222	206	204	190	190
20 d		151	....	} 088	{ 090	} 073	{ 068	090	054
21 c		102	112					044	054
22 c	29	973	972	946	943	910	908	984	984
23 c		850	846	838	839	794	780	800	798
24 c		715	709	714	....	673	669	671	667
25 d	28	472	460	482	418	453	385	362	....
26 d ?	25	....	616	....	640	....	644	....	665
27 d ?	24	....	845	....	825	....	816	....	....
28 d ?		....	589	....	621	....	573	....	596
29 d ?		....	322	....	306	....	282	....	250
30 c	23	980	000	....	....	994	987	944	968
31 c		....	212	....	194	....	196	....	180
32 d ?		....	066	....	078	....	984	....	046
33 d ?	22	....	680	....	620	....	620	....	....
34 d		....	488	....	514	....	490	....	478
35 c		298	300	316	292	316	313	261	283
36 d ?	21	....	146	....	120	....	....	....	114
37 d ?		....	066	....	038	....	....	....	032
38 d	20	....	972	....	934	....	014	....	996
39 d		....	804	....	778	....	822	....	780
40 d ??		....	....	....	688	....	642	....	....
41 c		596	620	592	594	554	556	575	571
42 c		440	440	448	449	456	453	418	412
43 c		314	312	294	296	320	317	293	286
44 c		198	197	180	180	210	209	219	211
45 c b		034	085	055	054	086	032	034	028
46 c	19	908	911	916	920	904	906	911	908
47 c		762	760	747	750	748	748	756	754
48 d		648	....	655	....	650	....	674	....
49 d		550	....	536	538	538	535	579	578
50 d	19	452	....	447	446	474	470	....	....
51 d	19	380	....	388	392	376	372	385	384
52 d		310	....	318	....	....	....	319	....
*53 d		....	....	....	....	230	230	210	210
54 d ?		196	....	168	....	....	....	....	....
55 d		109	....	065	086	110	105	094	098
56 d b	18	960	....	960	960	....	964	951	943
57 d ?		906	....	880	....	872	....	....	....
58 b		824	824	818	819	818	814	811	806

*Note by F. E. F.: Lines 53 and 54 possibly same line.

TABLE 19.—*Bolographs of glass prismatic solar spectrum—Observed positions of minima*—Continued.

[Initials refer to observers.]

Designation.	Distance.	May 16, 1898. Bolograph II.		May 17, 1898. Bolograph VIII.		May 9, 1898. Bolograph X.		May 13, 1898. Bolograph I.	
		F.E.F.	C.G.A.	F.E.F.	C.G.A.	F.E.F.	C.G.A.	F.E.F.	C.G.A.
	cm.								
59 d		066	....	672	....	672	....	669	....
60 c		552	554	354	556	536	557	540	544
61 d		375	376	442	444	372	....	388	380
62 c		300	300	321	320	296	288	291	290
63 e		182	186	160	162	179	174	174	176
*64 d }c	17	994	982	994	994	....	965	004	950
65 d }		960	....	930	....	968	....	950	....
66 d		855	850	849	838	836	885	852	856
67 c		722	723	716	718	740	735	726	728
68 c		612	612	606	609	606	606	630	630
69 d }c		496	} 490 {	493	} 438 {	....	} 434 {	510	} 444
70 d }		434		435		429		443	
71 d ?		302	306	250	287	....	....	....	336
72 c		205	204	186	188	134	130	135	136
73 d		009	005	014	014	....	954	....	982
74 c	16	914	914	937	940	896	804	921	926
75 d ?	14	....	585	....	564	....	486	....	526
76 d ?		....	388	....	341	....	368	....	....
77 d }b		143	198	190	....	200	190	168	....
78 b }		056	058	064	062	060	058	059	056
79 c	13	793	800	766	760	792	788	797	800
80 d ?		652	656	654	....	....	670	630	....
81 d		563	568	565	559	566	562	511	507
82 d }b		456	....	452	....	456	....	425	....
83 b		332	335	325	326	328	324	312	311
84 d ?		192	....	188	....	206	....	132	....
85 d		102	106	095	103	065	090	057	....
86 d ?	12	....	836	....	....	....	804	....	832
87 d ?		....	645	....	646	....	614	....	681
88 d ?		....	382	....	440	....	....	....	393
89 d ?	11	....	802	....	775	....	798	....	824
90 d }		544	542	548	544	518	521	536	534
91 d ? }b		438	....	438	....	....	....	460	....
92 b }		352	354	340	340	338	335	342	341
93 d		220	....	219	....	....	....	235	....
94 d		142	149	124	125	150	150	158	134
95 d		026	032	023	018	066	068	....	....
96 d	10	904	905	914	911	928	926	934	970
97 d ? ?		....	550	....	506	....	....	....	520
98 d	9	898	900	940	939	895	896	882	884
99 c		734	738	686	688	714	714	726	729
100 d		572	583	577	576	570	568	556	556
101 d ?		....	430	....	389	....	404	....	....
102 d		286	290	277	276	256	254	266	273
103 d ?		134	....	156	150	....	188	072	069
104 c		056	058	046	049	008	008	092	093
105 d ?	8	880	872	918	912	....	....	....	....
106 d ?		776	784	785	780	786	784	....	750
107 d		574	581	....	556	582	580	....	....
108 d		480	473	432	431	448	446	....	....
109 c		235	237	206	204	188	190	202	....
110 c		138	140	110	110	108	107	132	160
111 d	7	912	914	....	948	891	894	....	950
112 d ?		662	651	....	662	592	591	....	630
113 d ?		463	460	....	436	446	438	500	....
114 d		361	360	....	358	374	368	....	....
115 d ?		290	291	276	268	....	....	....	....
116 c		184	186	171	168	174	144	138	144

*Note by F. E. F.: Lines 64 and 65 possibly a single line.

ANNALS OF ASTROPHYSICAL OBSERVATORY.

TABLE 19.—*Bolographs of glass prismatic solar spectrum—Observed positions of minima*—Continued.

[Initials refer to observers.]

Designation.	Distance.	May 16, 1898. Bolograph II.		May 17, 1898. Bolograph VIII.		May 9, 1898. Bolograph X.		May 13, 1898. Bolograph I.	
		F.E.F.	C.G.A.	F.E.F.	C.G.A.	F.E.F.	C.G.A.	F.E.F.	C.G.A.
	cm.								
117 d ⎫	6	939	935	969	966	951	956	916	913
118 d ⎬		881	880	913	....	910	910	836	832
119 c ⎬ b		799	798	847	850	808	810	727	726
120 c ⎭		621	619	692	692	666	664	590	585
121 c		475	477	546	548	542	542	489	484
122 d		315	....	396	396	416	416	381	380
123 b		262	259	300	303	293	292	268	212
124 c ⎫ b		105	110	118	120	096	100	065	....
125 c ⎭		010	012	006	007	021	024	983	094
126 c ⎫	5	874	877	874	871	846	850	902	897
127 c ⎭		768	773	748	720	716	716	785	786
128 d f		664	676	628	....	634	....	681	687
129 c		518	518	495	490	477	474	523	528
130 c		359	362	338	342	351	351	366	366
131 c		223	224	228	228	222	220	183	180
132 b ⎫ b		099	099	100	102	099	099	043	042
133 c ⎭	4	959	965	964	965	976	975	911	908
134 c		848	850	822	822	866	867	....	....
135 d		752	756	758	....	786			
*136 d		655	654	696	703	710	713	602	604
137 c		522	524	616	615	596	596	518	518
138 b ⎫		352	358	428	425	436	434	378	378
139 b ⎭		208	210	248	250	194	192	194	193
140 d f		100	....	181	....				
141 d ⎫	3	997	000	002	983	053	068	040	035
142 b ⎬ Group.		918	919	895	892	843	844	924	921
143 d f ⎬		741		721		704		791	
144 b ⎬		620	620	574	576	548	550	622	618
145 d ⎬		479	430	452	448	440	443	514	514
146 c ⎭		333	338	320	320	316	316	347	350
147 d ⎫		158	162	154	149	160	161	109	105
148 c ⎬		042	044	046	042	046	045	008	006
149 d ⎬	2	839	840	892	881	898	900	821	....
150 d ⎬ Distinct group.		718	718	732	732	766	766	738	731
151 c ⎬		606	606	592	590	644	645	666	664
152 d f ⎬		488	....	493	483	521	522	463	....
153 c ⎭		410	411	412	410	456	460	391	388
154 d f		260	....	236	....	2 9	....	225	....
155 d		130	128	142	140	154	156	142	135
156 d		036	030	042	038	082	081	047	043
157 d	1	900	904	914	908	930	934	953	953
158 d		838	841	862	856	832	834	862	858
159 d		721	716	723	722	692	690	746	742
160 d		628	628	612	614	564	563	643	642
161 c b p		534	531	520	524	497	506	557	555
162 c		396	394	391	391	360	358	395	396
163 c		324	328	286	296	294	294	318	316
164 b		131	134	122	150	136	134	132	128
165 b	0	960	962	952	952	954	954	930	927
166 c		830	831	809	808	826	824	777	776
167 d		700	....	718	715	737	....	702	....
168 c		505	505	516	514	523	522	481	479
169 c		384	382	396	394	405	402	383	380
170 d		244	244	271	268	336	338	278	278
171 b		160	160	164	164	174	174	170	168
172 b	0	000	000	000	000	000	000	000	000
173 b		214	216	215	212	258	257	221	224

*Note by F. E. F. Lines 136-139 difficult to identify on account of change in character.

TABLE 19.—*Bolographs of glass prismatic solar spectrum—Observed positions of minima.*—Continued.

[Initials refer to observers.]

Designation	Distance.	May 16, 1898. Bolograph II.		May 17, 1898. Bolograph VIII.		May 9, 1898. Bolograph X.		May 13, 1898. Bolograph I.	
		F.E.F.	C.G.A.	F.E.F.	C.G.A.	F.E.F.	C.G.A.	F.E.F.	C.G.A.
	cm.								
174 d ?		267	....	262	....				
175 c		377	374	379	383	386	384	341	341
176 o		494	488	526	528	546	548	512	514
177 b	a	714	716	728	726	748	746	725	730
178 b	1	....	....	021	022	022	025	015	020
179 d ?		....	....	124	....	131	....	147	....
180 d		....	....	266	266	226	....	262	272
181 b		345	350	340	344	327	327	366	368
182 b } b σ		488	494	473	477	460	462	499	500
183 b		620	626	612	613	585	588	616	620
184 c		820	822	813	814	794	796	804	808
185 b		936	940	934	936	912	914	920	924
186 d	2	034	....	049	....	010	....	012	....
187 o		136	134	138	134	114	119	122	124
188 b		318	322	320	320	348	348	309	311
189 b		528	528	526	530	524	528	520	520
190 c		634	636	643	642	602	598	615	616
191 d		764	766	792	....	727	731	742	744
192 c		825	826	838	835	818	820	834	831
193 d ?		913	....	934	....			916	....
194 o	3	070	069	069	071	....	....	061	060
195 c		208	208	200	202	....	....	226	213
196 c		316	314	309	312	....	....	310	309
197 c }		512	512	470	478	456	456	492	494
198 c } b τ		636	636	626	632	604	606	631	629
199 c }		763	763	762	766	736	736	754	754
200 d		878	870	850	....	836	832	882	858
201 c	4	012	008	984	982	978	978	998	993
202 d		078	082	082	080	048	052	056	....
203 d		255	248	244	250	196	206	247	....
204 c		316	316	326	332	294	293	315	316
205 c		446	445	432	440	422	421	434	436
206 c		551	548	565	569	528	524	550	546
207 c		701	899	722	726	....	764	715	718
208 d ?		751	....	....	....	760	....		
209 c		838	836	834	837	847	848	827	828
210 d ? ?		899	....	872	....				
211 d		958	948	949	950	020	022	926	920
212 c	5	069	062	065	072	112	111	082	082
213 d ? ?		204	....	....	....	232	....	188	....
214 c		410	404	364	356	404	400	395	404
215 c		626	626	624	626	637	616	....	....
216 d		728	722	736	750				
217 o	6	068	072	....	....	026	050	055	037	036
218 d		....	....	262	....	256	....	201	....	234
219 d ? ?		....	....	877	....	862	....		
220 b	7	145	136	114	137	196	216	163	165
221 d		378	356	329	....	336	342	328	326
222 o		478	472	464	472	474	475	480	478
223 d		534	524	554	....	....	519	535	....
224 c		668	660	648	656	645	642	654	644
225 d		870	858	845	850	....	....	794	....
226 d		....	981	....	023	....	914	....	984
227 d ?	8	....	303	....	352	....	302	....	....
228 d ? ?		....	557	....	006	....	596	....	....
229 c	9	....	122	....	101	....	103	....	....
230 o		392	387	398	400	380	389	....	....

ANNALS OF ASTROPHYSICAL OBSERVATORY. 153

TABLE 19.—*Bolographs of glass prismatic solar spectrum—Observed positions of minima*—Continued.

[Initials refer to observers.]

Designation.	Distance.	May 16, 1896. Bolograph II.		May 17, 1896. Bolograph VIII.		May 9, 1896. Bolograph X.		May 13, 1896. Bolograph I.	
		F.E.F.	C.G.A.	F.E.F.	C.G.A.	F.E.F.	C.G.A.	F.E.F.	C.G.A.
	cm.								
231 d ††	11	....	228	....	226	....	172	....	....
*232 c	12	....	026	983	982	982	984	....	....
233 d ††	13	....	....	....	055	....	062	....	....
234 c		292	286	303	350	359	356	....	....
235 c		553	568	570	568	520	522	....	....
236 d ††		....	....	....	717	....	784	....	....
237 d	14	060	....	044	046	....	046	....	....
238 c		179	174	170	175	156	156	....	....
239 d		296	297	348	346	364	368	....	....
240 c		429	421	432	434	447	448	....	....
241 d		634	630	635	644	642	618	....	....
242 c		820	816	....	840	816	816	....	....
243 c	15	036	026	074	076	026	060	....	....
244 d		180	178	124	....	169	166	....	....
245 c		300	298	226	229	272	274	....	....
246 c		450	444	425	430	400	400	....	....
247 d		616	610	578	586	547	532	....	....
248 c		700	694	681	682	678	674	....	....
249 c		896	892	862	862	854	853	....	....
250 d		968	....	952	....	945	942	....	....
251 d	16	085	086	071	064	026	024	....	....
252 d		194	190	204	198	152	164	....	....
253 c		282	270	279	280	248	248	....	....
254 c		454	456	434	439	422	420	....	....
255 c		594	586	606	612	577	574	....	....
256 d †		750	....	720	....	756	....	....	....
257 c		824	822	832	831	806	811	....	....
258 c		978	977	968	966	946	945	....	....
259 c	17	152	150	144	144	118	120	....	....
260 c		356	355	352	345	356	354	386	378
261 d		470	....	448	....	....	....	....	....
262 c		522	523	498	497	470	472	525	518
263 d		676	679	612	614	578	584	646	644
264 c		776	777	740	744	724	724	766	757
265 d		907	912	846	840	822	815	845	838
266 c		994	994	980	984	945	946	982	974
267 b	18	152	150	136	138	112	116	139	136
268 c		343	340	342	343	329	332	340	338
269 d		488	498	468	464	493	492	482	489
270 d †		604	....	604	....	....	....	....	....
271 c		640	607	654	650	608	607	628	633
272 b		836	835	832	832	814	816	835	835
273 d	19	047	054	037	036	016	016	041	042
274 b		150	154	140	140	140	140	143	146
275 c		409	408	382	378	365	366	398	396
276 d		490	496	488	484	452	451	483	481
277 b		668	662	671	667	656	656	699	694
278 c		832	833	824	820	797	796	826	820
279 b	20	004	005	987	987	978	980	009	006
280 c		211	214	198	191	178	179	204	200
281 c	φ	318	320	321	338	354	352	320	324
282 d †		359	....	371	....	....	....	373	....
283 d †		539	....	546	....	....	....	561	....
284 d †		638	....	670	....	608	....	....	....
285 d †		769	....	....	....	756	....	783	....
†286 b	a	856	776	812	808	824	824	843	832
287 d		927	....	968	....	944	942	960	954

* Note by F. E. F.: Line 232 double in II, May 16, and VIII, May 17, at 12.029 and 11.950.
† Note by F. E. F.: Lines 285 and 286 off plate in VIII, May 17.

TABLE 19.—*Bolographs of glass prismatic solar spectrum—Observed positions of minima*—Continued.

[Initials refer to observers.]

Designation.	Distance.	May 16, 1898. Bolograph II.		May 17, 1898. Bolograph VIII.		May 9, 1898. Bolograph X.		May 13, 1898. Bolograph I.	
		F.E.F.	C.G.A.	F.E.F.	C.G.A.	F.E.F.	C.G.A.	F.E.F.	C.G.A.
	cm.								
288 c ⎫c⎫	21	143	146	129	124	⎰	163	158	202
289 d ⎬ ⎬		202	204	190	190	⎱163⎰	....	201	
290 d ⎪a⎪		374	375	340	340		356	377	378
291 c ⎪ ⎪		460	462	434	430	412	414	479	475
292 c ⎪ ⎪		746	750	728	726	728	731	....	762
293 d ? ⎪ ⎪		776	....	782	....	780		767	....
294 d ⎬b⎬		854	850	868	865	830	832	830	826
295 c ⎪ ⎪		950	952	950	948	961	962	989	986
296 d ⎪ ⎪	22	200	206	194	176	135	161	188	190
297 b ⎪ ⎪		260	260	247	244	224	222	258	254
298 d ⎪ ⎪		377	382	388	385	340	338	377	380
299 b ⎪ ⎪		476	480	484	479	457	458	489	486
300 b ⎪ ⎪		620	620	628	621	615	616	637	632
301 c ⎭ ⎭		766	768	774	772	750	750	785	783
302 c		957	956	954	956	932	934	975	972
303 c	23	094	094	084	084	074	073	137	
304 c		178	176	179	174	144	148	192	180
305 d		....	298	262	310	....	308	309	....
306 d ?		306	....	314	....	....	....	349	....
307 d		367	358	370	362	315	....	396	348
308 c		526	532	504	506	482	484	575	578
309 b		652	652	629	631	645	646	711	710
310 d		792	786	783	786	799	749	798	794
311 d		910	908	⎰ 957 ⎰	912	946	947	930	922
312 d ?	24	018	013	⎱ ⎱	004	076		028	
313 c		123	123	110	112	148	060	148	148
314 d		262	255	218	219	213	278	284	284
315 d		334	340	328	332	312		363	376
316 b		418	418	410	412	404	406	452	452
317 c		544	540	546	545	554	556	571	570
318 b		740	738	750	749	773	773	783	784
319 c		924	924	914	915	952	955	991	997
320 b	25	118	117	104	105	188	188	205	205
321 d ?		285	....	228	....	316		334	
322 c		343	345	330	331	444	442	474	469
323 d		550	550	535	539	....	639	....	....
324 d ?		646		626					
325 b	25	755	644	726	630	....	....	....	843

Designation.	Distance.	May 17, 1898. Bolograph XI.		May 17, 1898. Bolograph V.		April 30, 1898. Bolograph VIII.		May 14, 1898. Bolograph VIII.	
		F.E.F.	C.G.A.	F.E.F.	C.G.A.	F.E.F.	C.G.A.	F.E.F.	C.G.A.
	cm.								
318 b	21	410	416	558	562	....	....	402	406
319 c		244	244	362	339	270	....	215	220
320 b		036	034	122	124	052	....	033	....
321									
322 c	20	816	815	866	880	834	....	801	806
323 d ?		....	626		664				
324									
325 b		502	503	550	554	529	....	491	492
326 d		290	286	285	292	304	....	274	285
327 c		040	037	988	990	010	....	025	025
328 c	19	865	867	878	880	895	....	864	864

TABLE 19.—*Bolographs of glass prismatic solar spectrum—Observed positions of minima*—Continued.

[Initials refer to observers.]

Designation.	Distance.	May 17, 1898. Bolograph XI.		May 17, 1898. Bolograph V.		April 30, 1898. Bolograph VIII.		May 14, 1898. Bolograph VIII.	
		F.E.F.	C.G.A.	F.E.F.	C.G.A.	F.E.F.	C.G.A.	F.E.F.	C.G.A.
	cm.								
329 d ?		....	....	770	774	....	....	785	....
330 d		665	660	660	664	682	....	672	673
331 b		548	545	564	566	588	....	558	560
332 d		....	410	....	401	....	....	....	....
333 c		292	282	308	302	352	....	336	338
334 b		146	145	152	158	169	....	141	142
335 d		000	002	038	050	042	048	036	040
336 c	18	863	864	886	887	880	....	800	796
337 c		706	712	703	710	782	789	727	730
338 ?		496	502	483	490	509	507	486	488
339 d ?		....	382	....	410	....	....	....	....
340 c		240	232	258	264	268	266	236	240
341 d		010	....	086	....	013	....	027	035
342 d	17	892	....	....	....	856	....	826	836
343 c		636	680	655	662	643	....	616	618
344 c ?		....	....	472	....	483	....	....	....
345 c		159	159	154	160	202	196	140	142
346 d	16	988	988	964	970	986	970	966	968
347 d		} 649	{ 756	} 755	{ 760	}	....	688	{ 684
348 d			652		686				
349 d ?	14	480	....	....	554	536	....	498	497
350 d ?		372	....	....	....	438	....	372	....
351 d ?		323	....	....	....	390	....	322	....
352 d		240	....	274	271	268	266	233	230
353 d		090	088	079	136	106	....	127	130
354 d	13	944	946	974	986	001	074	976	....
355 d ?		822	....	....	....	....	....	858	....
356 b		745	748	744	750	778	777	760	762
357 d		628	....	606	....	650	....	624	....
358 d ?		431	....	422	....	....	....	....	....
359 c		282	274	296	304	....	....	....	....
360 d		158	160	145	144	....	....	....	....
361 d		066	059	008	004	....	....	....	....
362 d	12	877	874	878	878	....	....	826	....
363 d		768	764	816	822	....	....	760	....
364 b		662	664	700	702	697	696	679	678
365 d		....	475	486	493	524	530	480	490
366 d		....	093	076	085	....	086	042	045
367 d	11	....	707	....	759	....	....	656	660
368 d		....	563	....	592	....	....	552	554
369 c		....	460	453	472	498	497	428	485
370 d		216	....	252	....	202	....	166	....
371 c		154	152	174	....	....	190	083	082
372 d	10	996	....	024	....	....	....	960	....
373 b		916	914	954	962	920	918	897	898
374 c		710	712	733	740	742	738	710	716
375 d		....	....	634	....	650	....	615	....
376 c		474	473	464	460	492	492	479	476
377 d		....	262	216	219	300	....	229	....
378 c		211	194	141	150	218	220	179	179
379 d		076	084	035	044	116	117	102	105
380 d	9	906	907	914	915	980	....	934	935
381 d		788	780	798	802	828	831	801	802
382 c		700	702	697	706	740	739	717	716
*383 d		566	....	571	....	642	....	586	....

*Note by F. E. F.: Lines 383 to 391 displaced in V, May 17.

TABLE 19.—*Bolographs of glass prismatic solar spectrum—Observed positions of minima*—Continued.

[Initials refer to observers.]

Designation.	Distance.	May 17, 1898. Bolograph XI.		May 17, 1898. Bolograph V.		April 30, 1898. Bolograph VIII.		May 14, 1898. Bolograph VIII.	
		F.E.F.	C.G.A.	F.E.F.	C.G.A.	F.E.F.	C.G.A.	F.E.F.	C.G.A.
	cm.								
384 b		457	462	477	481	474	478	428	432
385 d		374	....	396				338	346
386 d		182	188	236	242	168	154	160	164
387 d		079	076	140	140	....	....	070	072
388 c	8	982	980	038	042	960	957	954	954
389 b		806	814	824	830	784	782	800	799
390 d }b		720		742				732	....
391 c		635	636	653	654	656	656	644	644
392 c		457	460	470	474	499	498	480	478
393 c		353	350	347	350	379	375	370	370
394 d		226	....	241		237		251	
395 c		142	144	130	135	158	154	146	146
396 c	7	842	842	846	844	865	862	856	856
397 d		681		689		712		692	
398 c		544	542	562	560	568	565	546	543
399 d		....	337	356	358				327
400 d ?		....	....	017	014	....	....	....	972
401 c ?	5			418	419				
402 c ?	4	165	218	220	226				
403 d ?		....	102	028	016				
404 c	3	655	662	668	678	694	690	....	666
405 c		355	350	351	366	300	300	....	362
406 c		....	....	170	174	172	161	....	122
407 c	2	....	....	960	961	979	982	922	924
408 d		....	854	884	884	888	....	....	830
409 d		801	....	836	....	816	....	834	
410 b		625	626	623	625	617	614	616	625
411 c		376	380	379	374	406	390	394	388
412 c		260	262	260	266	262	264	274	274
413 d	ψ	174	....	115	....	170		176	....
414 b	1	956	958	968	970	986	984	972	973
415 c		797	800	790	791	824	828	804	804
416 b	a	638	636	638	640	656	654	640	642
417 d	1	....	....	494	501	518	....	499	503
418 b		286	288	351	352	358	354	333	340
419 c		156	160	172	172	188	166	162	162
420 d ?	0	994	022	....	966			012	026
421 c		924	928	936	940	956	950	936	934
422 c		768	768	773	776	799	796	782	782
423 c		641	642	654	634	647	650	661	660
424 d		557	558	596	600	562	440	591	596
425 b		358	358	344	350	347	345	356	355
426 d		251	....	249	....	....	....	276	....
427 b	0	000	000	000	000	000	000	000	000
428 d		106	100	126	114	063	....	132	126
429 b		342	342	349	344	331	336	343	344
430 d		533	534	524	520	470	471	492	482
431 d ?		....	....	610		576		598	
432 c		731	734	716	712	746	746	750	744
433 c }b		936	934	919	915	905	908	914	915
434 b	1	173	186	122	128	136	136	163	164
435 d ? }b		454	....	418	....	450	....	421	
436 b		502	454	464	450	486	457	460	460
437 d		599	....	584	572	586	589	596	556
438 d ?		662	....	....	....	679		650	
439 c		855	856	868	866	858	862	866	866
440 c }b		952	957	970	964	918	919	926	926
441 b	2	130	133	146	143	130	132	126	129
442 d		256	256	274	263	229	230	237	232

TABLE 19.—*Bolographs of glass prismatic solar spectrum—Observed positions of minima*—Continued.

[Initials refer to observers.]

Designation.	Distance.	May 17, 1898. Bolograph XI.		May 17, 1898. Bolograph V.		April 30, 1898. Bolograph VIII.		May 14, 1898. Bolograph VIII.	
		F.E.F.	C.G.A.	F.E.F.	C.G.A.	F.E.F.	C.G.A.	F.E.F.	C.G.A.
	cm.								
443 c } b		480	471	486	488	454	455	452	446
444 b		624	624	629	624	614	612	605	604
445 d		828	822	805	796	796	....	800	792
446 b		994	991	968	964	960	958	964	960
447 d ?	3	068	083	062	....	....	....	058	....
448 c		231	232	223	236	204	213	196	192
449 c		450	453	430	427	391	397	406	410
450 d ?		530	....	490	....	548	....	526	....
451 d		700	704	712	718	675	....	678	688
452 b		814	817	824	820	810	813	836	829
453 d ?	4	....	104	....	121	....	....	....	....
454 b		198	196	194	190	166	167	166	159
455 c		529	526	525	510	....	....	514	499
456 c		776	783	766	758	732	733	766	760
457 d		906	908	873	863	875	874	952	944
458 d	5	068	036	128	122	060	056	131	128
459 c		....	338	396	364	325	328	408	345
460 c		677	686	714	698	668	670	714	710
461 c	6	044	044	004	002	990	006	001	998
462 c ?	7	....	170	....	194	....	....	....	146
463 d ?		....	427	....	402	....	....	....	454
464 d	8	502	528	526	526	496	....	502	500
465 d ?		729	....	725	....	684	....	744	....
466 c		894	892	909	898	830	831	866	....
467 c	9	044	044	026	....	006	014	033	030
468 d		....	....	424	413	396	396	447	445
469 b		902	902	914	909	882	879	898	914
470 c	10	395	404	402	380	354	362	427	431
471 d		532	636	570	580	....	....	590	593
472 c		747	750	749	743	806	812	792	796
473 d		931	948	856	860	....	....	948	934
474 c	11	205	204	200	226	250	200	212	216
475 c		323	322	370	365	441	406	374	....
476 d	12	090	088	198	106	115	078	074	080
477 d		887	959	914	922	946	942	914	912
478 d	13	420	448	433	352	507	....	496	500
479 d ?		....	547	....	510	....	....	....	616
480 d ?	14	....	....	308	300	319	....	287	289
481 d ?		....	....	406	400	428	....	388	380
482 d ?		....	....	675	668	635	....	643	633
483 c	15	264	264	272	270	281	316	292	290
484 d ?		....	....	470	....	486	....	475	....
485 d		....	716	705	706	709	741	750	748
486 d ?	16	....	....	000	....	978	....	010	....
487 d ?		....	....	280	....	323	....	301	....
488 c	17	462	458	477	464	490	....	482	477
489 d ?		662	....	658	....	....	....	692	692
490 d		751	752	771	764	768	....	781	778
491 d	18	296	300	332	368	265	....	320	316
492 c		489	484	478	482	458	....	464	464
493 c		713	716	702	695	688	....	730	728
494 d		840	842	848	842	....	....	852	854
495 c	19	204	202	217	209	188	194	217	212
496 c		385	386	372	362	386	397	397	392
497 d		574	572	562	556	594	....	620	616
498 d		694	694	700	694	731	758	722	726
499 d		....	768	814	806	876	....	874	877
500 c		945	953	974	970	990	997	978	975

TABLE 19.—*Bolographs of glass prismatic solar spectrum—Observed positions of minima—*Continued.

[Initials refer to observers.]

Designation.	Distance.	May 17, 1898. Bolograph XI.		May 17, 1898. Bolograph V.		April 30, 1898. Bolograph VIII.		May 14, 1898. Bolograph VIII.	
		F.E.F.	C.G.A.	F.E.F.	C.G.A.	F.E.F.	C.G.A.	F.E.F.	C.G.A.
	cm.								
501 c	20	236	233	275	270	250	266	266	264
502 d		328	....	337	....	354	....	362	....
503 d		542	544	575	461	486	....	539	536
504 c		632	626	636	630	570	574	632	628
505 d		902	908	888	889	832	816	891	898
506 b	21	012	010	002	000	993	998	028	028
507 d		188	186	202	196	205	218	201	201
508 c		324	314	330	326	347	344	368	362
509 c		596	596	606	611	612	566	618	617
510 c		719	712	732	730	736	740	764	756
511 d?	22	....	....	036	....	006	....	030	....
512 c		098	129	....	....	087	088	135	....
513 b		337	334	....	362	349	348	364	362
514 d		537	536	494	494	553	543	568	573
515 c		655	653	598	598	672	678	686	686
516 c		750	749	687	684	854	858	770	762
517 c		946	947	940	938	947	949	966	963
518 c	23	062	062	072	072	104	107	106	104
519 c		266	264	257	258	259	260	288	290
520 c		454	452	441	437	451	451	462	460
521 c		700	694	728	723	728	724	734	726
522 d		810	806	850	842	854	852	842	839
523 d		958	....	972	....	970	....	990	....
524 d?	24	....	144	132	140	....	155	148	....
525 c		301	290	298	297	328	333	312	320
526 d?		....	....	434	....	438	....	419	....
527 c		509	516	530	526	569	570	568	568
528 b		850	849	830	822	893	899	938	938
529 d		916	....	920	....	963	....	986	....
530 d	25	044	....	017	....	027	....	078	079
531 c		251	246	211	207	258	260	296	285
532 d?		396	....	445	....	424	....	478	458
533 d		588	586	603	590	603	596	594	596
534 d		708	699	706	701	698	699	715	702
535 c		894	890	916	916	924	922	948	956
536 c	26	200	} 212	236		311	317	219	....
537 c		310						350	....
538 c?	33	....	....	426	....	481	....	406	....
539 c		....	....	922	....	910	....	026	....
540 c	34	....	....	192	....	274	....	194	....
541 c		....	....	425	....	444	....	422	....
542 c	Ω	....	....	634 } 716 {		670 } 768 {		661 } 790 {	....
543 c		....	....	780		844			
544 c	35	....	....	202	200	264	236	242	246
545 c		....	....	486	470	452	450	535	475
546 b	36	....	....	031	029	014	070	080	079
547 c		....	....	554	600	614	612	498	659
548 b	37	....	....	....	....	....	....	....	238
549 b	37	....	....	....	....	....	....	....	850

Table 20 contains the ordinates of the minima in the rock-salt bolographs. Naturally the ordinates of minima of classes *a* and *b* are of little value as showing the trend of the curve, but those for the slighter deflections of classes *c* and *d* will be useful for this purpose, and it will be interesting to compare the relative intensity of energy in the different bolographs. It has been found more convenient to include in this table a second part, in which a very important reduction, described on a later page, has been introduced, but attention is now called only to the first part, to which the following description applies:

The measures in ordinates are, of course, subject to far greater error from "drift" than would arise from imperfect observations of positions upon the comparator. Accordingly, the observations of ordinates were recorded only to the nearest millimeter, and only one observer's results are included in the tables, as the differences between the two observers' measurements were nowhere appreciable. The results here recorded are the vertical distances between the minima of the bolograph and a base line determined as well as possible for each bolograph separately by the position of start and close, the bottom of the great absorption bands, and the drift observed in the battery records taken just before and after the bolograph itself, and allowing, finally, for the discontinuities in the curve. If the "drift" were zero, or throughout uniform, there would, of course, be no inaccuracy in the position of the base line as thus determined, and the vertical distances between it and the curve would give the true ordinates of the latter. But as the drift is not necessarily constant, it is presumable that at some places on the curve, notably at $\rho\ \sigma\ \tau$, which is a long way from any point of reference, errors as great as 5 mm may occur. There are included in the table of ordinates indications from two bolographs whose abscissæ are not recorded.

TABLE 20.—*Relative intensities of radiation from bolographs of rock-salt prismatic solar spectrum.*

[Initials refer to observers.]

Designation	Observed ordinates of minima.								Reduced ordinates of minima.							
	Apr. 1, 1898.	Apr. 2, 1898.	Apr. 6, 1898.	May 18, 1898.		Mar. 1, 1898.	Apr. 1, 1898.		Apr. 1, 1898.	Apr. 2, 1898.	Apr. 6, 1898.	May 18, 1898.		Mar. 1, 1898.	Apr. 1, 1898.	
	Bolograph II.	Bolograph XI.	Bolograph II.	Bolograph II.	Bolograph II.	Bolograph IV.	Bolograph V.	Bolograph V.	Bolograph II.	Bolograph XI.	Bolograph II.	Bolograph II.	Bolograph II.	Bolograph IV.	Bolograph V.	Bolograph V.
	Corrected ordinate measures.								Corrected ordinates multiplied by factor—							
	C.G.A.	C.G.A.	C.G.A.	C.G.A.	C.G.A.	C.G.A.	F.E.F.	F.E.F.	7.58	8.20	6.25	7.30	6.90	6.94	5.78	7.14
1 d†	2.2	2.4	1.5	2.1	2.0	1.7	......	......	16.7	19.7	9.4	15.3	13.8	11.8	......	10.0
2 d	1.8	1.8	......	......	1.4	2.0	1.4	......	13.6	14.8	......	......	9.7	11.6	10.0	......
3 d } b	......	......	1.1	1.5	1.4	1.3	2.1	1.3	......	......	6.9	11.0	9.7	9.0	12.1	9.3
4 d †	2.2	2.2	1.2	......	1.3	1.4	......	1.6	16.7	18.0	7.5	......	9.0	9.7	......	11.4
5 d	......	......	......	1.8	1.4	......	2.8	......	......	......	......	13.1	9.7	......	16.2	......
6 d	......	2.7	1.8	......	1.7	1.7	......	......	......	22.1	11.2	......	11.7	11.8	......	......
7 d	......	2.6	......	2.3	......	2.0	3.0	2.2	......	21.3	......	16.8	......	13.9	17.3	15.7
8 d	2.6	......	2.1	2.2	2.0	2.0	......	1.9	19.7	......	13.1	16.1	13.8	13.9	......	13.6
9 d	2.6	2.4	......	......	......	1.9	2.9	......	19.7	19.7	......	......	......	13.2	16.8	......
10 d	......	......	2.1	2.4	1.9	1.9	3.1	1.0	......	......	13.1	17.5	13.1	13.2	17.9	13.6
11 d A	......	......	......	......	......	......	......	......	......	......	......	......	......	......	......	......
12 d a	......	......	......	......	......	......	2.2	......	......	......	......	......	......	......	15.7	......
13 d } b	3.0	3.0	2.5	2.7	2.1	2.1	3.7	2.3	22.7	24.6	15.6	19.7	14.5	14.6	21.4	16.4
14 d	3.5	3.5	3.1	......	2.3	2.3	4.2	2.7	26.5	28.7	19.4	......	15.9	16.0	24.3	19.3
15 d	3.7	......	......	3.3	2.7	2.5	4.5	2.9	28.0	......	......	24.1	18.6	17.4	26.0	20.7
16 d	4.0	3.9	3.6	3.5	3.0	2.6	5.1	3.3	30.3	32.0	22.5	25.6	20.7	18.0	29.5	23.6
17 d	......	......	......	......	......	......	5.5	......	......	......	......	......	......	......	31.8	......
18 d	4.4	4.3	4.0	4.0	......	3.0	......	3.6	33.4	35.3	25.0	29.2	......	20.8	......	25.7
19 d	......	......	......	......	......	......	......	3.7	......	......	......	......	......	......	......	26.4
20 d	......	......	......	......	......	6.1	......	3.8	......	......	......	......	......	......	35.3	27.1

TABLE 20.—*Relative intensities of radiation from bolographs of rock-salt prismatic solar spectrum*—Continued.

[Initials refer to observers.]

Designation.	Observed ordinates of minima.								Reduced ordinates of minima.						
	Apr. 1, 1898.	Apr. 2, 1898.	Apr. 6, 1898.	May 18, 1898.	Mar. 1, 1898.	Apr. 1, 1898.	Apr. 1, 1898.		Apr. 1, 1898.	Apr. 2, 1898.	Apr. 6, 1898.	May 18, 1898.		Mar. 1, 1898.	Apr. 1, 1898.
	Bolograph II.	Bolograph XI.	Bolograph II.	Bolograph II.	Bolograph II.	Bolograph IV.	Bolograph V.	Bolograph V.	Bolograph II.	Bolograph XI.	Bolograph II.	Bolograph II.	Bolograph IV.	Bolograph V.	Bolograph V.
	Corrected ordinate measures.								Corrected ordinates multiplied by factor—						
	C.G.A.	C.G.A.	C.G.A.	C.G.A.	C.G.A.	C.G.A.	F.E.F.	F.E.F.	7.58	8.20	6.25	7.30	6.90	6.94	5.78	7.14
21 d ?	......	5.5	5.1	4.7	4.3	3.8	6.3	......	45.1	31.9	34.2	29.7	26.4	36.4	......
22 d	5.8	5.9	5.2	5.1	4.5				44.0	48.4	32.5	37.2	31.0		
23 d ?	6.3	6.6	5.5	......	5.2	4.5			47.8	54.1	34.4	......	35.9	31.2	
24 d ?	......	6.8	5.7	6.2						55.8	35.6	45.3			
25 d ?	......	6.8	5.7	......	5.5	4.6				55.8	35.6	......	38.0	31.9	
26 d	6.5	6.7	5.7	6.2	5.4	4.5		4.9	49.3	54.9	35.6	45.3	37.3	31.2	......	35.0
27 d	} 6.5	6.6	5.3	6.1	5.3	4.6	{ 8.1	4.9	} 49.3	54.1	33.1	44.5	36.6	31.9	{ 46.8	35.0
28 d							7.9	4.9							45.7	35.0
29 d	6.4	......	5.3	......	5.2	4.5		4.8	48.5	......	33.1	......	35.9	31.2	......	34.3
30 c	6.3	6.5	5.1	5.9	5.1	4.4	7.8	4.7	47.8	53.3	31.9	43.1	35.2	30.5	45.1	33.6
31 d	6.5	......	5.3	5.8	5.2	4.5	7.8	4.7	49.3	......	33.1	42.3	35.9	31.2	45.1	33.6
32 d ? } b							7.9	4.7							45.7	33.6
33 c	} 6.5	6.6	5.3	5.8	5.1	4.4	{ 7.8		} 49.3	54.1	33.1	42.3	35.2	30.5	{ 45.1	
34 d ?							4.7									33.6
35 d	} 6.5	6.7	......	5.8			{ 7.9		} 49.3	54.9		42.3			{ 45.7	
36 d	} 6.6	6.7	5.4	5.9	5.4	4.6	{ 8.0	4.8	} 50.0	54.9	33.8	43.1	37.3	31.9	{ 46.2	34.3
37 c							8.0	4.9							46.2	35.0
38 d							8.1	5.1							46.8	36.4
39 d							7.9	5.0							45.7	35.7
40 c	6.3	6.4	5.3	5.9	5.1	4.3	7.6	4.7	47.8	52.5	33.1	43.1	35.2	29.8	43.9	33.6
41 d	......	6.0	......	5.9			8.3	5.3		37.5		40.7			48.0	37.8
42 d	7.0	7.2	6.0	6.5	5.7	5.0			53.1	59.0	37.5	47.4	39.3	34.7		
43 d	} 7.0	7.2	6.0	......	5.9	4.9	{ 5.2		} 53.1	59.0	37.5	......	40.7	34.0	{	37.1
44 d																
45 d	7.1	7.2	6.0				5.2		53.8	59.0	37.5				37.1	
46 d	7.2	7.3	6.0	5.8			5.3		54.6	59.9	37.5	42.3			37.8	
47 d	7.2	7.4	......	5.8	6.1	5.1			54.6	60.7		42.3	42.1	35.4		
48 d ?				5.9	6.2							43.1	42.8			
49 d ?			......	5.9	6.3	5.3						43.1	43.5	36.8		
50 d ?	......	7.7	6.3	......	6.6	5.7				63.1	39.4	......	45.5	39.6		
51 d	7.8	7.9	8.5	6.2	6.8	5.8			59.1	64.8	40.6	45.3	46.0	40.3		
52 d ?																
53 d ?	8.1	8.3	6.9	......	7.3	6.2			61.4	68.1	43.1	......	50.4	43.0		
54 d ?	8.2	8.2	7.0	6.3	7.3	6.3			62.2	67.2	43.8	46.0	50.4	43.7		
55 c	7.8	7.8	6.7	6.1	7.0	6.0	9.2	5.7	59.1	64.0	41.9	44.5	48.3	41.6	53.2	40.7
56 d	8.1	8.1	7.0	6.3	7.2	6.2	......	6.0	61.4	66.4	43.8	46.0	49.7	43.0	......	42.8
57 d	8.0	7.8	6.9	6.3	7.3	6.2	......	6.1	60.6	64.0	43.1	46.0	50.4	43.0	......	43.6
58 d	......	7.7	......	......	6.0	9.7		5.6	......	63.1			41.6	56.1		40.0
59 b	7.4	7.5	6.0	5.5	6.6	5.6			56.1	61.5	37.5	40.2	45.5	38.9		
60 d ?							8.9	5.4							51.4	38.6
61 d	8.2	8.1	6.9	5.5	7.2	6.2	9.5	6.0	62.2	66.4	43.1	40.2	49.7	43.0	54.9	42.8
62 d ?																
63 d ?	......	8.4	7.4	6.6	......	6.5				68.9	46.2	48.2	......	45.1		
64 d ?	......	8.5	7.4	......	7.7	6.5				69.7	46.2	......	53.1	45.1		
65 d	8.3	8.5	7.4	......	......			62.9	69.7	46.2						
66 d ?	......	......	7.5	6.9	......	6.8					46.9	50.4	......	47.2		
67 d ?	} 8.5	8.5	7.3	6.8	7.8	6.7	{ 9.9	6.6	} 64.4	69.7	45.6	49.7	53.8	46.5	{ 57.2	47.1
68 d								6.4								45.7
69 b	7.8	7.9	6.7	6.0	7.1	6.2	8.9	5.7	59.1	64.8	41.9	43.8	49.0	43.0	51.4	40.7
70 d ?	......	8.5	7.1	6.9	7.0	6.6	8.8	......		69.7	44.4	50.4	52.4	45.8	50.9	......
71 d							9.1	6.4							52.6	45.7

ANNALS OF ASTROPHYSICAL OBSERVATORY. 161

TABLE 20.—*Relative intensities of radiation from bolographs of rock-salt prismatic solar spectrum*—Continued.

[Initials refer to observers.]

Designation.	Observed ordinates of minima.								Reduced ordinates of minima.							
	Apr. 1, 1896.			Apr. 2, 1898.	Apr. 6, 1898.	May 18, 1898.		Mar. 1, 1898.	Apr. 1, 1898.	Apr. 1, 1898.	Apr. 2, 1898	Apr. 6, 1898	May 18, 1898.		Mar. 1, 1898.	Apr. 1, 1898.
	Bolograph II.	Bolograph XI.	Bolograph II.	Bolograph II.	Bolograph II.	Bolograph II.	Bolograph IV.	Bolograph V.	Bolograph V.	Bolograph II.	Bolograph XI.	Bolograph II.	Bolograph II.	Bolograph II.	Bolograph IV.	Bolograph V.	Bolograph V.
	Corrected ordinate measures.									Corrected ordinates multiplied by factor—						
	C.G.A.	C.G.A.	C.G.A.	C.G.A.	C.G.A.	C.G.A.	C.G.A.	F.E.F.	F.E.F.	7.58	8.20	6.25	7.30	6.90	6.94	5.78	7.14
72 d	8.5	8.5	7.5	6.9	8.0	6.7	10.1			64.4	69.7	46.9	50.4	55.2	46.5	58.4	
73 d ?	8.6	8.5	7.6		8.1	7.0				65.2	69.7	47.5		55.9	48.6		
74 d	8.6	8.4	7.8	7.1	8.2	7.0				65.2	68.9	46.8	51.8	56.6	48.6		
75 d	8.6	8.8	7.9	7.1	8.2	6.9				65.2	72.2	49.4	51.8	56.6	47.9		
76 d	8.7	8.9	7.9	7.0						65.9	73.0	49.4	51.1				
77 d	8.8		8.0	7.0						66.7		50.0	51.1				
78 d	8.8	8.8	8.0	7.0	8.3	7.1				66.7	72.2	50.0	51.1	57.3	49.3		
79 d		8.9	8.1	7.0	8.3	7.3					73.0	50.6	51.1	57.3	50.7		
80 c	8.9	8.9	8.0	7.1	8.3	7.2		6.5		67.5	73.0	50.0	51.8	57.3	50.0		46.6
81 d ?																	
82 d ?		9.0	8.2		8.5	7.5					73.8	51.2		58.6	52.0		
83 d ?		9.0	8.1			7.5					73.8	50.6			52.0		
84 d ?		9.1	8.0		8.4	7.4					74.6	50.0		58.0	51.4		
85 d					8.4	7.4	11.0							58.0	51.4	63.6	
86 d ?																	
87 d		9.2	8.4		8.5	7.7		7.0			75.4	52.5		58.6	53.4		50.0
88 d ?	9.3	9.2	8.4		8.5					70.5	75.4	52.5		58.6			
89 d	9.2	9.0	8.3	6.8	8.4	7.4		8.9		69.7	73.8	51.9	49.7	58.0	51.4		49.3
90 d ?																	
91 d	9.1	8.6		6.4	7.9	7.0	10.7	6.8		69.0	70.5		46.7	54.5	48.6	61.8	48.6
92 d	8.8	8.4		6.3	7.6	6.9	10.6	6.6		66.7	68.9		46.0	52.4	47.9	61.3	47.1
93 c } b	8.7	8.3		6.4	7.3	6.8	10.3	8.5		65.9	68.1		46.7	50.4	47.2	59.5	48.6
94 c	8.3	7.9		6.3	7.1	8.3	9.8	6.2		62.9	64.8		46.0	49.0	43.7	56.6	44.3
95 d	8.3			6.4	7.2	6.5	9.9	6.2		62.9			46.7	43.0	45.1	57.2	44.3
96 d	8.3	8.1		6.3			9.7	6.3		62.9	66.4		46.0			56.1	45.0
97 d		8.2			7.3	6.6		6.2			67.2			50.4	45.8		44.3
98 d	8.6			6.6	7.4	6.8	9.8	6.3		65.2			48.2	51.0	47.2	56.6	45.0
99 d	9.0	9.0	9.5	7.0	8.3	7.5	10.8			68.2	73.8	59.4	51.1	57.3	52.0	62.4	
100 d		7.8	9.4		8.3	7.5	10.8	6.9			64.0	58.8		57.3	52.0	62.4	49.3
101 d	8.7	8.5	9.0	6.7	7.9	7.1		6.8		65.9	69.7	56.2	48.9	54.5	49.3		48.6
102 d ?					7.6	6.9	10.4	6.5						52.4	47.9	60.1	46.4
103 d		8.3	8.8	6.3	7.4	6.7		6.4			68.0	55.0	46.0	51.1	46.5		45.7
104 d	8.6	8.2	8.5	6.5	7.4		10.3	6.3		65.2	67.2	53.1	47.4	51.1		59.5	45.0
105 d	8.6	8.3	8.7	6.5	7.5	6.7	10.3	6.2		65.2	68.1	54.4	47.4	51.8	46.5	59.5	44.3
106 d	8.9	8.3	8.8	6.7	7.7	6.8	10.5	6.3		67.5	68.1	55.0	48.9	53.1	47.2	60.7	45.0
107 d ? } b							10.7	6.4								61.8	45.7
108 c	8.9	8.5	9.0	6.8	7.7	6.9	10.7	6.5		67.5	69.7	56.2	49.6	53.1	47.9	61.8	46.4
109 c	8.7	8.2	8.8	6.6	7.3	6.6	10.5	6.4		65.9	67.2	55.0	48.2	50.4	45.8	60.7	45.7
110 d								6.6									47.1
111 c	8.8	8.3	8.8	6.5	7.4	6.7	10.7	6.3		66.7	68.1	55.0	47.4	51.1	46.5	61.8	45.0
112 d								6.6									47.1
113 c	8.9	8.6	9.1	6.6	7.6	6.9		6.5		67.5	70.5	56.9	48.2	52.4	47.9		46.4
114 d }	9.1			6.8	8.1	7.2	{ 10.3			} 69.0			49.6	55.9	50.0	{ 59.5	
115 d								6.9									49.3
116 c	9.2	9.0	9.6	6.9	8.3	7.5	11.3	8.9		69.7	73.8	60.0	50.4	57.3	52.0	65.3	49.3
117 c	9.1	9.0	9.5	6.8	8.3	7.5	10.8	6.8		69.0	73.8	59.4	49.6	57.3	52.0	62.4	48.6
118 d							11.0	6.8								68.6	48.6
119 d	8.9	8.8	9.5	6.9	8.3	7.4	11.1	6.7		67.5	72.2	59.4	50.4	57.3	51.4	64.2	47.8
120 d		8.8	9.5		8.3		11.2	6.7			72.2	59.4		57.3		64.7	47.8
121 d	9.1	8.9	9.6	7.1	8.3		11.0	6.8		69.0	72.0	60.0	51.8	57.3		63.6	48.6

TABLE 20.—*Relative intensities of radiation from bolographs of rock-salt prismatic solar spectrum*—Continued.

[Initials refer to observers.]

Designation.	Observed ordinates of minima.								Reduced ordinates of minima.							
	Apr. 1, 1898.	Apr. 2, 1898.	Apr. 6, 1898.	May 18, 1898.		Mar. 1, 1898.	Apr. 1, 1898.		Apr. 1, 1898.	Apr. 2, 1898	Apr. 6, 1898	May 18, 1898.		Mar. 1, 1898.	Apr. 1, 1898.	
	Bolograph II.	Bolograph XI.	Bolograph II.	Bolograph II.	Bolograph II.	Bolograph IV.	Bolograph V.	Bolograph V.	Bolograph II.	Bolograph XI.	Bolograph II.	Bolograph II.	Bolograph II.	Bolograph IV.	Bolograph V.	Bolograph V.
	Corrected ordinate measures.								Corrected ordinates multiplied by factor—							
	C.G.A.	C.G.A.	C.G.A.	C.G.A.	C.G.A.	C.G.A.	F.E.F.	F.E.F.	7.58	8.20	6.25	7.80	6.90	6.94	5.78	7.14
122 d	9.2	8.8	9.7	7.0	8.4	7.6	11.1	6.7	69.7	72.2	60.6	51.1	58.0	52.7	64.2	47.8
123 d	8.9	8.2	8.9	......	7.5	6.9	11.2	6.5	67.5	67.2	55.6	......	51.8	47.9	64.7	46.4
124 d	......	7.8	8.6	6.4	......	......	10.3	6.3	......	64.0	53.8	46.7	......	......	59.5	45.0
125 d	8.3	......	......	5.8	6.5	5.8	9.4	6.0	62.9	......	......	42.3	44.8	40.2	54.3	42.8
126 d	7.8	......	7.6	......	6.1	5.3	......	5.5	59.1	......	47.5	......	42.1	36.8	......	39.3
127 d }b	6.9	6.3	7.0	4.7	5.2	4.7	8.5	4.8	52.3	51.7	43.8	34.3	35.9	32.6	49.1	34.3
128 d	6.6	5.8	6.0	4.5	4.9	4.4	7.7	4.4	50.0	47.6	37.5	32.8	33.8	30.5	44.5	31.4
129 d ? p							7.1								41.0	
130 c	5.9	5.3	5.4	3.9	4.2	3.8	6.5	3.9	44.7	43.5	33.8	28.5	29.0	26.4	37.6	27.8
131 c	6.1	5.3	5.7	4.0	4.6	4.1	......	4.0	46.2	43.5	35.6	29.2	31.7	28.4	......	28.6
132 c	6.1	5.3	5.7	4.3	4.6	4.1	6.8	4.0	46.2	43.5	35.6	31.4	31.7	28.4	39.3	28.6
133 d	6.4	......	......	4.5	4.8	4.4	7.2	4.3	48.5	......	......	32.8	33.1	30.5	41.6	30.7
134 d	7.4	6.6	7.3	5.8	5.1	5.5	......	5.1	56.1	54.1	45.6	42.3	35.2	38.2	......	36.4
135 d	} 7.2	6.4	7.0	5.4	6 1	5.4	{ 8.5 5.0	5.7 4.8	}54.6	52.5	43.8	39.4	42.1	37.5	{49.1 47.4	35.7 34.3
136 d							8.2									
137 c	6.6	6.0	6.3	5.0	5.5	4.8	7.9	4.4	50.0	49.2	39.4	36.5	38.0	33.3	45.7	31.4
138 c	6.5	5.8	6.1	4.6	5.0	4.5	7.7	4.3	49.3	47.6	38.1	33.6	34.5	31.2	44.5	30.7
139 e	6.7	6.1	6.4	5.1	5.3	4.8	7.9	4.8	50.8	50.0	40.0	37.2	36.6	33.3	45.7	32.5
140 d }b	7.1	6.5	......	5.4	......	......	8.4	3.9	53.8	53.3	......	39.4	......	......	48.6	27.8
141 c	7.1	6.4	6.7	5.2	5.8	5.0	8.4	3.6	53.8	52.5	41.9	38.0	40.0	34.7	48.6	25.7
142 b σ	7.0	6.3	6.6	5.1	5.7	4.8	8.2	4.8	53.0	51.7	41.2	37.2	39.3	33.3	47.4	34.3
143 b }	7.3	6.6	6.8	5.2	5.8	5.2	8.6	5.3	55.3	54.1	42.5	38.0	40.0	36.1	49.7	37.8
144 c	7.4	6.9	7.2	5.3	6.2	5.6	8.9	5.7	56.1	56.6	45.0	38.7	42.8	38.9	51.4	40.7
145 c	7.7	6.9	7.3	5.5	6.3	5.7	9.0	5.7	58.4	56.6	45.6	40.2	43.5	39.6	52.0	40.7
146 c a	......	7.1	7.5	5.4	6.4	5.7	9.1	5.9	......	58.2	46.9	39.4	44.2	39.6	52.6	42.1
147 d	......	......	7.8	......	7.0	6.3	9.8	......	......	......	48.8	......	48.3	43.7	56.6	......
148 c	7.8	7.1	7.9	5.7	6 6	5.9	9.5	6.3	59.1	58.2	49.4	41.6	45.5	40.9	54.9	45.0
149 d	8.4	7.6	8.3	5.8	7.2	6.4	11.0	6.5	63.7	62.3	51.9	42.3	49.7	44.4	63.6	48.6
150 c	8.4	7.7	8.5	5.8	7.4	6.6	10.3	6.4	63.7	63.1	53.1	42.3	51.1	45.8	59.5	45.7
151 c	8.6	8.0	8.7	6.1	7.8	7.1	10.7	6.7	65.1	65.6	54.4	44.5	53.8	49.3	61.8	47.8
152 d ?							10.9								63.0	
153 c	9.0	8.4	9.3	6.4	7.2	7.4	11.0	6.9	68.2	68.9	58.1	46.7	49.7	51.4	63.6	49.3
154 d	9.6	9.1	10.3	6.8	9.3	8.4	11.9	7.6	72.8	74.6	64.4	49.6	64.2	58.3	68.8	54.3
155 c	9.9	9.3	10.4	7.5	9.6	8.9	12.3	7.8	75.0	76.3	65.0	54.8	66.2	61.8	71.1	55.7
156 d ?	}......	9.3	10.3	7.4	9.6	8.9	{ 12.1	8.0	}......	76.3	64.4	54.0	66.2	61.8	{69.9	57.1
157 d							......	7.7							55.0	
158 c	9.7	9.0	10.1	7.0	9.3	8.6	11.8	7.5	73.5	73.8	63.1	51.1	64.2	59.7	68.2	53.6
159 c	9.7	9.1	10.3	7.3	9.4	8.6	11.7	7.6	73.5	74.6	64.4	53.3	64.9	59.7	67.6	54.3
160 d }b							12.0	7.6							69.4	54.3
161 d }c	} 10.1	9.5	10.6	7.6	9.7	9.1	{ 12.3	7.8	}76.6	77.9	66.2	55.5	66.9	63.2	{71.1	55.7
162 d }							12.3								71.1	
163 d							12.6	7.8							72.8	55.7
164 d	10.6	9.9	11.5	8.1	9.7	9.7	12.9	8.1	80.3	81.2	71.9	59.1	66.9	67.3	74.6	57.8
165 d	10.7	10.2	11.6	8.2	9.9	10.1	......	8.3	81.1	83.6	72.5	59.9	68.3	70.1	......	59.3
166 d	10.7	10.3	......	8.2	9.9	10.3	......	......	81.1	84.5	......	59.9	68.3	71.5	......	......
167 d	......	......	......	8.3	10.1	10.3	......	......	......	......	......	60.6	69.7	71.5	......	......
168 d	10.9	10.6	12.0	8.4	10.4	10.7	13.3	......	82.6	86.9	75.0	61.3	71.8	74.3	76.9	......
169 d	11.0	10.6	......	8.6	......	......	13.3	......	83.4	86.9	......	62.8	......	......	76.9	......
170 d ?	......	10.7	......	......	10.4	10.7	......	......	......	87.7	......	......	71.8	74.3	......	......
171 d	11.0	10.6	......	8.7	......	10.6	......	9.2	83.4	86.9	......	63.5	......	73.6	......	65.7
172 c	10.6	10.3	11.8	8.5	10.3	10.5	13.3	9.1	80.3	84.5	73.7	62.0	71.1	72.9	76.9	65.0
173 d	10.9	10.6	......	......	10.5	10.8	......	9.3	82.6	86.9	......	......	72.4	75.0	......	66.4

ANNALS OF ASTROPHYSICAL OBSERVATORY. 163

TABLE 20.—*Relative intensities of radiation from bolographs of rock-salt prismatic solar spectrum*—Continued.

[Initials refer to observers.]

Designation.	Observed ordinates of minima.									Reduced ordinates of minima.							
	Apr. 1, 1898.	Apr. 2, 1898.	Apr. 6, 1898.	May 18, 1898.		Mar. 1, 1898.	Apr. 1, 1898.		Apr. 1, 1898.	Apr. 2, 1898	Apr. 6, 1898	May 18, 1898.		Mar. 1, 1898	Apr. 1, 1898.		
	Bolo-graph II.	Bolo-graph XI.	Bolo-graph II.	Bolo-graph II.	Bolo-graph IV.	Bolo-graph II.	Bolo-graph V.	Bolo-graph V.	Bolo-graph II.	Bolo-graph XI.	Bolo-graph II.	Bolo-graph II.	Bolo-graph IV.	Bolo-graph V.	Bolo-graph V.		
	Corrected ordinate measures.								Corrected ordinates multiplied by factor—								
	C.G.A.	C.G.A.	C.G.A.	C.G.A.	C.G.A.	C.G.A.	F.E.F.	F.E.F.	7.58	8.20	6.25	7.80	6.90	6.94	5.78	7.14	
174 d	.....	10.6	12.3	8.7	10.5	10.9	.....	.....	86.9	76.9	63.5	72.4	75.6	.....	.....	
175 d	11.3	10.5	12.3	.....	10.6	10.9	14.4	9.5	85.6	86.1	76.9	.....	73.1	75.6	83.2	67.8
176 d ?	.....	10.6	12.4	.....	10.7	11.0	.....	.....	86.9	77.5	.....	73.8	76.3	.....	.....	
177 d	.....	10.6	12.5	.....	10.7	11.1	.....	.....	86.9	78.1	.....	73.8	77.0	.....	.....	
178 d ?	.....	10.8	12.6	.....	.....	11.1	.....	.....	86.6	78.7	.....	.....	77.0	.....	.....	
179 d ?	.....	10.5	12.6	9.4	11.1	11.3	.....	.....	86.1	78.7	68.6	76.6	78.4	.....	.....	
180 d	11.8	10.6	12.6	9.6	11.1	11.3	.....	.....	89.4	86.9	78.7	70.1	76.6	78.4	.....	
181 c	.....	10.4	12.8	9.5	11.1	11.4	.....	.....	85.3	80.0	69.4	76.6	79.1	.....	.....	
182 c	.....	10.6	12.8	9.5	11.2	11.6	.....	.....	86.9	80.0	69.4	77.3	80.5	.....	.....	
183 d	.....	10.8	12.7	9.5	11.3	11.5	.....	.....	88.6	79.4	69.4	78.0	79.8	.....	.....	
184 d	11.2	10.7	12.9	9.1	11.3	.....	.....	.....	84.9	87.7	80.6	66.4	78.0	.....	.....	
185 d	.....	10.8	12.7	.....	11.4	11.6	.....	.....	.....	88.0	79.4	.....	78.7	80.5	.....	
186 c	11.1	10.7	12.7	9.3	11.2	11.4	.....	.....	84.1	87.7	79.4	67.9	77.3	79.1	.....	
187 c	11.2	10.8	12.7	9.4	11.3	11.5	.....	9.5	84.9	88.6	79.4	68.6	78.0	79.8	.....	67.8
188 d	.....	.....	.....	9.3	11.4	11.6	.....	.....	.....	.....	.....	67.9	78.7	80.5	.....	
189 d	11.3	10.8	12.6	9.4	11.5	.....	.....	.....	85.6	88.6	78.7	68.6	79.4	.....	.....	
190 d	11.2	10.8	12.6	9.4	11.4	11.5	.....	9.5	84.9	88.6	78.7	68.6	78.7	79.8	.....	67.8
191 d ?	.....	.....	.....	.....	.....	.....	.....	9.5	.....	.....	.....	.....	.....	.....	.....	67.8
192 c	11.0	10.6	12.4	9.4	11.2	11.4	14.1	9.2	83.4	86.9	77.5	68.6	77.3	79.1	81.5	65.7
193 c	11.3	10.7	14.7	9.7	11.4	11.7	.....	9.3	85.6	87.7	79.4	70.8	78.7	81.2	.....	66.4
194 ?	11.4	10.7	12.6	9.8	11.5	11.7	14.5	9.6	86.4	87.7	78.7	71.5	79.4	81.2	83.8	68.5
195 c	.....	10.7	12.6	9.9	11.4	11.8	.....	.....	.....	87.7	78.7	72.3	78.7	81.9	.....	.....
196 c	11.4	10.7	12.5	9.8	11.3	11.6	.....	9.4	86.4	87.7	78.1	71.5	78.0	80.5	.....	67.1
197 c	11.3	10.6	12.4	9.7	11.4	11.5	14.4	9.6	85.6	86.9	77.5	70.8	78.7	79.8	83.2	68.5
198 c	11.8	10.6	12.3	9.8	11.3	11.4	.....	9.7	85.6	86.9	76.9	71.5	78.0	79.1	.....	69.3
199 c	11.5	10.8	12.4	10.0	11.4	11.6	14.4	.....	87.2	88.6	77.5	73.0	78.7	80.5	83.2	.....
200 d	.....	.....	.....	9.9	11.1	11.1	.....	10.1	.....	.....	.....	72.3	76.6	77.0	.....	72.1
201 c	11.0	10.2	11.8	9.4	10.8	11.0	13.9	9.5	83.4	83.6	73.7	68.6	74.5	76.3	80.3	67.8
202 ?	11.1	.....	11.8	9.4	10.9	14.0	.....	9.6	84.1	.....	73.7	68.6	75.2	75.6	80.9	68.5
203 d	.....	10.3	11.7	9.5	.....	10.8	14.0	9.6	.....	84.5	73.1	69.4	.....	75.0	80.9	68.5
204 c	11.0	10.2	11.6	9.5	10.7	10.6	13.7	9.4	83.4	83.6	72.5	69.4	78.8	73.6	79.2	67.1
205 c	10.8	10.0	11.4	9.4	10.5	10.4	13.6	9.4	81.9	82.0	71.2	68.6	72.4	72.2	78.6	67.1
206 ?	10.7	10.1	11.2	9.4	10.5	10.4	13.6	9.4	81.1	82.8	70.0	68.6	72.4	72.2	78.6	67.1
207 c	10.6	9.9	10.9	9.3	10.1	10.1	13.6	9.3	80.3	81.2	68.1	67.9	70.0	70.1	78.6	66.4
208 d ?	.....	.....	.....	.....	.....	.....	13.4	.....	.....	.....	.....	.....	.....	.....	77.5	.....
209 d	10.3	9.5	10.5	9.1	9.8	9.7	12.9	9.2	78.1	77.9	65.6	66.4	67.6	67.3	74.6	65.7
210 d	9.5	9.0	9.7	8.8	9.0	9.0	11.9	8.7	72.0	73.8	60.6	64.2	62.1	62.5	68.8	62.1
211 d	8.5	7.9	8.5	7.7	7.8	7.5	10.3	.....	64.4	64.8	53.1	56.2	53.8	52.0	59.5	.....
212 d	.....	.....	.....	.....	.....	.....	7.6	.....	.....	.....	.....	.....	.....	.....	54.3	.....
213 c	6.9	6.2	6.2	5.9	5.8	5.4	8.4	6.0	52.3	50.8	38.7	43.1	40.0	37.5	48.6	42.8
214 c	6.6	5.8	5.6	5.6	5.3	4.9	7.5	5.6	50.0	47.6	35.0	40.9	36.6	34.0	43.4	40.0
215 b b	5.9	5.1	4.6	4.8	4.6	4.3	6.4	4.6	44.7	41.8	28.7	35.0	31.7	29.8	37.0	32.8
216 c	5.9	5.1	4.6	4.8	4.7	4.4	.....	4.6	44.7	41.8	28.7	35.0	32.4	30.5	.....	32.8
217 c	6.2	5.2	5.0	5.1	4.9	4.7	6.9	5.0	47.0	42.6	31.2	37.2	33.8	32.6	39.9	35.7
218 c	.....	.....	.....	5.1	4.8	4.6	.....	.....	.....	.....	.....	37.2	33.1	31.9	.....	.....
219 b φ	5.6	4.7	4.2	4.4	4.4	4.0	6.3	4.0	42.4	38.5	26.2	32.1	30.4	27.8	36.4	28.6
220 d	8.2	7.0	7.2	6.7	6.7	6.4	9.2	6.7	62.2	57.4	45.0	48.9	46.2	44.4	53.2	47.8
221 c a	8.1	7.0	6.9	6.7	6.7	6.4	9.3	6.4	61.4	57.4	43.1	48.9	46.2	44.4	53.8	45.7
222 c	7.0	6.0	5.5	5.8	5.7	5.4	7.9	5.5	53.1	49.2	34.4	42.3	39.3	37.5	45.7	39.3
223 c b	7.1	6.0	5.4	5.5	5.4	4.9	.....	5.5	53.8	49.2	33.7	40.2	37.3	34.0	.....	39.3
224 c	7.9	6.9	6.5	6.3	6.5	6.0	9.3	6.2	59.9	56.6	40.6	46.0	44.8	41.6	53.8	44.3
225 c	8.3	7.2	7.2	6.9	7.1	6.6	10.0	6.5	62.9	59.0	45.0	50.4	49.0	45.8	57.8	46.4

TABLE 20.—*Relative intensities of radiation from bolographs of rock-salt prismatic solar spectrum*—Continued.

[Initials refer to observers.]

Designation.	Observed ordinates of minima.								Reduced ordinates of minima.						
	Apr. 1, 1898.	Apr. 2, 1898.	Apr. 6, 1898.	May 18, 1898.	Mar. 1, 1898.	Apr. 1, 1898.			Apr. 1, 1898.	Apr. 2, 1898.	Apr. 6, 1898.	May 18, 1898.	Mar. 1, 1898.	Apr. 1, 1898.	
	Bolograph II.	Bolograph XI.	Bolograph II.	Bolograph II.	Bolograph IV.	Bolograph V.	Bolograph V.		Bolograph II.	Bolograph XI.	Bolograph II.	Bolograph II.	Bolograph IV.	Bolograph V.	Bolograph V.
	Corrected ordinate measures.								Corrected ordinates multiplied by factor—						
	C.G.A.	C.G.A.	C.G.A.	C.G.A.	C.G.A.	C.G.A.	F.E.F.	F.E.F.	7.58	8.20	6.25	7.30	6.90	6.94	5.78	7.14
226 d							6.9									49.3
227 d	9.2	8.0	7.8	8.2	8.8	8.3	11.4	7.9	69.7	65.6	48.7	59.9	60.7	57.6	65.9	54.4
228 d	10.0	8.7	9.2	8.5	9.1	8.8	12.1	8.0	75.8	71.3	57.5	62.0	62.8	61.1	69.9	57.1
229 d	10.2	8.9	9.6	8.8	9.2	9.0	12.5	8.1	77.3	73.0	60.0	64.2	63.5	62.5	72.2	57.8
230 d	11.1	10.0					13.6	9.0	84.1	82.0					78.6	64.3
231 c	11.2	10.0	10.8	9.6	10.4	10.1	13.8	8.8	84.9	82.0	67.5	70.1	71.8	70.1	79.8	62.8
232 d							14.5								83.8	
233 d	11.6	10.5	11.4	10.0	11.3	10.9		9.3	87.9	86.1	71.2	73.0	78.0	75.6		66.4
234 d								9.4								67.1
235 c	11.5	10.6	11.3		11.1	10.7		9.3	87.2	86.9	70.6		76.6	74.2		66.4
236 c	11.6	10.7	11.5	9.9	11.2	11.0	15.9	9.4	87.9	87.7	71.9	72.3	77.3	76.2	91.9	67.1
237 c	11.3	10.5	11.2	9.7	11.1	10.7	15.5	9.9	85.6	86.1	70.0	70.8	76.6	74.3	89.6	70.7
238 c	11.6	10.7	11.5	9.8	11.4	11.1	15.9	10.1	87.9	87.7	71.9	71.5	78.7	77.0	91.9	72.1
239 c	11.5	10.7	11.4	9.8	11.4	11.1	15.8	10.1	87.2	87.7	71.2	71.5	78.7	77.0	91.3	72.1
240 c	12.1	11.3	12.2	10.3	12.0	11.7	16.3	10.6	91.7	92.7	76.2	75.2	82.8	81.2	94.2	75.7
241 d							16.6								95.9	
242 d	12.2	11.4	12.3	10.3	12.3	11.9	16.8	10.7	92.5	93.5	76.9	75.2	84.9	82.6	95.9	76.4
243 c	12.1	11.2	12.2	10.3	12.2	11.9	16.8	10.7	91.7	91.8	76.2	75.2	84.2	82.6	94.2	76.4
244 c	11.8	10.8	11.7	10.0	11.7	11.5	15.7	10.3		88.6	73.1	73.0	80.7	79.8	90.7	73.5
245 d							15.7	10.5							90.7	75.0
246 c	11.8	11.0	11.8	9.9	11.5	11.5	16.2	10.2	89.4	90.2	73.7	72.3	79.4	79.8	89.0	72.8
247 c	12.1	11.2	12.2	10.2	12.1	12.0	16.0	11.9	91.7	91.8	76.2	74.5	83.5	83.3	92.4	85.0
248 d	12.3		12.6	10.4	12.2	12.0	16.3	12.3	93.2		78.7	75.9	84.2	83.3	94.2	87.8
249 d	12.5		12.8	10.8	12.5	12.3	16.6	12.5	94.8		80.0	78.8	86.2	85.4	95.9	89.2
250 d							12.6									90.0
251 d	12.7	12.0	13.1	10.9		12.9	16.8	12.6	96.3	98.4	81.9	79.6			89.5	97.1	90.0
252 d	12.7	12.0	13.0	10.7	12.9	13.1	17.1	12.7	96.3	98.4	81.2	78.1	89.0	90.9	98.8	90.7
253 d																
254 c	12.9		13.1	11.0	13.5	13.3	17.1	12.9	97.8		81.9	80.3	89.7	91.6	98.8	92.1
255 d f			13.4		13.3	13.3	17.4				83.7		91.8	92.3	100.6	
256 d	12.9	12.4	13.4	11.1	13.5	12.3		13.1	97.8	101.7	83.7	81.0	93.2	92.3		93.5
257 d	12.9	12.4	13.5	11.3	13.5	13.4	17.9	13.2	97.8	101.7	84.4	82.5	93.2	93.0	103.5	94.2
258 d	12.9	12.4	13.4	11.3	13.6	13.5	17.9	13.1	97.8	101.7	83.7	82.5	93.8	93.7	103.5	93.5
259 d	12.8	12.4	13.2	11.1	13.5	13.7	18.0	12.8	97.0	101.7	82.5	81.0	93.2	95.1	104.0	91.4
260 d f	12.8	12.4	13.2		13.5			12.5	97.0	101.7	82.5		93.2			89.2
261 d f	12.8	12.3		11.0	13.6	13.8	17.9	12.2	97.0	100.9		80.3	93.8	95.8	103.5	87.1
262 d	12.7	12.1	13.0		13.5	13.6	17.6	13.3	96.3	99.2	81.2		93.2	94.4	101.7	95.0
263 d	12.5	12.1	12.8	10.4			17.4	12.8	94.8	99.2	80.0	75.9			100.6	91.4
264 d							16.9	12.6							97.7	90.0
265 d	} 12.3	11.5	12.4	10.3	13.1	13.0	{	12.3	} 93.2	94.3	77.5	75.2	90.4	90.2	{	87.8
266 d							16.3								94.2	
267 d	11.5		11.5	9.7	12.4	12.3	15.5	11.7	87.2		71.9	70.8	85.6	85.4	89.6	83.5
268 c	11.1	10.2	11.2	9.0	12.1	12.0	15.3	11.4	84.1	83.6	70.0	65.7	82.5	83.3	88.4	81.4
269 d	10.8	9.9		8.1			15.2	11.2	81.9	81.2					87.9	80.0
270 b	10.4	9.6	10.5	8.7	11.4	11.3	14.7	10.6	78.8	78.7	65.6	63.5	78.7	78.4	85.0	75.7
271 d	11.5	11.0		9.6	12.5	12.6	16.2	11.8	87.2	90.2		70.1	86.2	87.4	93.6	84.3
272 d f	12.0	11.5		10.3		13.1	16.8		91.0	94.3		75.2		90.9	97.1	
273 d							16.9	12.5							97.7	89.2
274 c	11.9	11.5	14.3	10.2	12.6	12.7	16.8	12.3	90.2	94.3	89.4	74.5	86.9	88.1	97.1	87.8
275 d	12.5	12.3	15.0	10.7	13.3	13.4	17.9	13.1	94.8	100.9	93.7	78.1	91.8	93.0	103.5	93.5
276 d	12.7	12.4	15.1	10.7	13.4	13.4	18.0	13.3	96.3	101.7	94.4	78.1	92.5	93.0	104.0	95.0

ANNALS OF ASTROPHYSICAL OBSERVATORY.   165

TABLE 20.—*Relative intensities of radiation from bolographs of rock-salt prismatic solar spectrum*—Continued.

[Initials refer to observers.]

Designation.		Observed ordinates of minima.								Reduced ordinates of minima.							
		Apr. 1, 1898.	Apr. 2, 1898.	Apr. 6, 1898.	May 18, 1898.		Mar. 1, 1898.	Apr. 1, 1898.	Apr. 1, 1898.	Apr. 2, 1898	Apr. 6, 1898	May 18, 1898.		Mar. 1, 1898	Apr. 1, 1898.
		Bolograph II.	Bolograph XI.	Bolograph II.	Bolograph II.	Bolograph IV.	Bolograph V.	Bolograph V.	Bolograph II.	Bolograph XI.	Bolograph II.	Bolograph II.	Bolograph IV.	Bolograph V.	Bolograph V.
		Corrected ordinate measures.							Corrected ordinates multiplied by factor—							
		C.G.A.	C.G.A.	C.G.A.	C.G.A.	C.G.A.	C.G.A.	F.E.F.	F.E.F.	7.58	8.20	6.25	7.30	6.90	6.94	5.78	7.14
277 d !		} 12.8	12.2	14.9	10.6	13.1	13.1	17.9	13.2	} 97.0	100.0	93.1	77.4	90.4	90.9	} 102.3	94.2
278 d								17.8	13.1							102.9	90.5
279 d		} 12.9	....	14.7	10.6	12.9	12.7	17.7	....	} 97.8	....	91.9	77.4	89.0	86.1	} 102.3	
280 d								17.5	13.1							101.2	93.5
281 d		....	....	14.1	10.3	12.5	12.3	17.2	12.8	....	....	88.1	75.2	86.2	85.4	99.4	91.4
282 d		12.4	11.8	14.1	10.3	12.3	12.2	16.9	12.8	94.0	96.8	88.1	75.2	84.9	84.7	97.7	91.4
283 d		12.1	11.4	13.7	10.2	12.0	11.8	16.5	12.8	91.7	93.5	85.6	74.5	82.8	81.9	95.4	91.4
284 d !		11.8	....	....	....	....	11.4	....	12.4	89.4	....	....	....	....	79.1	....	88.5
285 c		11.7	10.8	13.2	9.5	11.3	11.0	15.9	12.4	88.7	88.6	82.5	69.4	78.0	76.3	91.9	88.5
286 d																	
287 c		11.4	10.4	12.7	9.1	10.9	10.7	15.4	12.1	86.4	85.3	79.4	66.4	75.2	74.3	89.0	86.4
288 c		11.3	10.3	12.5	9.0	10.8	10.3	14.1	12.2	85.7	84.5	78.1	65.7	74.5	71.5	81.5	87.1
289 d		10.3	9.0	10.9	8.1	9.1	8.5	13.3	10.8	78.1	73.8	68.1	59.1	62.8	59.0	76.9	77.1
290 c		9.5	8.2	10.0	7.2	8.3	7.6	12.5	11.1	72.0	67.2	62.5	53.3	57.8	52.7	72.2	79.2
291 d		9.6	8.3	10.1	7.1	8.6	7.8	12.5	10.2	72.8	68.1	63.1	51.8	59.3	54.1	72.2	72.8
292 d		9.3	7.8	9.3	6.8	8.1	7.3	12.0	9.6	70.5	64.0	58.1	49.6	55.9	50.7	69.4	68.5
293 d		7.0	5.6	6.7	5.4	6.0	5.2	8.8	6.8	53.0	45.9	41.9	39.4	41.4	36.1	50.9	48.6
294 d		4.7	3.4	4.0	3.1	4.0	3.3	5.1	4.6	35.6	27.9	25.0	22.6	27.6	22.9	29.5	32.8
295 d		3.3	1.9	....	....	2.5	1.9	....	3.2	25.0	15.6	....	....	17.2	13.2	....	22.8
296 d		2.0	1.3	1.5	....	2.0	1.5	1.9	1.6	15.6	10.7	9.4	....	13.8	10.4	11.0	11.4
297 d		1.4	1.2	1.1	0.9	1.6	....	1.3	1.1	10.6	9.8	6.9	6.6	11.0	....	7.5	7.9
298 d		1.3	1.1	1.1	0.8	....	....	1.2	0.9	9.9	9.0	6.9	5.8	....	....	6.9	6.4
299 d		1.2	0.9	0.9	0.7	1.6	....	1.1	0.9	9.1	7.4	5.6	5.1	11.0	....	6.4	6.4
300 d		....	....	....	....	....	....	1.1	0.7	....	....	....	....	....	....	6.4	5.0
301 d		1.1	0.9	0.8	0.8	1.5	1.1	1.1	0.9	8.3	7.4	5.0	5.8	10.4	7.6	6.4	6.4
302 d		1.6	1.0	0.9	1.4	1.6	1.1	1.4	1.3	12.1	8.2	5.6	10.2	11.0	7.6	8.1	9.3
303 d		1.3	0.9	0.9	....	1.5	....	1.4	0.8	9.9	7.4	5.6	....	10.4	....	8.1	5.7
304 d		1.1	0.8	0.6	1.3	1.5	1.0	1.1	0.7	8.3	6.6	3.7	9.5	10.4	6.9	6.4	5.0
305 d		1.5	0.8	0.8	1.6	1.4	0.9	1.4	1.3	11.4	6.6	5.0	11.7	9.7	6.2	8.1	9.3
306 d		....	0.8	0.8	1.9	1.6	0.9	1.5	1.1	....	6.6	5.0	13.9	11.0	6.2	8.7	7.9
307 c		1.4	0.7	0.8	1.7	1.6	0.9	1.5	1.1	10.6	5.7	5.0	12.4	11.0	6.2	8.7	7.9
308 c		1.6	0.8	0.9	1.9	1.6	1.0	1.7	1.3	12.1	6.6	5.6	13.9	11.0	6.9	9.8	9.3
309 c	♯	1.8	0.9	1.2	2.3	1.8	1.2	2.0	1.6	13.6	7.4	7.5	16.8	12.4	8.3	11.6	11.4
310 o		2.2	1.0	1.3	2.4	1.8	1.3	2.3	2.0	16.7	8.2	8.1	17.5	12.4	9.0	13.3	14.3
311 c		3.0	1.6	2.0	3.3	2.3	1.5	3.2	2.9	22.7	13.1	12.5	24.1	15.9	10.4	19.1	20.7
312 o	a	3.3	1.8	2.1	3.7	2.5	1.7	3.9	3.2	25.0	14.8	13.1	27.0	17.2	11.8	22.5	22.8
313 d		4.2	2.6	3.2	4.7	3.0	2.3	4.9	4.3	31.8	21.3	20.0	34.3	20.7	16.0	28.3	30.7
314 c		4.4	2.7	3.4	5.0	3.3	2.2	5.1	4.6	33.4	22.1	21.2	36.5	22.8	15.3	29.5	32.8
315 c		4.3	2.8	3.4	4.9	3.4	2.6	5.0	4.6	32.6	21.3	21.2	35.8	23.5	18.0	28.9	32.8
316 c		5.7	3.0	3.6	5.1	3.8	2.8	5.5	5.0	43.2	24.6	22.7	37.2	26.2	19.4	31.8	35.7
317 d		5.1	3.3	4.2	5.5	4.0	2.9	6.3	5.3	38.7	27.1	26.2	40.2	27.6	20.1	36.4	37.8
318 c		5.5	3.7	4.6	5.9	4.2	3.2	6.8	5.9	41.7	30.3	28.7	43.1	29.0	22.2	39.3	42.1
319 o		5.9	4.0	4.9	6.3	4.7	3.7	7.5	6.2	44.7	32.8	30.6	46.0	32.4	25.7	43.4	44.3
320 d		8.0	5.7	7.3	8.0	6.4	5.2	9.8	8.6	60.6	46.7	45.6	58.4	44.2	36.1	56.6	61.4
321 o		7.8	5.6	7.4	8.2	6.7	5.9	9.2	8.1	59.1	45.9	46.2	59.9	46.2	40.9	53.2	57.8
322 c		7.0	4.6	6.3	7.3	5.9	5.0	8.3	7.3	53.1	37.7	39.4	53.3	40.7	34.7	48.0	52.1
323 o	b ψ¹	6.7	4.4	6.0	7.0	5.2	4.3	8.4	6.9	50.8	36.1	37.5	51.1	35.9	29.8	48.6	49.3
324 c		7.8	5.6	7.5	8.2	6.3	5.7	9.7	8.1	59.1	45.9	46.9	59.9	43.5	39.6	56.1	57.8
325 d		....	....	....	....	....	....	10.1	....	....	....	....	....	....	....	58.4	....
326 o		9.0	6.9	9.1	9.3	7.7	6.9	11.5	9.4	68.2	56.6	56.9	67.9	53.1	47.9	60.5	67.1
327 d		10.6	8.9	11.6	10.7	10.0	9.1	....	11.0	80.3	73.0	72.5	78.1	69.0	63.2	....	78.5
328 d		10.9	9.3	12.0	11.1	10.4	9.8	13.7	11.6	82.6	76.3	75.0	81.0	71.8	68.0	79.2	82.8

TABLE 20.—*Relative intensities of radiation from bolographs of rock-salt prismatic solar spectrum*—Continued.

[Initials refer to observers.]

Designation.	Observed ordinates of minima.									Reduced ordinates of minima.							
	Apr. 1, 1898.	Apr. 2, 1898.	Apr. 6, 1898.	May 18, 1898.		Mar. 1, 1898.	Apr. 1, 1898			Apr. 1, 1898.	Apr. 2, 1898	Apr. 6, 1898	May 18, 1898.		Mar. 1, 1898	Apr. 1, 1898.	
	Bolograph II.	Bolograph XI.	Bolograph II.	Bolograph II.	Bolograph II.	Bolograph IV.	Bolograph V.	Bolograph V.		Bolograph II.	Bolograph XI.	Bolograph II.	Bolograph II.	Bolograph II.	Bolograph IV.	Bolograph V.	Bolograph V.
	Corrected ordinate measures.									Corrected ordinates multiplied by factor—							
	C.G.A.	C.G.A.	C.G.A.	C.G.A.	C.G.A.	C.G.A.	F.E.F.	F.E.F.		7.58	8.20	6.25	7.30	6.90	6.94	5.78	7.14
329 b	11.2	9.7	12.5	11.5	12.0	10.5	14.8	11.7		84.9	79.5	78.1	84.0	82.8	72.9	85.5	83.5
330 d	12.3	11.1	14.0	12.5	12.1	12.0	16.2	12.9		93.2	91.0	87.5	91.2	83.5	83.3	93.6	92.1
331 d		11.4		12.8	12.8	12.5	16.5				93.5		93.4	88.3	86.8	95.4	
332 d	12.6	11.4	14.6	12.9	13.1	12.7	16.9	13.4		95.5	93.5	91.2	94.2	90.4	88.1	97.7	95.7
333 d							17.0	13.6								98.3	97.1
334 d ?	13.0	12.0		13.2	13.8	12.4	17.2	13.7		98.5	98.4		66.4	95.2	86.1	99.4	97.8
335 d	13.0	12.1	15.3	13.1			17.2	13.9		98.5	99.2	95.6	95.6			99.4	99.2
336 d ?	13.1		15.4		14.1	13.9	17.3	13.9		99.3		96.2		97.3	96.5	100.0	99.2
337 d	13.0	12.2	15.6	13.2	14.1	14.0	17.3	13.9		98.5	100.0	97.5	96.4	97.3	97.2	100.0	99.2
338 d ?	13.2	12.3			14.3		17.3	14.0		100.1	100.9			98.7		100.0	100.0
339 d ?	13.2	12.3	15.7	13.4	14.3	14.2	16.1	14.0		100.1	100.9	98.1	97.8	98.7	9 4.5	93.1	100.0
340 d	13.0	12.3	15.8	13.4	14.2	14.2	17.1	13.9		98.5	100.9	98.7	97.8	98.0	98.5	98.8	99.2
341 d	12.9		15.6	13.4	14.0	13.9	16.6	13.8		97.7		97.5	97.8	96.6	96.5	95.9	98.5
342 d ?							16.2									93.6	
343 d	12.4	11.6	14.7	12.5	13.5	13.5	15.9	13.2		94.0	95.1	91.9	91.2	93.2	93.7	91.9	94.2
344 d ?					13.4	13.2		12.9						92.5	91.6		92.1
345 c	12.1	11.4	14.4	12.2	13.2	13.2	15.6	12.9		91.7	93.5	90.0	89.1	91.1	91.6	90.2	92.1
346 d ? } b	} 12.6	11.7	15.0	12.5	13.5	13.6	{ 13.9	13.8		}95.5	95.9	93.7	91.2	93.2	94.4	{	92.8
347 d ?							{ 15.5									89.6	
348 c	12.5	11.7	14.7	12.3	13.4	13.5	15.5	13.2		94.8	95.9	91.9	89.8	92.5	93.7	89.6	94.2
349 d ?																	
350 d		11.7	14.8	12.3	13.7	13.8	15.5	13.5			95.9	92.5	89.8	94.5	95.8	89.6	96.4
351 c	12.2	11.3	14.5	11.7	13.2	13.3	15.5	13.2		92.5	92.7	90.6	85.4	91.1	92.3	89.6	94.2
352 c } b	12.2	11.4	14.6	11.7	13.2	13.3	15.4	13.1		92.5	93.5	91.2	85.4	91.1	92.3	89.0	93.5
353 d	12.6	11.8	15.1	11.9	13.5		15.9	13.4		95.5	96.8	94.4	86.9	93.2		91.9	95.7
354 d	12.6		15.4			13.7	16.0	13.5		95.5		96.2			95.1	92.5	96.4
355 d		12.0			13.7			13.6			98.4			94.5			97.1
356 d							16.2	13.8								93.6	98.5
357 d	12.6		15.1	12.3	13.6	13 8	15.9	13.5		95.5		94.4	89.8	93.8	95.8	91.9	97.1
358 d	12.3	11.5	14.8	12.1		13.4	15.8	13.2		93.2	94.3	92.5	88.3		93.0	91.3	94.2
359 d		11.5		11.7	13.2		15.6	13.2			94.3		85.4	91.1		90.2	94.2
360 d ?		11.6	14.8	11.7	13.2	13.5	15.7	13.3			95.1	92.5	85.4	91.1	93.7	90.7	95.0
361 d ?		11.5	14.9	11.6	13.2	13.5	15.8	13.2			9 : .3	93.1	84.7	91.1	93.7	91.3	94.2
362 d ?		11.7	14.9	11.9	13.3	13.7	15.8	13.5			95.9	93.1	86.9	91.8	95.1	91.3	96.4
363 c	12.1	11.4	14.6	11.7	12.0	13.4	15.5	13.1		91.7	93.5	91.2	85.4	89.0	93.0	89.6	93.5
364 d								13.3									95.0
365 c	12.0	11.4	14.4	11.4	12.8	13.1	15.4	13.2		91.0	93.5	90.0	83.2	88.3	90.9	89.0	94.2
366 c	11.8	11.3	14.4	11.3	12.7	12.9	15.1	13.1		89.4	92.7	90.0	82.4	87.6	89.5	87.3	93.5
367 d	11.8	11.2	14.3	11.3	12.6	12.9	14.1	13.1		80.4	91.8	89.4	82.4	86.9	89.5	81.5	93.5
368 d	11.7	11.1	14.0	11.3	12.5	12.8	14.9	13.1		88.7	91.0	87.5	82.4	86.2	88.8	86.1	93.5
369 d							14.9	13.0								86.1	92.8
370 d		10.8	13.7	11.1	12.2	12.3	14.5	13.0			88.6	85.6	81.0	84.2	85.4	83.8	92.8
371 d	11.3	10.8	13.6	11.1	12.0	12.2	14.3	12.6		85.6	88.6	85.0	81.0	82.8	84.7	82.7	91.4
372 d ?																	
373 c	11.0	10.4	13.0	10.6	11.5	11.6	13.8	12.4		83.4	85.3	81.2	77.3	79.4	80.5	79.8	88.5
374 d							13.5	12.2								78.0	87.1
375 d	10.6	9.9	12.4	10.2	11.0	11.1	13.2	11.9		80.3	81.2	77.5	74.4	75.9	77.0	76.3	85.0
376 d	} 10.6	9.8	12.5	10.1	10.8	10.9	{ 13.0	12.0		}80.3	80.4	78.1	73.7	74.5	73.6	{ 75.1	85.7
377 d							{ 13.0	11.7								75.1	83.5
378 c	} 10.4	9.6	12.1	10.0	10.4	10.6	{ 13.0			}78.8	78.7	75.6	73.0	71 8	73.6	{ 75.1	
379 d							{ 12.6									72.8	

ANNALS OF ASTROPHYSICAL OBSERVATORY.

TABLE 20.—*Relative intensities of radiation from bolographs of rock-salt prismatic solar spectrum*—Continued.

[Initials refer to observers.]

Designation.		Observed ordinates of minima.								Reduced ordinates of minima.							
		Apr. 1, 1898.	Apr. 2, 1898.	Apr. 6, 1898.	May 18, 1898.			Mar. 1, 1898.	Apr. 1, 1898.	Apr. 1, 1898.	Apr. 2, 1898	Apr. 6, 1898	May 18, 1898.		Mar. 1, 1898	Apr. 1, 1898.
		Bolograph II.	Bolograph XI.	Bolograph II.	Bolograph II.	Bolograph II.	Bolograph IV.	Bolograph V.	Bolograph V.	Bolograph II.	Bolograph XI.	Bolograph II.	Bolograph II.	Bolograph II.	Bolograph IV.	Bolograph V.	Bolograph V.
		Corrected ordinate measures.								Corrected ordinates multiplied by factor—							
		C.G.A.	C.G.A.	C.G.A.	C.G.A.	C.G.A.	C.G.A.	F.E.F.	F.E.F.	7.58	8.20	6.25	7.80	6.90	6.94	5.78	7.14
380 d } ₀ 381 d }		} 10.2	9.4	11.7	9.5	10.1	10.4	{...... 12.3	11.8 11.6	}77.3	77.1	73.1	69.3	69.7	72.2	{...... 71.1	84.3 82.8
382 d ?		......	......	......	......	......	......	......	11.2	......	......	......	......	......	......	......	80.0
383 c		9.7	8.7	11.0	9.4	9.5	9.5	11.6	11.0	73.5	71.3	68.7	68.6	65.6	65.9	67.0	78.5
384 d		9.1	8.0	9.8	......	......	8.9	11.5	......	69.0	65.6	61.2	......	......	61.8	66.5	......
385 d		8.5	7.5	9.3	8.1	8.0	7.8	10.1	8.9	64.4	61.5	58.1	59.1	52.2	54.1	58.4	63.5
386 c		7.9	6.8	8.3	7.3	7.0	6.7	9.0	9.2	59.9	55.8	51.9	53.3	48.3	46.5	52.0	65.7
387 d		6.5	5.4	6.5	6.2	5.5	5.2	6.8	7.8	49.3	44.3	40.6	46.0	38.0	36.1	39.3	55.7
388 d		4.5	3.4	3.9	4.2	3.4	3.1	4.5	5.4	34.1	27.9	24.4	30.7	23.5	21.5	26.0	38.6
389 d		3.3	2.5	2.4	2.8	2.0	1.9	2.9	3.8	25.0	20.5	15.0	20.4	13.8	13.2	16.8	27.1
390 c		2.2	1.7	1.7	1.9	1.6	1.4	1.9	3.0	16.7	13.9	10.6	13.9	11.0	9.7	11.0	21.4
391 d		1.1	1.1	0.7	1.0	1.1	1.1	2.0	8.3	9.0	4.4	7.3	7.6	7.6	6.4	14.3
392 d ? 393 d ?		} 0.9	1.1	0.6	0.6	......	0.9	{...... 0.9	1.7 1.7	}6.8	9.0	3.7	4.4	......	6.2	{...... 5.2	12.1 12.1
394 d		0.8	1.0	0.6	0.7	......	0.9	0.8	1.7	6.1	8.2	3.7	5.1	......	6.2	4.6	12.1
395 d	Ω	0.7	0.7	0.4	0.5	......	0.7	0.7	1.5	5.3	5.7	2.5	3.6	......	4.9	4.0	10.7
396 d		0.6	0.6	0.4	0.5	......	0.6	0.8	1.4	4.5	4.9	2.5	3.6	......	4.2	4.6	10.0
397 d		0.5	0.7	0.3	0.3	0.6	0.6	0.8	1.5	3.8	5.7	1.9	2.2	4.1	4.2	4.6	10.7
398 d	a	0.9	0.6	0.5	0.5	0.5	0.6	1.0	1.7	6.8	4.9	3.1	3.6	3.4	4.2	5.8	12.1
399 d		0.8	0.6	0.5	0.5	0.7	0.6	0.9	1.6	6.1	4.9	3.1	3.6	4.8	4.2	5.2	11.4
400 d		0.5	0.6	0.3	0.3	0.5	......	0.6	1.4	3.8	4.9	1.9	2.2	3.4	......	3.5	10.0
401 d		0.7	0.7	0.4	0.4	0.5	0.6	0.9	1.6	5.3	5.7	2.5	2.9	3.4	4.2	5.2	11.4
402 d		1.3	0.8	0.8	0.9	......	0.7	1.3	2.2	9.9	6.6	5.0	6.6	......	4.9	7.5	15.7
403 d ?		......	......	......	......	......	......	......	1.5	......	......	......	......	......	......	......	8.7
404 d		1.5	1.0	1.0	1.3	0.8	0.8	1.9	2.4	11.4	8.2	6.2	9.5	5.5	5.6	11.0	17.1
405 c		2.1	1.2	1.6	1.7	1.0	1.2	2.6	3.1	15.9	9.8	10.0	12.4	6.9	8.3	15.0	22.1
406 d		2.8	1.8	2.4	2.2	1.5	1.6	3.2	3.6	21.2	14.8	15.0	16.1	10.4	11.1	18.5	25.7
407 c		2.5	1.8	2.5	2.2	2.1	3.0	......	3.7	18.9	14.8	15.6	16.1	14.5	20.8	......	26.4
408 d		......	2.7	......	2.3	2.5	2.9	4.8	4.9	......	22.1	......	16.8	17.2	20.1	27.7	35.0
409 d		4.9	4.0	5.9	3.7	4.3	4.6	5.9	6.2	37.1	32.8	36.9	27.0	29.7	31.9	34.1	44.3
410 d		5.8	4.9	7.0	5.1	5.4	5.6	7.0	7.2	44.0	40.2	43.7	37.2	37.3	38.9	40.5	51.4
411 d		6.1	5.4	7.1	5.3	5.6	6.1	6.9	7.3	46.2	44.3	44.4	38.7	38.6	42.3	39.9	52.1
412 b	ω₁	2.8	2.2	3.2	2.3	2.9	3.3	3.4	3.9	21.2	18.0	20.0	16.8	20.0	22.9	19.7	27.8
413 c		3.1	2.5	3.8	2.7	3.3	4.0	3.7	4.4	23.5	20.5	23.7	19.7	22.8	27.8	21.4	31.4
414 d		6.1	6.1	7.7	6.0	6.1	6.9	7.4	7.9	46.2	50.0	48.1	43.8	42.1	47.9	42.8	56.4
415 d		6.5	6.5	7.7	5.9	6.5	7.4	7.4	7.8	49.3	53.3	48.1	43.1	44.8	51.4	42.8	55.7
416 b	ω₂	4.9	5.0	6.2	4.7	5.2	5.1	5.9	6.4	37.1	41.0	38.7	34.3	35.9	35.4	34.1	45.7
417 c		5.3	5.4	6.5	4.9	5.4	4.9	6.0	6.6	40.2	44.3	40.63	5.8	37.3	34.0	34.7	47.1
418 d		5.7	6.0	7.3	5.5	6.0	5.5	6.7	7.3	43.2	49.2	45.6	40.2	41.4	38.2	38.7	52.1

TABLE 20.—*Relative intensities of radiation from bolographs of rock-salt prismatic solar spectrum*—Continued.

[Initials refer to observers.]

Designation.	Dec. 7, 1897. Bolograph V.	Dec. 9, 1897. Bolograph II.	Mar. 14, 1896. Bolograph IV.	Apr. 6, 1896. Bolograph II.	Dec. 7, 1897. Bolograph VIII.	Dec. 9, 1897. Bolograph V.	Mar. 14, 1896. Bolograph XI.	Dec. 7, 1897. Bolograph V.	Dec. 9, 1897. Bolograph II.	Mar. 14, 1896. Bolograph IV.	Apr. 6, 1896. Bolograph II.	Dec. 7, 1897. Bolograph VIII.	Dec. 9, 1897. Bolograph V.	Mar. 14, 1896. Bolograph XI.
	C.G.A.	C.G.A.	C.G.A.	C.G.A.	C.G.A.	C.G.A.	C.G.A.	\multicolumn{7}{c}{Corrected ordinates multiplied by a factor.}						
412 b ω_1	1.8	3.4	4.3	4.0	4.8	3.8	2.4	11.8	9.5	11.6	14.9	18.1	22.0	18.5
413 d	2.2		4.8	4.8	5.5	4.5	3.6	14.4		13.0	17.9	20.6	26.1	27.7
414 d														
415 d	5.8		13.1	11.1	10.9	8.8	6.2	37.9		35.4	41.4	41.0	51.0	47.7
416 b ω_2	4.3	9.1	9.8	8.5	8.7	7.0	4.7	28.1	25.4	26.5	31.7	32.6	40.5	36.1
417 c	4.5	9.7	10.4	8.9	9.2	7.3	4.9	29.4	27.1	28.1	33.2	34.6	42.3	37.7
418 d				9.5	9.7	7.7	5.4			35.4	36.5	44.6	41.5	
419 d	6.1		14.0	11.0	11.1	9.1	6.6	39.9		37.8	41.0	41.8	52.7	50.8
420 d	5.3	13.8	14.4	11.3	11.4	9.3	6.7	41.2	38.5	38.9	42.1	42.9	53.8	51.5
421 d ?	6.4	14.2		11.6	11.7	9.4	6.7	41.9	39.6		42.3	44.0	54.4	51.5
422 d ?	6.4	14.2	14.5	11.7	11.6	9.3		41.9	39.6	39.2	42.6	43.7	53.8	
423 c	6.3	14.0	14.4	11.7	11.7	9.2	6.8	41.2	39.1	38.9	42.6	44.0	53.3	52.3
424 d		14.2	14.6	11.8	11.5	9.2	6.7		39.6	39.4	44.0	43.2	53.3	51.5
425 d		13.4	13.8	11.1	10.7	8.9	6.5		37.4	37.3	41.4	40.2	51.5	50.0
426 c	5.7	12.9	13.3	10.7	10.3	8.4	6.1	37.8	36.0	35.6	39.9	38.7	48.6	48.9
427 d	5.6	12.7	13.2	10.6	10.3	8.6		36.6	35.4	35.6	39.5	38.7	49.8	
428 d		12.5	12.8	10.5	10.2	8.5	6.1		34.9	34.6	39.2	38.4	49.2	45.9
429 d	5.4	12.2		10.3	10.0	8.2		35.3	34.0		38.4	37.6	47.5	
430 d			12.4		9.9		6.1			33.5		35.7		46.9
431 d ?		12.7	12.8	10.4		8.2			35.4	34.6	38.8		47.5	
432 d ?	5.3	12.3	12.5			7.8	5.9	34.7	34.3	33.8			45.2	45.4
433 d			11.1	10.2	8.9	7.1				30.0	38.0	33.4	41.1	
434 d		10.4	10.8	8.8	8.8	7.0	5.4		29.0	29.2	32.8	33.1	40.5	41.5
435 d ?			10.6		8.6	6.8				28.6		32.3	39.4	
436 d ?	4.4	9.9	10.1					28.8	27.6	27.3				
437 d	4.4		10.2	8.5	8.4	6.6	5.1	28.8		27.5	31.7	31.5	38.2	39.2
438 c	4.1	9.1	8.9	9.9	8.0	6.3	4.8	26.6	25.4	24.0	36.9	30.1	36.5	36.9
439 d		9.1		9.9	7.9	6.3	4.8		25.4		36.9	29.6	36.5	36.9
440 d	3.7	7.8	7.8	7.4	7.4	6.0	4.5	24.2	21.8	21.1	27.6	27.9	34.7	34.6
441 d c	3.5			7.0	7.0	5.7	4.3	22.9			26.1	26.2	33.3	33.1
442 d		7.4	7.4		7.0	5.7			20.6	20.0		26.2	33.3	
443 d	3.0	6.1	6.4	6.3	6.3	5.2	4.0	19.6	17.0	17.3	23.5	23.7	30.1	30.8
444 d c														
445 d c	2.8	5.7	6.0	6.0	6.0	5.1	3.9	18.3	15.9	16.2	22.4	22.6	29.5	30.0
446 d	2.4	4.8		5.5	5.3	4.6	3.6	15.7	13.4		20.5	20.0	26.6	27.7
447 c	2.1	3.9	4.3	4.9	4.8	4.2	3.3	13.7	10.9	11.6	18.3	18.1	24.3	25.4
448 d	2.1		4.1	4.6	4.5	4.1	3.2	13.7		11.1	17.2	17.0	23.7	24.6
449 c	1.5	2.9	3.1	3.7	3.8	3.4	2.6	9.8	8.1	8.4	13.8	14.2	19.7	20.0
450 d ?				3.4	3.8	3.5					12.7	14.2	20.3	
451 d ?														
452 d ?	2.1	1.1		2.9	3.0	2.0	2.1	7.8	3.1		10.8	11.1	17.4	16.1
453 d ?														
454 d	0.6	1.4		1.8	1.9	2.4	1.6	3.9	3.9		6.7	7.2	13.9	12.3
455 c	0.5	1.2	1.1	1.6	1.8	1.9	1.4	3.3	3.3	3.0	6.0	6.7	11.0	10.8
456 d	0.4	1.1		1.2	1.3	1.6	1.2	2.6	3.1		4.5	5.0	9.3	9.2
457 d	0.2			0.7	0.8	1.9	0.8	1.3			2.6	3.0	11.0	6.2
458 d ?														
459 a X		1.0		0.3							0.3			
			0.3	0.5	0.1					0.5	0.5	0.2		
460 d				0.5	0.4	0.8	0.6				0.5	0.4	1.4	1.5
461 d		0.2	1.1	0.9	1.0	1.4	1.1		0.2	0.6	0.9	1.0	2.2	2.1
462 d	1.0	0.4	1.1	1.3	1.3	2.1	1.6	1.0	0.4	0.6	1.3	1.3	3.7	3.1
463 c	1.3	0.7	1.4	1.6	1.8	2.4	1.9	1.3	0.6	0.7	1.6	1.8	4.2	3.6
464 d	2.0	0.8	1.7	2.5	2.7	3.2	2.5	2.0	0.7	0.9	2.5	2.7	5.6	4.8
465 d				3.0		3.1					3.0		6.0	
466 c	2.2	0.8	2.1	2.8	3.1	3.6	2.9	2.2	0.7	1.1	2.8	3.1	6.3	5.6
467 d		1.0	2.5	3.2	3.4	3.6	3.0		0.8	1.3	3.2	3.5	6.3	5.8

TABLE 20.—*Relative intensities of radiation from bolographs of rock-salt prismatic solar spectrum*—Continued.

[Initials refer to observer.]

Designation.	Dec. 7, 1897. Bolograph V.	Dec. 9, 1897. Bolograph II.	Dec. 9, 1897. Bolograph IV.	Mar. 14, 1898. Bolograph II.	Mar. 14, 1898. Bolograph VIII.	Apr. 6, 1898. Bolograph V.	Apr. 6, 1898. Bolograph XI.	Dec. 7, 1897. Bolograph V.	Dec. 9, 1897. Bolograph II.	Dec. 9, 1897. Bolograph IV.	Mar. 14, 1898. Bolograph II.	Mar. 14, 1898. Bolograph VIII.	Apr. 6, 1898. Bolograph V.	Apr. 6, 1898. Bolograph XI.
	C.G.A.	C.G.A.	C.G.A.	C.G.A.	C.G.A.	C.G.A.	C.G.A.	\multicolumn{7}{c}{Corrected ordinates multiplied by a factor.}						
468 d?		1.1			2.6		2.7		0.9			2.6		5.2
469 b	1.8	0.9	2.3	2.3	2.4	2.8	2.3	1.7	0.8	1.3	2.3	2.4	4.9	4.4
470 d		1.4		4.0	4.2				1.2		4.0	4.2		
471 d?				4.0	4.3	4.2	3.7				4.0	4.4	7.3	7.1
472 c	3.6	1.6	3.9	3.8	3.9	4.0	3.4	2.5	1.3	2.1	3.8	3.8	7.0	6.5
473 d														
474 c	3.2	1.4	3.7	3.5	3.5	3.8	3.2	3.1	1.2	2.0	3.5	3.5	6.6	6.1
475 c	2.2	1.0	2.4	2.7	2.8	3.1	2.7	2.2	0.8	1.3	2.7	2.8	5.4	5.2
476 c	1.9	0.9	2.2	2.4	2.4	2.8	2.3	1.9	0.7	1.2	2.4	2.4	4.9	4.4
477 c	2.3	1.1	2.5	2.9	3.0	3.3	2.8	2.2	0.9	1.3	2.9	2.9	5.7	5.4
478 d } x_1				2.6	2.6	3.0	2.5				2.6	2.6	5.2	4.8
479 c } a	1.7	0.8	2.0	2.4	2.5	2.9	2.5	1.7	0.7	1.1	2.4	2.6	5.0	4.8
480 d?				2.5	2.6						2.5	2.6		
481 c	1.8	0.9	2.2	2.4	2.5	2.8	2.5	1.7	0.8	1.2	2.4	2.6	4.9	4.8
482 c	3.7	1.4	4.2	3.9	3.9	3.9	3.4	3.6	1.2	2.3	3.9	3.8	6.8	6.5
483 c	1.3	1.6	3.8	4.0	3.9	4.1	3.6	1.3	1.2	2.0	4.0	3.8	7.1	6.9
484 d		2.0		5.8	5.5	4.6	3.8		1.7		5.8	5.5	8.0	7.3
485 d }	7.0	3.2		5.7	5.8	5.3	4.6	6.8	2.7		5.7	5.8	9.2	8.8
486 d }														
487 d		2.7		4.4	4.6				2.3		5.2	4.4	8.0	
488 d			2.3	2.1	1.5	2.1	2.1			1.3	2.1	1.5	3.7	4.0
489 c } x_2	1.1		1.7	1.6		1.9	1.7	1.1		0.9	1.6		3.3	3.3
490 d														
491 d	1.3		1.7	2.0	1.7	2.0	1.8	1.3		0.9	2.0	1.7	3.5	3.5
492 d?		0.9				2.2	2.0		0.8				3.8	3.8
493 c	2.7	1.5	3.4	2.9	2.7	2.8	2.5	2.6	1.3	1.9	2.9	2.7	4.9	4.8
494 b } a	2.1	1.1	2.9	2.8	2.6	2.9	2.6	2.0	0.9	1.6	2.8	2.6	5.0	5.0
495 d		1.7		3.1	3.3	3.3	3.0		1.4		3.1	3.3	5.7	5.8
496 d? }	2.3	1.1	3.0	2.7	2.5	2.7	2.4	2.2	0.9	1.6	2.7	2.5	4.7	4.6
497 c }														
498 b	1.8	1.2	2.7	2.1	2.0	2.4	2.0	1.7	1.0	1.4	2.1	2.0	4.2	3.8
499 c	3.8	1.9	5.0	3.7	3.4	3.8	3.2	3.7	1.6	2.7	3.7	3.4	6.6	6.1
500 c	3.9	1.9	5.1	3.9	3.5	2.9	3.2	3.8	1.6	2.8	3.9	3.5	5.0	6.1
501 d				4.5	4.1	4.3					4.5	4.1	7.5	
502 d	5.4	3.0	7.3	4.8	4.4	4.4	3.7	5.3	2.5	4.0	4.8	4.4	7.7	7.1
503 b	5.4	3.2	7.9	4.8	4.3	4.3	3.7	5.3	2.7	4.3	4.8	4.3	7.5	7.1
504 c		4.3	10.8						3.6	5.8				
505 c	7.3	4.4	10.8					7.1	3.7	5.8				
506 d	7.6	4.5	11.2					7.4	3.8	6.1				
507 d		4.7	11.4						4.0	6.2				
508 d	6.6	4.0	10.0					6.5	3.4	5.4				
509 c	6.6	3.9	9.5	5.4		4.7	4.0	6.5	3.3	5.2	5.4		8.2	7.7
510 d?		3.9	9.8		5.0				3.3	5.3		5.0		
511 c	6.3	3.7	9.2	5.2	5.0			6.2	3.1	5.0	5.2	5.0		
512 c	6.3	3.8	9.2					6.2	3.2	5.0				
513 c	6.6	4.0	9.6	5.3	4.9			6.5	3.4	5.2	5.3	4.9		
514 c	6.7	4.0	9.8					6.5	3.4	5.3				
515 d?														
516 d?														
517 b	5.2	2.9	7.0	4.5	4.2	4.1	3.5	5.1	2.5	3.8	4.5	4.2	7.1	6.7
518 d?	6.1	3.7	9.0	4.8	4.5	4.2		5.9	3.1	4.9	4.8	4.5	7.3	
519 d } c	6.0	3.7	8.8	4.7		4.1	3.5	5.9	3.1	4.7	4.7		7.1	6.7
520 d }	6.0	3.7	8.8					5.9	3.1	4.7				
521 c	6.6	3.3	8.3	4.3	4.3	3.9		6.5	2.8	4.5	4.3	4.3	6.8	
522 c	5.5	3.3	7.9	4.1	4.0	3.7		5.4	2.8	4.3	4.1	4.0	6.4	
523 d?	5.6		8.1					5.5		4.4				
524 c	5.4	3.2	7.7					5.3	2.7	4.1				

TABLE 20.—*Relative intensities of radiation from bolographs of rock-salt prismatic solar spectrum*—Continued.

[Initials refer to observers.]

Designation.	Dec. 7, 1897. Bolograph V.	Dec. 9, 1897. Bolograph II.	Dec. 9, 1897. Bolograph IV.	Mar. 14, 1898. Bolograph II.	Mar. 14, 1898. Bolograph VIII.	Apr. 6, 1898. Bolograph V.	Apr. 6, 1898. Bolograph XI.	Dec. 7, 1897. Bolograph V.	Dec. 9, 1897. Bolograph II.	Dec. 9, 1897. Bolograph IV.	Mar. 14, 1898. Bolograph II.	Mar. 14, 1898. Bolograph VIII.	Apr. 6, 1898. Bolograph V.	Apr. 6, 1898. Bolograph XI.
	C.G.A.	C.G.A.	C.G.A.	C.G.A.	C.G.A.	C.G.A.	C.G.A.	\multicolumn{7}{c}{Corrected ordinates multiplied by a factor.}						
525 c	5.2	3.1	7.5	3.8		3.4		5.1	2.6	4.1	3.8		5.9	
526 d	5.1	3.2	7.5	3.7	3.4	3.3		5.0	2.7	4.1	3.7	3.4	5.7	
527 d?														
528 d	4.6	2.9	7.1	3.4	3.1			4.5	2.5	3.8	3.4	3.1		
529 d?														
530 d?														
531 c	3.8	2.5	5.9	2.9		2.8	2.4	3.7	2.1	3.2	2.9		4.9	4.6
532 d?														
533 d?														
534 c	3.8	2.5	5.9	2.7	2.8	2.7	2.2	3.7	2.1	3.2	2.7	2.8	4.7	4.2
535 d	3.9	2.6	6.0	2.7	2.6			3.8	2.2	3.2	2.7	2.6		
536 d	3.1	1.9	4.6					3.1	1.6	2.5				
537 a Y				0.3							0.3			
				0.6							0.3			
538 c		1.0	0.6	0.6	0.4	0.8	0.7		0.4	0.2	0.3	0.2	0.7	0.7
539 d		1.5	1.1	1.4		1.4	1.2		0.6	0.4	0.7		1.2	1.2
540 d		2.6	2.6	2.2	1.9	2.0			1.1	1.1	1.1	1.0	1.7	
541 d		2.8	2.6			2.2			1.2	1.1			1.9	
542 c		2.3	2.1	2.2	1.4	1.9	1.8		1.0	0.8	1.1	0.6	1.6	1.7
543 c		2.0	1.6	1.9	1.0				0.8	0.7	0.9	0.5		
544 d } c		2.1	1.9	1.9	0.9				0.9	0.8	0.9	0.5		
545 d		2.2		2.0	1.0				0.9		1.0	0.5		
546 d			2.2	2.0	1.0		1.5			0.9	1.0	0.5		1.4
547 d		2.2	2.0	2.0	1.0	1.7			0.9	0.8	1.0	0.5	1.5	
548 d		2.3	1.9						1.0	0.8				
549 c		1.8	1.2	1.5	0.5	1.5	1.4		0.8	0.5	0.7	0.3	1.3	1.3
550 c		1.6	1.3	1.4	0.2	1.3	1.2		0.7	0.5	0.7	0.2	1.1	1.2
551 d		1.7	1.3			1.3	1.3		0.7	0.5			1.1	1.3
552 d } c		1.2	1.1	1.3	0.2	1.1	1.0		0.5	0.4	0.6	0.2	1.0	1.0
553 d		1.4				1.2	1.0		0.6				1.0	1.0
554 d		1.8			0.6	1.4	1.2		0.8			0.3	1.2	1.2
555 b		1.3	0.7	1.3	0.1	1.2	1.0		9.5	0.3	0.6	0.05	1.1	1.0
556 d		1.8	1.1	1.6	0.4				0.7	0.4	0.8	0.2		
557 b		0.8	0.7	1.1	0.0	1.1	1.0		0.4	0.3	0.5	0.05	1.0	1.0
558 d?		0.9	0.5	1.2	0.2	1.1	1.0		0.4	0.2	0.6	0.2	1.0	1.0
559 d?		1.1	0.7	1.6	0.5	1.4	1.2		0.5	0.3	0.8	0.3	1.2	1.2
560 c		1.2	0.7	1.5	0.4	1.3	1.1		0.5	0.3	0.7	0.2	1.1	1.1
561 d		1.1	0.7	1.7	0.5	1.4			0 5	0.3	0.8	0.3	1.2	
562 d?		1.2	1.0	1.4	0.5	1.2	1.1		0.5	0.4	0.7	0.3	1.0	1.1
563 c		1.0	0.6	1.1	0.3	1.2			0.4	0.2	0.5	0.2	1.0	
564 d?		0.8	0.5	0.9	0.0	0.7	0.7		0.4	0.2	0.4	0.05	0.6	0.7
565 d		0.8	0.5	0.8	0.0	0.7	0.6		0.4	0.2	0.4	0.05	0.6	0.6
566 d } b		0.7	0.4	1.2	0.6	0.8	0.7		0.3	0.2	0.6	0.3	0.7	0.7
567 d		1.0	0.6	1.6		1.2			0.4	0.2	0.8		1.0	
568 d?		1.1	0.9		0.7	1.3	1.1		0.5	0.4		0.3	1.1	1.1
569 d			0.9	1.7		1.3				0.4	0.8		1.1	
570 d?		0.8		1.2	0.4	1.0			0.4		0.6	0.2	0.9	
571 c		0.8	0.5	1.0	0.1	0.8	0.7		0.4	0.2	0.5	0.05	0.7	0.7
572 d		0.5		1.0	0.3	0.8	0.6		0.2		0.5	0.2	0.7	0.6
573 c		0.5	0.4	0.5	0.0	0.4	0.3		0.2	0.2	0.2	0.05	0.3	0.3
574 d				0.7	0.2						0.3			
575 d														
576 c		} 0.1		0.5	0.0	0.4	0.1		0.1		0.2	0.05	0.3	0.1
577 c				0.6	0.0	0.2	0.1				0.3	0.0	0.2	0.1
578 d				0.3	0.0		0.0				0.1	0.0		0.0
579 d		0.1	0.2	0.5	0.0	0.2	0.0		0.1	0.1	0.2	0.0	0.2	0.0

REDUCTION OF OBSERVATIONS.

1. REDUCTION OF MEASUREMENTS IN ABSCISSÆ.

(a) *Temperature correction.*—The temperatures at which the bolographs were taken were so constant and so near 20° C. that no correction has been necessary for either of the four classes of errors which, as has been shown on page 105, are introduced by changes of temperature.

(b) *Elimination of bolographic or speed error.*—In taking the mean of the measurements on the various bolographs, attention was paid to the elimination of displacements of deflections introduced by the fact that the spectrum steadily moved across the bolometer strip, instead of remaining stationary throughout the time during which the galvanometer needle is adjusting itself. This error is nearly eliminated, as has been shown on page 120, by taking some bolographs with motion in one direction and others in the reverse. Accordingly, this was done. In reducing the observations all the measurements of both observers on plates taken from short to long wave-lengths were placed in one group, and all those in the contrary direction in another, and the two corresponding means taken. The final mean of these two preliminary means gave the most probable value for the position of the lines. In determining its probable error it is obvious that the separate observations should be subtracted from the appropriate preliminary means to obtain the residuals, and not from the final mean. Otherwise there would be introduced in the probable error (which is a measure of accidental or unknown errors) that known error in the separate bolographs which is eliminated by taking the mean of the two preliminary means, and which therefore no longer affects the result. In this procedure it is assumed that the speed error is entirely eliminated by the device of taking half the bolographs in one direction and the remainder in the opposite. This, as has been shown, is true only where the approaches to a deflection are alike, whether one goes from short wave-lengths to long or vice versa. But there are comparatively few deflections where this is not the case, and the error remaining for the unsymmetrical deflections will rarely amount to 0.1 mm., or 0.6″ arc.

(c) *Reduction to a constant reference point.*—As has been stated, several different sharp deflections, convenient to the purpose, were chosen on the several different types of bolographs as zero points for the measurements, and have been preserved as such as far as reduction (b). The results in the final mean obtained in (b) were next reduced to the middle of the head of the A line as a common zero point, so that they might be joined onto visual observations of the absolute minimum deviation of A.

(d) *Reduction to angular distance from A.*—It has been shown that 1 cm. in abscissa of the bolographs corresponds at the speed ratio employed to 59.998″ ± 0.003″ of spectrum. Accordingly, the final mean distances from A as obtained in (c) were multiplied by this number and reduced to degrees, minutes, and seconds. The probable error in the speed ratio is so small as to be negligible in the 100 cm. of abscissæ covered by the observations.

(e) *Reduction of angular distances from A to absolute deviations and refractive indices.*—For this reduction we require to know the minimum deviation of A for the prism. The measurements of the deviation of A took place when the prism had a slightly different angle from that which it had when the bolographs were made, owing to its having been "fogged" and therefore repolished. Nevertheless, as its refracting angle at the time the bolographs were taken was known by measurements accurate to within 10″ of arc causing in any case an uncertainty of less than ½″ of arc in the greatest angular distances from A, no error of consequence is thereby introduced. The deviation of A at a different well-determined angle being accurately known, its deviation has been calculated for the angle at which the bolographs were made. From this value were subtracted the angular distances obtained in (d) giving the absolute minimum deviations of the lines determined, and from these and the angle at which the bolographs were assumed to be taken, the refractive indices were computed.

The three angles used in these computations were determined as follows:

Angle of the salt prism as used for bolographs from November 18, 1897, to July, 1898. Date of measurement, July 9, 1898. Observers, C. G. A. and O. E. M.

Method of measurement.—Image of slit reflected alternately from two faces of movable prism to cross hairs of fixed 3-inch telescope of 3 feet focal length. Angle given by subtracting the difference of vernier readings from 180°. Four verniers read at each setting.

Observations of vernier differences.

No.	Measurement.
	° ′ ″
1	120 6 18.2
2	120 6 16.8
3	120 6 17.8
Mean	120 6 17.6

Mean angle of prism 59° 53′ 42.4″.

REMARKS.—The axis of rotation of the prism passed through the center of its rear face. Hence the angle was measured, not at the center of prism faces, but halfway between the center and the back edges of the faces. As noted on page 103, rock-salt prisms are particularly likely to become concave, owing to their large temperature coefficient of expansion, and had the measurement of the angle been made at constant temperature it is probable that, having been taken thus 3.3 cm. below the center of the prism faces, it would have been larger than the mean angle by perhaps 10″ or 15″. But the temperature was rising during the observations at a rate, as given by the thermograph near the prism, of about 0.2° per hour. Consequently the prism faces tended to approach convexity, but probably not to such a degree as that the effect just mentioned should be wholly counteracted. Hence we may suppose the observations to have been probably too large, but by an amount of not more than 10″ of arc. However, the angle was assumed to be 59° 53′ 30″ in the reductions. Should the angle really have been greater or less than 59° 53′ 30″ by as much as 10″ an error ranging gradually from 0.0″ of arc at A to ± 0.6″ at the extremity of the region of the spectrum examined would be introduced in the deviations.

OBSERVATIONS FOR DETERMINING THE REFRACTIVE INDEX OF A.

To determine the absolute deviation and refractive index of the A line in the salt prism the following observations were made, in which special precautions were taken to eliminate error either from defective collimation or incomplete flatness of prism faces. The nature of these precautions has been fully explained at a previous page, but a brief restatement may not be out of place here. In each of the two methods employed the prism was provided with a pair of diaphragms as narrow as consistent with suitable visual resolution of A and adjusted symmetrically upon the two faces of the prism. By this means the refractive angle was necessarily measured for the same place upon the prism as the deviation, and thus error from imperfect flatness of faces was avoided. In the first method the angle was measured with the prism fixed and telescope movable, and with a pair of diaphragms whose centers were 2 cm. from the apex. The deviation was measured twice in this method, once when the prism was so placed that the divergency of the beam decreased the deviation as much as it did the angle, and once when it increased the deviation as much as it decreased the angle. The half sum of these two results gave the proper deviation, and their half difference gave the proper correction to the angle. In the second method the angle was measured with telescope fixed and prism movable. The centers of the diaphragms were 6.5 cm. from the edge. The position of the prism was such that identically the same beam was reflected from each face, and thus the angle measured required no correction.[1] In measuring the deviation the same procedure was used as in the first method, except that the prism was so placed that identically the same beam was used on the one side as on the other, and thus the deviations as measured required no correction.

The table of measurements which follows, requires some explanation of the temperatures recorded. By "prism temperature" is indicated the temperature of a second large prism already mentioned, which has a thermometer bulb at its center, and gives as nearly as may be the temperature of the center of the similar large prism actually used optically. This temperature (corrected of course for thermometer error) was taken as the temperature of the prism used in the actual observations by the second method, and a temperature two-thirds the interval between this and the temperature of the air was taken as that of the observations by the first method. The latter procedure was followed on account of the nearness of the diaphragms to the edge of the prism in the first method and the nearer approach of that part of the prism to the air temperature. The deviations were corrected to 20° C. and a barometric pressure of 760 mm.

Measurements of the refractive index for A in the rock-salt prism "R. B. I." Reduced to temperature 20° C., barometric pressure 76 cm.

No. of observation.	Date.	Time of day.	Observed.		Corrected.		Assumed temperature of prism.	Barometer.	Observed deviation of A.	Corrections to deviation.		Corrected deviation of A.	Residuals.	Distance of axis of rotation from apex.	Remarks.
			Prism temperature.	Air temperature.	Prism temperature.	Air temperature.				Temperature.	Barometer.				
	1898.	h. m.	°	°	°	°	°	cm.	° ′ ″	″	″	° ′ ″	″	cm.	
1	Oct. 19	9 40	19.4	19.6	19.18	19.56	19.44	76.3	40 21 57.1	−6.7	+0.6	40 21 51.0	−2.7	0.5	First method. Measured with diaphragms 4 cm. high, 2 cm. wide, center being 2.0 cm. from refracting angle.
2		50	19.39	19.76	19.17	19.72	19.53	76.4	58.2	−5.6	+0.8	53.4	−0.3		
3		12 0	19.70	19.84	19.48	19.80	19.70	76.4	56.6	−3.6	+0.8	53.8	+0.1		
4		0 20	19.74	19.96	19.52	19.92	19.78	76.4	58.4	−2.6	+0.8	56.6	+2.9		

θ' = mean deviation A, = 40° 21′ 53.7″, = p.e. = 0.7″.

5	Oct. 19	2 15	19.76	19.82	19.54	19.78	19.70	76.4	40 21 16.7	−3.6	+0.8	40 21 13.9	−2.7	2.5	
6		30	19.77	19.82	19.55	19.78	19.71	76.4	18.1	−3.5	+0.8	15.4	−1.2		Do.
7		3 25	19.67	19.60	19.45	19.56	19.53	76.4	24.5	−5.6	+0.8	19.7	+3.1		
8		35	19.68	19.80	19.46	19.76	19.66	76.4	20.7	−4.1	+0.8	17.4	+0.8		

θ'' = mean deviation A, = 40° 21′ 16.6″, = p.e. = 0.9″.

$$\frac{\theta' + \theta''}{2} = \qquad 40° 21' 35.1'',\ \text{p. e.} = 1.1.$$

$$\frac{\theta' - \theta''}{2} = \qquad 0° 0' 18.6''.$$

[1] The prism was specially polished before the measurements with a design to avoid the usual concavity. Both surfaces were afterwards found to be somewhat convex, owing to overcorrection, and especially near the edges.

Measurements of the refractive index for A in the rock-salt prism "R. B. I."—Continued.

No. of obser-vation.	Date.	Time of day.	Observed.		Angle of prism.	Residu-als.	Distance of axis of rota-tion from apex.	Remarks.
			Prism tempera-ture.	Air tempera-ture.				
	1896.	A. m.	°	°	° ′ ″	″	″	
9	Oct. 19	10 20	19.50	19.14	59 56 12.0	−4.5		
10		40	19.60	20.00	16.5	−1.0		
11		11 25	19.65	19.74	17.6	+0.1		Same diaphragm used as in observations 1 to 8. Prism fixed. Telescope moved.
12		30	19.69	19.80	18.1	+0.6		
13		3 45	19.68	19.84	23.2	+4.7		

Mean angle 59° 56′ 17.5″, p.e. = 0.9″.

$$18.6 = \frac{\theta' - \theta''}{2} = \text{correction for divergence of light, p.e.} = 1.1''.$$

α = prism angle = 59° 56′ 36.1″, p.e. = 1.4″.

No. of obser-vation.	Date.	Time of day.	Observed.		Corrected.		Assumed temperature of prism.	Barometer.	Observed deviation of A.	Corrections to devia-tion.		Corrected deviation of A.	Residuals.	Distance of axis of rotation from apex.	Remarks.
			Prism tem-perature.	Air tempera-ture.	Prism tem-perature.	Air tempera-ture.				Temperature.	Barometer.				
	1896.	A. m.	°	°	°	°	°	cm.	° ′ ″	″	″	° ′ ″	″	cm.	Second method. Measured with diaphragms 4 cm. high, 2 cm. wide, center being 6.5 cm. from re-fracting angle.
14	Oct. 20	9 50	19.26	19.56	19.04	19.52	19.04	77.0	40 21 31.6	−11.5	+1.9	40 21 22.0	+1.3	4.5	
15		10 00	19.29	19.62	19.07	19.58	19.07	77.0	31.9	−11.2	+1.9	22.6	+1.9		
16		40	19.37	19.72	19.15	19.68	19.15	77.0	29.2	−10.2	+1.9	20.9	+0.2		
17		50	19.42	19.94	19.20	19.90	19.20	77.0	25.0	−9.6	+1.9	17.3	−3.4		

Mean deviation 40° 21′ 20.7″, p.e. = 0.8″.

No. of obser-vation.	Date.	Time of day.	Observed.		Angle of prism.	Residu-als.	Distance of axis of rota-tion from apex.	Remarks.
			Prism tempera-ture.	Air tempera-ture.				
	1896.	A. m.	°	°	° ′ ″	″	cm.	
18	Oct. 20	1 10	19.96	20.06	59 56 15.5	+0.6	5.7	
19		25	19.77	20.16	17.1	+2.3		Same diaphragm used as in observation 14 to 17. Prism moved. Telescope fixed.
20		33	19.72	20.24	14.6	−0.3		
21		2 00	19.86	20.32	12.4	−2.5		

α = mean angle = 59° 56′ 14.9″, p.e. = 0.7″.

First method.	Second method.
θ = 40° 21′ 35.1″ p.e. = 1.1.	40° 21′ 20.7″ p.e. = 0.8″.
α = 59° 56′ 36.1″ p.e. = 1.4.	59° 56′ 14.9″ p.e. = 0.7″.
n = 1.536800 p.e. 0.0000056, w't = 1.	1.536827, p.e. = 0.0000033, w't 2.
n (weighted mean) = 1.536818 ± 0.0000037.	

Computing from the value of n just obtained the deviation of A for a prism of refracting angle 59° 53′ 30″, as used in obtaining the bolographs, we have designating the absolute deviation by θ_A, $\theta_A = 40° 18′ 20.6″$.

This absolute deviation is used in connection with the results taken from the bolographs to obtain the deviations of the infra-red spectral lines and their refractive indices.

(*f*) *Reduction of angular deviations to 60° prism angle for the salt prism.*—While the refractive indices are, in default of wave-lengths, naturally the standard references for the lines determined, since they are independent of the prism angle and depend only on the substance of the prism and the temperature, yet, as they are the result of a somewhat tedious calculation, it has been thought best to give also the position of the lines in terms of minimum deviation of a 60° rock-salt prism at 20° C. to facilitate the use of the tables. To do this we require to correct the observed deviations for the slight difference in the prism angle. An approximate differential formula for this reduction has been derived,[1] but it has been thought better to use another method of approximation which is somewhat closer, but not so general. This is to compute from the refractive indices at A and at the end of the spectrum region covered by the bolographs, i. e. at about 5.3μ the deviations for a 60° prism. Subtracting from these values the deviations for a prism of 59° 53' 30" we have the corrections at A and at 5.3μ. Taking the second difference of the deviations and dividing it by the distance in centimeters of abscissæ between the two points selected we obtain a mean coefficient for the interval, from which, with the observed distances of the lines from A, the correction for the angular distance from A of each is to be determined. The coefficient as thus computed, corresponding to an angle of 59° 53' 30", was -0.236, so that if x be the abscissa of any line, its minimum deviation for a 60° prism is obtained by adding to the deviation given in (*e*) the correction computed for the A line and subtracting $0.236x$.

(*g*) *Reduction to wave-lengths.*—As stated on another page, it has been felt to be so desirable to obtain accurate wave-lengths for the infra-red lines, that a new determination of the dispersion curve for rock salt has been made with the great prism R. B. I. An account of this research will be found in Part II. As it was done by the bolographic method, the maxima due to the radiations of known wave-lengths were primarily found, not at such and such absolute deviations of the prism, as in the earlier investigation of the writer, and of Paschen, Rubens, and others, but at such and such distances from the A line, so that the wave-lengths of absorption lines may be determined wholly independently of their absolute deviation, and dependent only on differences of deviation from A. This eliminates a large source of error, and enables us to get the wave-lengths of the absorption lines in the rock-salt spectrum with an accuracy nearly corresponding to the accuracy of the determinations of their positions on the bolographs. To do this it is merely necessary to plot on a large scale the wave-lengths and corresponding distances from A and then to find upon the curve best representing these observations, the wave-lengths corresponding to the measured positions of the absorption lines as given in Column VI, Table 21. This procedure has been followed, and the wave-lengths given in Table 21 are the result.

---

[1] By J. E. Keeler. See Am. Jour. of Sci., third series, Vol. xxx, p. 450, 1885.

To obtain the wave-lengths of the lines discovered in the spectrum of the glass prism a number of the more prominent lines were identified between the rock-salt bolographs and glass bolographs. The wave-lengths being known from the former, it was merely necessary to make a plot in which the wave-lengths of these lines should determine ordinates, and their distances from A in the glass prism spectrum the abscissæ. By interpolation in this plot the wave-lengths of all the remaining absorption lines were determined.

2. REDUCTIONS OF MEASURES IN ORDINATES.

(a) *Reduction to a common scale throughout the region of spectrum.*—Since at several points in the curves the slit width was changed, and with it the scale of ordinates, it becomes necessary to introduce a multiplier to reduce to a constant slit width.[1] While the use of this multiplier is self-evident in different parts of the same bolograph, some explanation is necessary concerning the procedure in joining on the Ω to Y bolographs with the A to Ω bolographs. The ordinates found in the short region common to both sets of bolographs should be the same for curves of the same dates. Accordingly, the ordinates of the Ω to Y bolographs have all been multiplied by such a quantity as to bring these common ordinates in agreement with those of the A to Ω bolographs nearest to them in point of time.

(b) *Reduction of bolographs of different dates to a common scale.*—In order to compare the ordinates to detect the occurrence of relative variation in the energy at different parts of the spectrum it is necessary that the bolographs taken at different times shall all be reduced to the same scale of relative ordinates. This has been done by calling the energy at a point near Ω, usually of maximum energy, 100, and multiplying the the ordinates of every bolograph by a factor obtained by dividing 100 by the measured ordinate of this point on the bolograph in question.

TABLES OF FINAL VALUES.

In the following tables we have the final result of all these reductions. The absorption lines discovered in the rock-salt prismatic spectrum will be found in Table 21, while those from the glass prismatic spectrum appear in Table 22. Finally, in order to facilitate the use of the results by those who are concerned only with the wave-lengths and character of the lines, a third table is annexed from which all data relating to the prisms is eliminated, and in which all the lines discovered are arranged in regular order. In this table the results from the glass prism, so much fuller than those from the salt prism on account of greater dispersion, have been used exclusively as far as they go,[2] and the salt prism results begin at Ω.

---

[1] In a few bolographs it was necessary to take account of the loss of energy by diffraction owing to narrow slit widths.

[2] In Ψ and Ω some lines were inserted taken from the results with rock salt, for, as will be seen later, at certain seasons of the year the general absorption here is much increased, so as to obscure entirely minute lines, and this was the case when the glass bolographs were taken.

ANNALS OF ASTROPHYSICAL OBSERVATORY.

TABLE 21.—*Absorption lines in the infra-red solar spectrum, from bolographs of the rock-salt prismatic spectrum.*[1]

Designation. (Explained on p. 132.)	Number of observations by—		Ordinates.		Differences of deviation.	Whole deviation.	Whole deviation reduced to 60° prism.	Probable error. Cols. VI VII. VIII.	$n=\dfrac{\sin\frac{1}{2}(\theta+a)}{\sin\frac{1}{2}a}$	Probable error of n.	Wave-length.
	F. E. F.	C. G. A.	Mean of reduced.	Average deviation from mean.							
I.	II.	III.	IV.	V.	VI.	VII.	VIII.	IX.	X.	XI.	XII.	XIII.
					° ′ ″	° ′ ″	° ′ ″	″				μ
1 d ?	2	6	14.4	2.8	−0 0 7.1	40 18 27.7	40 25 28.0	0.12	2.40	1.536840	.0090086	0.7596
2 d	6	3	11.9	1.8	− 2.9	23.5	23.8	.24	2.41	827	086	.7600
3 d }b	6	4	9.7	1.3	±0 0 0.0	20.6	20.9	.12	2.40	818	086	.7604
4 d	4	5	12.0	3.5	+0 0 3.7	16.9	17.2	.12	2.40	807	086	.7609
5 d ?	2	2	12.0	2.2	6.8	13.8	14.1	.32	2.44	797	087	.7612
6 d	3	4	14.2	4.0	9.1	11.5	11.8	.18	2.41	790	086	.7614
7 d	3	3	17.0	1.8	15.0	4.7	4.9	.24	2.41	768	086	.7621
8 d	5	5	15.0	1.9	19.1	1.5	1.7	.30	2.42	752	086	.7624
9 d	6	3	17.4	2.4	22.3	17 58.3	24 58.5	.24	2.41	749	086	.7629
10 d	2	4	14.7	2.0	25.4	55.2	55.4	.24	2.41	739	086	.7631
11 d }A	5	0	....	....	27.1	53.5	53.7	.36	2.43	734	087	.7634
12 d a	6	0	15.7	0.0	30.4	50.2	50.4	.36	2.43	723	086	.7638
13 d	6	4	18.7	3.4	34.2	46.4	46.6	.30	2.42	711	086	.7641
14 d }b	6	5	21.4	4.3	40.8	39.8	39.9	.36	2.43	691	087	.7650
15 d	6	4	22.5	3.6	45.7	34.9	35.0	.30	2.42	676	086	.7653
16 d	5	6	25.3	4.1	50.9	29.7	29.8	.24	2.41	660	086	.7660
17 d	6	0	31.8	0.0	56.8	23.8	23.9	.42	2.44	641	087	.7666
18 d	4	5	28.2	4.4	59.5	21.1	21.2	.18	2.41	633	086	.7668
19 d	4	0	26.4	0.0	1 5.5	15.1	15.1	.36	2.43	614	087	.7676
20 d	4	0	31.2	4.1	10.9	9.7	9.7	.18	2.41	601	086	.7680
21 d ?	0	5	34.0	4.6	2 36.2	15 44.4	22 44.1	.54	2.46	332	088	.7772
22 d	3	5	38.6	6.1	3 22.0	14 58.6	21 58.1	.24	2.41	189	086	.7826
23 d ?	0	5	40.7	8.2	5 37.3	12 43.3	19 42.8	.60	2.47	1.535767	088	.7987
24 d ?	0	3	45.6	6.8	6 18.2	2.4	1.2	.18	2.41	689	086	.8039
25 d ?	0	4	40.3	7.7	7 18.4	11 2.2	18 0.8	.18	2.41	452	086	.8119
26 d	2	6	41.2	7.5	25.0	10 55.6	17 54.2	.54	2.46	431	088	.8126
27 d	5 }6		{ 40.9	5.9	31.4	49.2	47.7	.48	2.45	411	087	.8136
28 d	5		{ 40.4	5.4	36.4	44.2	42.7	.48	2.45	396	087	.8142
29 d	6	4	36.6	4.8	41.6	39.0	37.5	.18	2.41	379	086	.8150
30 c	6	6	40.1	7.3	47.0	33.6	32.1	.12	2.40	363	085	.8160
31 d }b	6	5	38.6	6.9	51.8	28.8	27.2	.12	2.40	347	086	.8164
32 d ?	6 }6		39.6	6.0	57.6	23.0	21.4	.24	2.41	329	086	.8173
33 c	4		{ 45.1	0.0	8 0.1	20.5	18.9	.24	2.41	322	086	.8178
34 d ?	5 }8		{ 33.6	0.0	3.2	17.4	15.8	.12	2.40	313	086	.8181
35 d	5		{ 45.7	0.0	5.3	15.3	13.7	.24	2.41	306	086	.8183
36 d	6 }6		{ 40.2	6.0	10.7	9.9	8.3	.24	2.41	289	086	.8190
37 c	6		{ 40.6	5.6	13.9	6.7	5.1	.24	2.41	279	086	.8196
38 d	4	0	41.6	5.2	25.5	9 55.1	16 53.4	.48	2.45	243	087	.8211
39 d	6	0	40.7	5.0	31.8	48.8	47.1	.30	2.42	223	086	.8221
40 c	6	6	42.4	7.1	35.0	44.6	42.9	.12	2.40	210	086	.8228
41 d	6	2	41.0	3.5	46.1	34.5	32.7	.18	2.41	178	086	.8240
42 d	2	6	45.2	8.0	56.7	23.9	22.1	.06	2.40	145	086	.8257
43 d	4 }5		37.1	0.0	9 9.1	11.5	9.6	.60	2.48	106	089	.8274
44 d	4		{........	....	13.1	7.5	5.6	.36	2.43	093	087	.8280
45 d	4	3	46.8	9.5	19.8	0.8	15 58.9	.12	2.40	073	086	.8290
46 d	5	4	46.4	8.7	32.9	8 47.7	45.7	.42	2.44	032	087	.8308
47 d	6	5	47.0	8.5	42.2	38.4	36.4	.18	2.41	003	086	.8321
48 d ?	2	2	43.0	0.2	48.5	32.1	30.1	.18	2.41	1.534984	086	.8331
49 d ?	0	3	41.1	2.9	54.3	26.3	24.3	.42	2.44	965	087	.8340
50 d ?	0	4	46.9	8.1	10 12.9	7.7	5.6	.18	2.41	907	086	.8366
51 d	0	6	49.5	8.3	29.6	7 51.0	14 48.8	.30	2.42	855	086	.8389
52 d ?	3	0	....	....	51.0	29.6	27.3	.78	2.52	788	090	.8420
53 d ?	0	5	53.2	9.2	11 25.0	6 55.6	13 53.2	.12	2.40	682	086	.8468
54 d ?	2	6	52.2	8.3	36.3	44.3	41.9	.24	2.41	647	086	.8484
55 c	6	6	49.2	7.2	44.6	36.0	33.5	.18	2.41	621	086	.8496

[1] Table 21 is represented in Plates XX and XX A, except that since these plates were printed about thirty lines belonging to Class D have been added to the table, which has 579 lines.

TABLE 21.—*Absorption lines in the infra-red solar spectrum, from bolographs of the rock-salt prismatic spectrum*—Cont'd.

Designation.	Number of observations by—		Ordinates.		Differences of deviation.	Whole deviation.	Whole deviation reduced to 60° prism.	Probable error. Cols. VI, VII. VIII.	$n=\frac{\sin\frac{1}{2}(\theta+a)}{\sin\frac{1}{2}a}$	Probable error of n.	Wave-length.
	F. E. F.	C. G. A.	Mean of reduced.	Average deviation from mean.							
I.	II.	III.	IV.	V.	VI.	VII.	VIII.	IX.	X.	XI.	XII.	XIII.
					° ′ ″	° ′ ″	° ′ ″	″	″			μ
56 d	6	6	50.4	7.7		54.7	25.9	23.4	0.24	2.41	590	086	0.8510
57 d	5	6	50.1	7.1	12 3.6	17.0	14.5	.18	2.41	562	086	.8522
58 d	4	2	50.2	9.4		8.0	12.6	10.0	.24	2.41	548	086	8530
59 b	6	6	46.6	8.1		13.3	7.3	4.7	.06	2.40	532	086	.8538
60 d?	2	0	45.0	6.4		16.7	3.9	1.3	.36	2.43	521	087	.8542
61 d	6	6	50.3	8.2		21.1	5 59.5	12 56.9	.30	2.42	507	086	.8549
62 d?	3	0	......	......		27.2	53.4	50.8	.84	2.54	488	091	.8558
63 d?	0	4	52.1	8.4		40.8	39.8	37.1	.48	2.45	446	087	.8577
64 d?	0	4	53.5	8.1		48.3	32.4	29.6	.48	2.45	422	087	.8588
65 d	3	3	59.6	8.9		56.9	23.7	20.9	.24	2.41	395	086	.8601
66 d?	0	3	48.2	1.5	13 14.7	5.9	3.1	.00	2.40	340	086	.8628
67 d?	3	6	52.2	5.0		19.7	0.9	11 58.1	.24	2.41	225	086	.8634
68 d	6		45.1	0.0		23.6	4 57.0	54.1	.30	2.42	312	086	.8640
69 b	6	6	49.2	6.9		30.6	50.0	47.1	.36	2.43	290	087	.8652
70 d?	4	5	52.3	5.9		37.3	43.3	40.4	.48	2.45	269	087	.8661
71 d	5	6	49.2	3.4		39.7	40.9	38.0	.24	2.41	262	086	.8666
72 d	5	6	55.9	7.1		46.4	34.2	31.3	.18	2.41	241	086	.8676
73 d?	0	5	57.4	8.1		53.1	27.5	24.5	.24	2.41	220	086	.8687
74 d	0	6	56.6	6.9	14 4.8	15.8	12.8	.66	2.49	183	089	.8705
75 d	2	6	57.2	7.7		17.2	3.4	0.3	.24	2.41	145	086	.8716
76 d	0	4	59.6	9.6		25.1	3 55.5	10 52.4	.36	2.43	120	087	.8739
77 d	0	3	55.9	7.2		32.7	47.9	44.8	.30	2.42	096	088	.8752
78 d	0	6	57.8	7.8		39.8	40.8	37.6	.30	2.42	074	088	.8764
79 d	0	5	56.5	6.9		49.9	30.7	27.5	.18	2.41	043	086	.8782
80 c	5	6	56.9	7.8		59.3	21.3	18.1	.06	2.40	013	086	.8799
81 d?	2	0			15 10.6	10.0	6.7	.30	2.42	1.583978	086	.8819
82 d?	2	4	58.9	7.4		18.6	2.0	9 58.7	.36	2.43	953	087	.8834
83 d?	0	3	58.8	10.0		28.4	2 52.2	48.8	.18	2.41	922	086	.8851
84 d?	0	4	58.5	8.0		34.5	46.1	42.7	.30	2.42	903	086	.8862
85 d	2	3	57.7	4.2		41.3	39.3	35.9	.24	2.41	882	086	.8874
86 d	2	0				44.7	35.9	32.5	.48	2.45	871	087	.8880
87 d	3	4	58.9	7.2		50.6	30.0	26.6	.18	2.41	853	086	.8891
88 d?	3	4	64.2	8.7	16 2.1	18.5	15.0	.24	2.41	817	086	.8913
89 d	4	6	57.7	8.1		10.5	10.1	6.6	.30	2.42	791	086	.8928
90 d?	2	0				19.6	1.0	8 57.4	.30	2.42	762	086	.8944
91 d	4	5	57.1	8.6		25.2	1 55.4	51.8	.30	2.42	745	086	.8954
92 d	5	5	53.8	8.5		31.4	49.2	45.6	.18	2.41	726	086	.8966
93 c	5	5	55.2	8.0		37.9	42.7	39.1	.30	2.42	705	086	.8977
94 c	5	5	52.5	7.7		45.1	35.5	31.8	.06	2.40	683	086	.8990
95 d	6	4	49.9	6.8		53.4	27.2	23.5	.12	2.41	657	086	.9006
96 d	5	3	55.3	7.8		56.7	23.9	20.2	.24	2.42	647	086	.9015
97 d	6	3	51.9	7.6	17 0.1	20.5	16.8	.18	2.41	636	086	.9021
98 d	6	4	52.2	5.8		4.8	15.8	12.1	.18	2.41	621	086	.9030
99 d	6	6	60.6	6.5		15.6	5.0	1.2	.18	2.41	587	086	.9050
100 d	6	4	57.3	4.4		21.9	0 58.7	7 54.9	.18	2.41	568	086	.9062
101 d	6	6	56.2	6.9		27.8	52.8	49.0	.12	2.41	550	086	.9072
102 d?	6	2	51.7	4.6		30.7	49.9	46.1	.06	2.40	540	086	.9078
103 d		5	52.0	6.3		33.7	46.9	43.1	.30	2.42	531	086	.9084
104 d	6	6	55.5	7.3		38.7	41.9	38.0	.12	2.41	516	086	.9093
105 d	6	6	54.6	7.2		43.5	37.1	33.2	.12	2.41	500	086	.9102
106 d	6	6	55.7	7.3		48.4	32.2	28.3	.18	2.41	485	086	.9111
107 d?	4	0	53.8	8.0		52.3	28.3	24.4	.30	2.42	473	086	.9119
108 c	6	6	56.5	7.4		56.2	24.4	20.5	.18	2.41	461	086	.9126
109 c	6	6	54.9	7.3	18 4.4	16.2	12.2	.12	2.41	435	086	.9141
110 d	4	0	47.1	0.0		12.1	8.5	4.5	.24	2.42	411	086	.9156

TABLE 21.—*Absorption lines in the infra-red solar spectrum, from bolographs of the rock-salt prismatic spectrum*—Cont'd.

Designation.	Number of observations by—		Ordinates.		Differences of deviation.	Whole deviation.	Whole deviation reduced to 60° prism.	Probable error. Cols. VI, VII, VIII.	$n=\frac{\sin\frac{1}{2}(\theta+a)}{\sin\frac{1}{2}a}$	Probable error of n.	Wavelength.
	F.E.F.	C.G.A.	Mean of reduced.	Average deviation from mean.							
I.	II.	III.	IV.	V.	VI.	VII.	VIII.	IX.	X.	XI.	XII.	XIII.
					° ′ ″	° ′ ″	° ′ ″	″			μ
111 c	6	6	55.2	7.8	16.3	4.3	0.3	0.12	2.41	398	086	0.9163
112 d	4	0	47.1	0.0	24.0	39 59 56.6	6 52.6	.12	2.41	374	086	.9178
113 c	6	6	55 7	8.0	26.6	54.0	49.9	.06	2.40	366	086	.9183
114 d	4	} 4	59.5	0.0	30.0	50.6	46.5	.12	2.41	355	086	.9190
115 d	3		49.3	0.0	33.7	46.9	42.8	.12	2.41	344	086	.9197
116 c	6	6	59.7	7.5	37.4	43.2	39.1	.18	2.41	332	086	.9203
117 c	4	6	59.6	7.1	47.4	33.2	29.1	.24	2.42	301	086	.9222
118 d	5	0	56.1	7.5	51.7	28.9	24.7	.24	2.42	287	086	.9230
119 d	6	6	58.8	7.9	56.8	23.8	19.6	.18	2.41	271	086	.9240
120 d	6	6	60.3	6.5	19 1.3	19.3	15.1	.24	2.42	257	086	.9250
121 d	6	5	60.3	6.7	5.7	14.9	10.7	.36	2.43	244	087	.9257
122 d	6	6	59.5	7.1	11.8	8.8	4.6	.18	2.41	225	086	.9269
123 d	6	5	57.3	7.9	22.8	58 57.8	5 53.5	.24	2.42	190	086	.9290
124 d	5	3	53.8	6.3	27.3	53.3	49 0	.24	2.42	176	086	.9299
125 d	4	4	47.9	7.2	31.8	48.8	44.5	.12	2.41	162	086	.9307
126 d	5	4	45.0	6.7	35.4	45.2	40.9	.06	2.40	151	086	.9314
127 d }b ρ	6	6	41.8	7.5	41.5	39.1	34.8	.06	2.40	132	086	.9326
128 d	6	6	38.5	6.6	45.6	35.0	30.6	.18	2.41	119	086	.9333
129 d ?	3	0	41.0	0.0	51.9	28.7	24.3	.00	2.40	099	086	.9347
130 c	6	6	33.9	6.0	53.4	27.2	22.8	.18	2.41	094	086	.9350
131 c	6	6	34.7	6.0	20 0.9	19.7	15.3	.06	2.40	071	086	.9363
132 c	5	6	35.6	5.6	7.4	13.2	8.8	.06	2.40	051	086	.9376
133 d	6	4	36.2	5.9	11.4	9.2	4.7	.06	2.41	038	086	.9384
134 d	6	6	44.0	6.8	24.2	57 56.4	4 51.9	.06	2.41	1.532996	086	.9410
135 d ?	4	} 6	42.4	6.7	27.9	52.7	48.2	.54	2.46	986	088	.9418
136 d	6		40.8	6.6	30.4	50.2	45.7	.36	2.43	979	087	.9422
137 c	6	6	40.4	5.9	35.6	44.0	39.4	.06	2.41	959	086	.9435
138 c	6	6	38.7	6.3	42.7	37.9	33.3	.06	2.41	940	086	.9448
139 c	6	6	40.8	6.0	50.8	29.8	25.2	.06	2.41	915	086	.9464
140 d	6	3	44.6	8.8	56.7	23.9	19.3	.24	2.42	896	086	.9477
141 c }b σ	6	6	41.9	7.3	21 1.4	19.2	14.5	.12	2.41	882	086	.9486
142 b	6	6	42.2	6.4	10.0	10.6	5.9	.06	2.41	855	086	.9504
143 b	6	6	44.2	6.6	21.1	56 59.5	3 54.8	.24	2.42	820	086	.9528
144 c	6	6	46.3	6.3	32.2	48.4	43.6	.18	2.41	786	086	.9550
145 c	6	6	47.1	6.4	37.5	43.1	38.3	.12	2.41	769	086	.9562
146 c a	6	6	46.1	5.5	41.8	38.8	34.0	.18	2.41	756	086	.9570
147 d	4	3	49.4	3.6	49.6	31.0	26.2	.06	2.41	731	086	.9588
148 c	6	5	49.3	6.1	53.8	26.8	21.9	.12	2.41	718	086	.9597
149 d	6	6	53.3	7.4	22 1.0	19.6	14.7	.36	2.43	696	087	.9612
150 c	6	6	53.0	6.8	5.5	12.1	7.2	.06	2.41	672	086	.9630
151 c	6	6	55.3	6.7	16.2	4.4	2 59.4	.12	2.41	648	086	.9646
152 d ?	0	0	63.0	0.0	20.1	0.5	55.5	.24	2.42	636	086	.9652
153 c	6	6	57.0	7.7	26.4	55 54.2	49.2	.12	2.41	616	086	.9668
154 d	6	6	63.4	7.0	34.5	46.1	41.1	.12	2.41	579	086	.9686
155 c	6	6	65.7	6.4	43.8	36.8	31.7	.12	2.41	562	086	.9706
156 d ?	3	} 5	63.5	6.4	54.1	26.5	21.4	.12	2.41	529	086	.9729
157 d	3		55.0	0.0	56.8	23.8	18.7	.12	2.41	521	086	.9734
158 c	6	6	63.4	6.5	23 1.3	19.3	14.2	.18	2.41	507	086	.9745
159 c	6	6	64.0	6.2	9.2	11.4	6.2	.18	2.41	482	086	.9762
160 d }b r	4	0	61.8	7.6	13.0	7.6	2 4	.30	2.42	470	086	.9771
161 d }c	5	} 6	63.4	7.7	20.8	54 59.8	1 54.6	.42	2.44	446	087	.9789
162 d ?	6		71.1	0.0	24.2	56.4	51.2	.30	2.42	435	086	.9796
163 d	5	6	64.2	8.6	28.9	51.7	46.5	.30	2.42	421	086	.9808
164 d	5	6	69.9	7.1	39.8	40.8	35.5	.30	2.42	387	086	.9832
165 d	5	6	70.7	7.2	47.2	33.4	28.1	.12	2.41	364	086	.9850
166 d	0	5	73.1	7.8	55.3	25.3	20.0	.66	2.49	338	089	.9869

TABLE 21.—*Absorption lines in the infra-red solar spectrum, from bolographs of the rock-salt prismatic spectrum*—Cont'd.

Designation.	Number of observations by—		Ordinates.		Differences of deviation.	Whole deviation.	Whole deviation reduced to 60° prism.	Probable error. Cols. VI, VII, VIII.	$n = \dfrac{\sin \frac{1}{2}(\theta + a)}{\sin \frac{1}{2} a}$	Probable error of n.	Wave-length.
	F. E. F.	C. G. A.	Mean of reduced.	Average deviation from mean.							
I.	II.	III.	IV.	V.	VI.	VII.	VIII.	IX.	X.	XI.	XII.	XIII.
					° ′ ″	° ′ ″	° ′ ″	″				μ
167 d	0	3	67.3	4.4	24 6.0	14.6	9.2	0.42	2.44	305	067	0.9893
168 d	3	6	75.5	5.6	39.0	53 41.6	0 36.1	.48	2.46	201	088	.9972
169 d	3	3	77.5	7.6	49.4	31.2	25.6	.06	2.41	169	086	.9996
170 d ?	2	3	77.9	6.5	59.0	20.7	15.1	.42	2.44	136	087	1.0021
171 d	4	4	74.6	8.4	25 4.7	15.9	10.3	.12	2.41	121	086	.0033
172 c	6	6	73.3	5.6	14.0	6.6	0.9	.12	2.41	092	086	.0056
173 d	3	4	76.7	5.5	24.7	52 55.9	39 59 50.2	.36	2.43	058	087	.0082
174 d	0	5	75.1	5.7	39.9	40.7	34.9	1.20	2.69	011	096	.0122
175 d	3	5	78.3	5.7	54.0	20.6	20.8	0.24	2.42	1.531967	086	.0150
176 d ?	0	4	78.6	4.1	26 2.6	18.0	12.2	.36	2.43	940	087	.0182
177 d	0	4	79.0	4.0	11.8	8.8	2.9	.06	2.41	911	086	.0205
178 d ?	0	2	80.8	3.9	36.5	51 44.1	58 38.1	.30	2.43	834	087	.0270
179 d ?	0	5	77.7	4.1	27 7.1	13.5	7.4	.60	2.48	738	089	.0351
180 d	0	6	80.0	5.4	19.3	1.3	57 55.2	.36	2.43	700	087	.0384
181 c	0	5	78.1	4.1	30.4	50 50.2	44.0	.24	2.42	665	086	.0414
182 d	0	5	78.8	4.4	39.2	41.4	35.2	.36	2.43	638	087	.0439
183 d	0	5	79.0	4.3	50.0	30.6	24.3	.60	2.48	604	089	.0468
184 d	0	5	79.5	5.9	28 19.5	1.1	56 54.7	.72	2.51	511	086	.0550
185 d	0	4	81.8	3.4	27.9	49 52.7	46.3	.48	2.46	485	088	.0578
186 c	5	6	79.2	4.5	35.8	44.8	38.4	.12	2.41	460	086	.0595
187 c	2	6	78.2	5.7	43.1	37.5	31.0	.18	2.42	438	086	.0616
188 d	2	3	75.7	5.2	49.7	30.9	24.4	.18	2.42	417	086	.0634
189 d	0	5	80.2	5.5	55.4	25.2	18.7	.36	2.44	399	087	.0651
190 d	5	6	78.2	7.1	29 2.6	18.0	11.4	.06	2.41	377	086	.0672
191 d ?	2	0	67.8	0.0	9.4	11.2	4.6	.66	2.50	356	089	.0690
192 c	6	6	77.5	5.2	11.8	8.8	2.2	.12	2.41	348	086	.0699
193 c	6	6	78.5	5.7	25.1	48 55.5	55 48.9	.18	2.42	306	086	.0736
194 c	6	6	79.6	5.1	32.3	48.3	41.6	.12	2.41	284	086	.0758
195 c	3	5	79.9	3.9	43.4	37.2	30.5	.24	2.42	249	086	.0791
196 c	6	6	78.5	5.5	53.0	27.6	20.8	.24	2.42	219	086	.0820
197 c	6	6	78.9	5.0	59.6	21.0	14.2	.18	2.42	198	086	.0839
198 c	6	6	78.2	4.9	30 10.4	10.2	3.4	.24	2.42	164	086	.0871
199 c	5	6	81.2	4.4	20.2	0.4	54 53.5	.30	2.43	134	087	.0901
200 d	0	3	74.5	1.3	28.6	47 52.0	45.1	.96	2.59	107	092	.0927
201 c	6	6	76.0	4.9	34.2	46.4	39.5	.18	2.42	090	086	.0944
202 d	6	5	75.2	4.3	39.2	41.4	34.5	.12	2.41	074	086	.0960
203 d	5	4	75.2	5.0	46.3	34.3	27.3	.24	2.42	052	087	.0981
204 c	6	6	75.3	5.0	51.6	29.0	22.0	.42	2.45	035	087	.0998
205 c	5	6	74.2	4.9	59.9	20.7	13.7	.18	2.42	'010	086	.1024
206 d	6	6	74.1	5.0	31 6.8	13.8	6.8	.30	2.43	1.530988	087	.1046
207 c	5	6	72.8	5.4	12.5	8.1	1.0	.24	2.42	970	086	.1063
208 d ?	2	0	77.5	0.0	17.8	2.8	53 55.7	.42	2.45	953	087	.1081
209 d	5	6	70.4	4.8	20.3	0.3	53.2	.18	2.42	945	086	.1089
210 d	6	6	65.8	4.3	27.7	46 52.9	45.8	.18	2.42	922	086	.1112
211 d	6	6	57.7	4.5	35.1	45.5	38.3	.12	2.41	899	086	.1136
212 d	2	0	54.3	0.0	41.7	38.9	31.7	.66	2.50	878	089	.1157
213 c	6	6	44.2	4.8	44.2	36.4	29.2	.12	2.41	871	086	.1165
214 c	6	6	40.9	4.5	52.6	28.0	20.8	.06	2.41	844	086	.1192
215 b	6	6	35.2	4.5	32 1.1	19.5	12.2	.12	2.41	818	086	.1221
216 c b	6	6	35.1	4.6	10.7	9.9	2.6	.06	2.41	787	086	.1252
217 c	6	6	37.5	4.2	17.7	2.9	52 55.6	.12	2.41	766	086	.1276
218 c ⊕	4	3	34.1	2.1	20.9	45 59.7	52.4	.06	2.41	756	086	.1287
219 b	6	6	32.8	4.7	36.8	43.8	36.4	.00	2.41	705	086	.1340
220 d a	6	6	50.6	5.2	46.6	34.0	26.6	.12	2.41	675	086	.1373

TABLE 21.—*Absorption lines in the infra-red solar spectrum, from bolographs of the rock-salt prismatic spectrum*—Cont'd.

Designation.	Number of observations by—			Ordinates.		Differences of deviation.	Whole deviation.	Whole deviation reduced to 60° prism.	Probable error. Cols. VI, VII, VIII.	$n=\frac{\sin\frac{1}{2}(\theta+a)}{\sin\frac{1}{2}a}$	Probable error of n.	Wave-length.
	F. E. F.	C. G. A.		Mean of reduced.	Average deviation from mean.							
I.	II.	III.		IV.	V.	VI.	VII.	VIII.	IX.	X.	XI.	XII.	XIII.
						° ′ ″	° ′ ″	° ′ ″	″				μ
221 c	6	6		50.1	5.6	55.9	24.7	17.2	0.06	2.41	646	086	1.1406
222 c	6	6		42.6	3.0	33 6.8	13.8	6.8	.18	2.42	612	086	.1443
223 c b	6	6		41.1	6.1	11.6	9.0	1.5	.12	2.41	597	086	.1460
224 c	6	6		48.4	6.2	19.8	0.8	51 53.2	.06	2.41	571	086	.1491
225 c	6	6		52.0	5.9	27.2	44 53.4	45.8	.18	2.42	548	086	.1516
226 d	6	0		49.3	0.0	31.0	49.6	42.0	.18	2.42	536	086	.1528
227 d	5	6		60.6	4.9	37.3	43.3	35.7	.24	2.42	516	086	.1551
228 d	6	6		64.7	5.7	44.1	36.5	28.8	.24	2.42	495	086	.1576
229 d	5	5		66.3	5.9	48.3	32.3	24.6	.24	2.42	482	086	.1592
230 d	4	2		77.2	6.5	59.0	21.6	13.9	.18	2.43	448	086	.1631
231 c	5	6		73.6	6.4	34 2.9	17.7	10.0	.18	2.42	436	086	.1645
232 d	6	0		83.8	0.0	10.7	9.9	2.1	.48	2.46	411	088	.1674
233 d	2	6		76.9	6.1	12.9	7.7	50 59.9	.24	2.43	405	086	.1683
234 d	5	0		67.1	0.0	15.9	4.7	56.9	.66	2.50	395	089	.1694
235 c	6	5		77.0	6.7	25.6	43 58.0	47.2	.12	2.42	365	086	.1731
236 c	6	6		79.0	7.6	35.3	45.3	37.4	.12	2.42	334	086	.1767
237 c	5	6		78.0	6.9	46.5	34.1	26.2	.06	2.41	299	086	.1809
238 c	6	6		79.8	7.0	52.4	26.2	20.3	.12	2.42	281	086	.1832
239 c	6	6		79.6	6.9	35 1.7	18.9	10.9	.06	2.41	252	086	.1868
240 d	3	6		83.7	6.9	7.8	12.8	4.8	.24	2.42	233	086	.1890
241 d	6	0		95.9	0.0	9.9	10.7	2.7	.18	2.42	226	086	.1900
242 d	4	6		84.7	7.0	15.7	4.9	49 56.9	.24	2.42	208	086	.1921
243 c	6	6		84.0	6.4	20.7	43 59.9	51.9	.12	2.42	192	086	.1942
244 c	6	6		79.9	5.8	29.2	51.4	43.3	.06	2.41	166	086	.1974
245 d	6	0		82.5	7.5	34.2	46.4	38.3	.48	2.46	150	088	.1995
246 c	6	6		80.8	6.5	40.7	39.9	31.8	.12	2.42	129	086	.2021
247 c	6	6		84.8	5.4	51.6	29.0	20.8	.06	2.41	095	086	.2065
248 d	6	5		85.3	5.5	58.4	22.2	14.0	.18	2.42	074	086	.2092
249 d	5	5		87.2	5.2	36 6.8	13.8	5.6	.18	2.42	048	086	.2125
250 d	2	*0		90.0	6.0	15.6	5.0	48 56.7	.72	2.52	020	090	.2161
251 d	4	5		90.4	3.9	18.8	1.8	53.5	.24	2.43	010	087	.2174
252 d	4	6		90.4	5.8	25.1	41 55.5	47.2	.12	2.42	1.529990	086	.2200
253 d	3	0				27.9	52.7	44.4	.24	2.43	982	087	.2212
254 c	6	5		90.3	5.4	36.9	43.7	35.4	.18	2.42	953	086	.2250
255 d?	0	3		92.1	4.4	42.8	37.8	29.4	.36	2.44	935	087	.2274
256 d	4	6		91.9	5.4	49.8	30.8	22.4	.24	2.43	913	087	.2303
257 d	4	6		93.8	5.5	58.8	21.8	13.4	.36	2.44	885	087	.2341
258 d	4	6		93.8	5.4	37 7.0	13.6	5.1	.48	2.46	859	088	.2378
259 d	5	6		93.2	6.2	14.4	6.2	47 57.7	.30	2.43	835	087	.2410
260 d?	4	4		92.7	5.5	23.7	40 56.9	48.4	.42	2.45	807	087	.2450
261 d?	0	5		93.9	5.9	27.5	53.1	44.6	.30'	2.43	795	087	.2466
262 d	6	5		94.4	4.1	34.0	46.6	38.0	.24	2.43	774	087	.2495
263 d	3	4		90.3	8.2	39.2	41.4	32.8	.30	2.43	758	087	.2518
264 d	4	6		93.8	3.8	44.1	36.5	27.9	.24	2.43	743	087	.2540
265 d	3			87.8	0.0	48.0	32.6	24.0	.30	2.43	730	087	.2559
266 d	4	0		94.2	0.0	51.4	29.2	20.6	.12	2.42	720	086	.2572
267 d	6	5		82.0	6.1	53.8	24.8	16.2	.30	2.43	706	087	.2594
268 c	6	6		80.0	6.1	38 0.4	20.2	11.5	.06	2.42	691	086	.2615
269 d b	3	2		82.8	2.6	5.6	15.0	6.3	.24	2.43	675	087	.2639
270 b	6	6		75.5	5.5	9.5	11.1	2.4	.12	2.42	663	086	.2657
271 d	6	5		85.6	4.8	17.9	2.7	46 54.0	.18	2.42	637	086	.2698
272 d?	5	4		89.7	5.8	28.8	39 51.8	43.0	.24	2.43	602	087	.2750
273 d	3	0		93.4	4.2	33.0	47.6	38.8	.60	2.42	589	086	.2770
274 c	6	6		88.5	4.2	37.7	42.9	34.1	.12	2.42	575	086	.2793
275 d	6	6		93.7	4.6	49.7	30.9	22.0	.36	2.44	537	087	.2850
276 d	6	6		94.4	4.9	57.5	23.1	14.2	.30	2.43	513	087	.2889

TABLE 21.—*Absorption lines in the infra-red solar spectrum, from bolographs of the rock-salt prismatic spectrum*—Cont'd.

Designation.	Number of observations by—		Ordinates.		Differences of deviation.	Whole deviation.	Whole deviation reduced to 60° prism.	Probable error. Cols. VI, VII, VIII.	$n=\frac{\sin \frac{1}{2}(\theta+a)}{\sin \frac{1}{2}a}$	Probable error of n.	Wave-length.
	F. E. F.	C. G. A.	Mean of reduced.	Average deviation from mean.							
I.	II.	III.	IV.	V.	VI.	VII.	VIII.	IX.	X.	XI.	XII.	XIII.
					° ′ ″	° ′	° ′ ″	″				μ
277 d 1	2	6	96.8	4.6	39 6.3	14.3	5.4	0.60	2.49	485	089	1.2932
278 d	6		98.2	4.7	8.9	11.7	2.8	.30	2.43	477	087	.2946
279 d	3	5	102.3	0.0	13.2	7.4	45 58.4	.12	2.42	463	086	.2968
280 d	3		97.4	3.8	16.5	4.1	55.1	.18	2.42	453	086	.2983
281 d	6	4	87.6	5.4	22.1	36 57.5	48.5	.18	2.42	432	086	.3016
282 d	4	6	89.1	5.9	27.3	53.3	44.3	.30	2.43	419	087	.3039
283 d	6	6	87.1	5.9	35.0	45.6	36.6	.24	2.43	395	087	.3079
284 d 1	2	2	85.7	4.4	40.0	40.6	31.5	.12	2.42	379	088	.3104
285 c	6	6	83.0	6.4	43.2	37.4	28.3	.24	2.43	369	087	.3120
286 d	2	0	......	......	44.6	36.0	26.9	.00	2.42	365	086	.3129
287 c	6	6	80.3	6.5	53.1	27.5	18.4	.18	2.42	338	086	.3172
288 c	4	6	78.6	6.1	40 0.9	19.7	10.6	.18	2.42	314	086	.3212
289 d	6	6	69.4	7.1	11.6	9.0	44 59.8	.18	2.42	280	086	.3269
290 c	6	6	64.6	8.1	17.0	3.6	54.4	.12	2.42	263	086	.3299
291 d	6	6	64.3	7.2	27.5	37 53.1	43.9	.18	2.42	230	086	.3354
292 d	6	6	60.8	7.3	33.9	46.7	37.4	.18	2.42	210	086	.3390
293 d	6	6	44.6	5.0	42.0	38.6	29.3	.12	2.42	185	086	.3434
294 d	6	6	28.0	3.5	48.5	32.1	22.8	.06	2.42	164	086	.3471
295 d	5	4	18.8	4.1	55.2	25.4	16.0	.30	2.44	143	087	.3510
296 d	5	5	11.8	1.7	41 1.4	19.2	9.8	.24	2.43	124	087	.3542
297 d	6	5	8.6	1.6	9.9	10.7	1.3	.24	2.43	097	087	.3590
298 d	4	4	7.5	1.3	15.0	5.6	43 56.2	.42	2.45	061	087	.3618
299 d	5	5	7.3	1.6	20.1	0.5	51.0	.42	2.45	065	087	.3647
300 d	4	0	5.7	0.7	25.5	36 55.1	45.6	.30	2.44	048	087	.3679
301 d	4	6	7.2	1.3	27.7	52.9	43.4	.18	2.42	041	086	.3691
302 d	5	4	9.0	1.6	35.3	45.3	35.8	.30	2.44	017	087	.3734
303 d	4	5	7.8	1.6	42.9	37.7	28.2	.36	2.44	1.528994		.3780
304 d	6	6	7.1	1.7	48.7	31.9	22.3	.18	2.43	976	087	.3815
305 d	6	6	8.5	2.0	56.9	23.7	14.1	.24	2.43	950	087	.3863
306 d a	3	5	8.5	2.3	42 5.8	14.8	5.2	.24	2.43	922	087	.3915
307 c	6	6	8.4	2.2	9.0	11.6	2.0	.24	2.43	912	087	.3935
308 c	6	6	9.4	2.3	18.0	2.6	42 52.9	.24	2.43	883	087	.3990
309 c	5	6	11.1	2.5	28.1	35 54.5	44.8	.12	2.42	858	086	.4038
310 c ↓	5	6	12.4	3.0	33.4	47.2	37.5	.12	2.42	835	086	.4082
311 c	6	6	17.3	4.3	41.7	38.9	29.1	.12	2.42	809	086	.4132
312 c	6	5	19.3	5.0	49.1	31.5	21.7	.12	2.42	786	086	.4180
313 d	6	6	25.4	5.9	56.4	24.2	14.4	.18	2.43	763	087	.4224
314 c	6	6	26.7	6.4	43 3.3	17.3	7.4	.18	2.43	741	087	.4268
315 c	6	6	26.8	5.8	11.0	9.6	41 59.7	.06	2.42	717	086	.4317
316 d	6	6	30.2	6.8	18.2	2.4	52.5	.12	2.42	695	086	.4362
317 d	6	6	31.8	6.5	23.9	34 56.7	46.8	.06	2.42	677	086	.4398
318 c	6	6	34.6	7.0	31.0	49.6	39.6	.18	2.43	654	087	.4444
319 c	6	6	37.5	7.1	39.1	41.5	31.5	.06	2.42	629	086	.4496
320 d	5	6	51.2	8.0	46.4	34.2	24.2	.18	2.43	606	087	.4545
321 c)	6	6	51.2	6.4	56.1	24.5	14.4	.18	2.43	576	087	.4608
322 c	6	6	44.9	6.8	44 4.2	16.4	6.3	.12	2.42	550	086	.4663
323 c)b ψ	6	6	42.4	7.6	10.2	10.4	0.3	.06	2.42	532	086	.4702
324 c	6	6	51.1	7.1	20.3	0.3	40 50.1	.18	2.43	500	087	.4770
325 d)	6	0	58.4	0.0	23.8	33 56.8	46.6	.24	2.43	489	087	.4792
326 c	6	6	60.5	6.9	33.0	47.6	37.4	.12	2.42	460	086	.4855
327 d	6	6	73.5	4.7	40.4	40.2	30.0	.24	2.43	437	087	.4905
328 d	5	6	77.1	4.3	45.0	35.6	25.3	.18	2.43	422	087	.4936
329 b	6	6	81.4	3.4	57.8	22.8	12.5	.06	2.42	382	086	.5024
330 d	6	6	80.4	3.5	45 4.8	15.8	5.5	.18	2.43	360	087	.5071
331 d	3	4	91.5	3.1	10.0	10.0	39 59.6	.18	2.43	342	087	.5111
332 d	6	6	93.3	2.5	16.3	4.3	53.9	.06	2.42	324	086	.5150

TABLE 21.—*Absorption lines in the infra-red solar spectrum, from bolographs of the rock-salt prismatic spectrum*—Cont'd.

Designation.	Number of observations by—		Ordinates.		Differences of deviation.	Whole deviation.	Whole deviation reduced to 60° prism.	Probable error. Cols. VI, VII, VIII.	$n=\frac{\sin\frac{1}{2}(\theta+a)}{\sin\frac{1}{2}a}$		Probable error of n.	Wavelength.
	F.E.F.	C.G.A.	Mean of reduced.	Average deviation from mean.								
I.	II.	III.	IV.	V.	VI.	VII.	VIII.	IX.	X.	XI.	XII.	XIII.
					° ′ ″	° ′ ″	° ′ ″	″				μ
333 d	3	0	97.7	0.6	26.5	0.1	49.7	0.12	2.42	311	086	1.5180
334 d ?	5	5	06.0	3.0	27.8	32 52.8	42.4	.30	2.44	288	087	.5230
335 d	4	4	97.9	1.6	35.3	45.3	34.8	.18	2.43	264	087	.5281
336 d ?	4	4	98.1	1.4	42.1	38.5	28.0	.48	2.47	243	082	.5330
337 d	5	6	98.3	1.2	47.5	33.1	22.6	.24	2.43	226	087	.5366
338 d ?	4	3	99.9	0.5	56.2	24.4	13.9	.48	2.47	199	088	.5426
339 d ?	5	6	98.4	1.6	50.5	21.1	10.5	.24	2.43	189	087	.5451
340 d	6	6	96.8	0.6	46 7.3	13.3	2.7	.24	2.43	164	087	.5506
341 d	5	5	97.2	0.7	16.6	4.0	38 53.4	.36	2.45	135	087	.5572
342 d ?	4	0	93.6	0.0	25.8	31 54.8	44.1	.24	2.43	106	087	.5640
343 d	6	6	93.2	1.1	28.8	51.8	41.1	.24	2.43	096	087	.5662
344 d ?	5	2	92.1	0.3	33.3	47.3	36.6	.24	2.43	082	087	.5696
345 c	6	6	91.2	1.1	35.9	44.7	34.0	.18	2.43	074	087	.5715
346 d ? \}b	2 \}	6	92.8	0.0	43.5	37.1	26.4	.54	2.48	050	089	.5771
347 d ?	6 \}		89.6	0.0	45.5	35.1	24.4	.24	2.43	044	087	.5786
348 c	6	6	92.8	1.8	52.3	26.3	17.5	.12	2.43	022	087	.5836
349 d ?	4	0			57.7	22.9	12.1	.06	2.42	006	086	.5879
350 d	5	5	93.5	2.5	47 2.2	18.4	7.6	.18	2.43	1.527991	087	.5910
351 c	6	6	91.0	1.9	8.4	12.2	1.4	.12	2.43	972	087	.5960
352 c \}b	6 \}	6	91.1	1.9	17.3	3.3	37 52.4	.06	2.42	944	086	.6028
353 d	6	5	93.5	2.7	27.6	30 53.0	42.1	.30	2.44	912	087	.6110
354 d ?	4	3	95.1	1.1	31.6	49.0	38.1	.06	2.42	899	086	.6141
355 d ?	6	2	96.7	1.4	34.8	45.8	34.9	.12	2.43	889	087	.6166
356 d	5	0	96.0	2.4	40.5	40.1	29.2	.18	2.43	871	087	.6211
357 d	6	5	94.0	1.9	46.6	34.0	23.0	.18	2.43	852	087	.6262
358 d	6	5	92.4	1.5	53.2	27.4	16.4	.24	2.44	832	087	.6316
359 d	3	3	91.0	2.6	57.0	23.6	12.6	.42	2.46	820	088	.6348
360 d ?	6	5	91.9	2.5	48 2.4	18.2	7.2	.12	2.43	803	087	.6391
361 d ?	6	5	91.8	2.3	8.3	12.3	1.2	.42	2.46	784	088	.6441
362 d ?	6	5	92.9	2.5	18.1	2.5	36 51.4	.30	2.44	753	087	.6522
363 c	6	6	90.9	2.1	21.8	29 58.8	47.7	.12	2.43	742	087	.6554
364 d	4	0	95.0	0.0	30.9	49.7	38.6	.54	2.48	713	089	.6633
365 c	6	6	90.0	2.4	36.4	44.2	33.0	.18	2.43	696	087	.6682
366 c	6	6	89.0	2.5	44.4	36.2	25.0	.12	2.43	671	087	.6752
367 d	6	6	88.0	3.4	52.2	27.4	16.2	.30	2.44	643	087	.6832
368 d	6	6	88.0	2.5	49 1.7	18.9	7.6	.24	2.44	616	087	.6910
369 d	4	0	89.4	3.4	5.5	15.1	3.8	.54	2.48	604	089	.6942
370 d	6	5	85.9	2.7	13.8	6.8	35 55.5	.18	2.43	578	087	.7021
371 d	6	6	85.2	2.5	20.1	0.5	49.2	.30	2.44	559	087	.7080
372 d ?	2	0			27.1	28 53.5	41.1	.90	2.59	536	092	.7147
373 c	6	6	81.9	2.8	29.5	51.1	39.7	.18	2.43	529	087	.7170
374 d	4	0	82.6	4.6	36.7	43.9	32.5	.30	2.44	507	087	.7241
375 d	6	6	78.4	2.8	41.5	39.1	27.7	.18	2.43	491	087	.7287
376 d	6 \}	6	80.4	5.3	45.1	35.5	24.1	.18	2.43	480	087	.7324
377 d	6 \}		79.3	4.2	49.1	31.5	20.0	.30	2.44	467	087	.7364
378 o	3 \}	6	75.1	0.0	53.5	27.1	15.6	.18	2.43	453	087	.7410
379 d	6 \}		72.8	0.0	56.1	24.5	13.0	.48	2.47	445	088	.7438
380 d ?\}c	3 \}	6	84.3	0.0	50 2.5	18.1	6.6	.18	2.43	425	087	.7505
381 d ?	6 \}		77.0	5.8	5.1	15.5	4.0	.30	2.44	417	087	.7533
382 d ?	2	0	80.0	0.0	10.0	10.6	34 59.1	.48	2.47	402	087	.7587
383 c	6	6	69.9	3.4	14.6	6.0	54.4	.18	2.43	387	087	.7637
384 d	5	4	64.8	2.7	23.3	27 57.3	45.7	.30	2.44	360	087	.7650
385 d	6	0	58.9	3.2	28.5	52.1	40.5	.06	2.43	344	087	.7709
386 c	6	6	54.2	4.7	34.7	45.9	34.8	.18	2.43	324	087	.7762
387 d	6	6	43.7	5.2	44.0	36.6	24.9	.12	2.43	295	087	.7854

TABLE 21.—*Absorption lines in the infra-red solar spectrum, from bolographs of the rock-salt prismatic spectrum*—Cont'd.

Designation.	Number of observations by—		Ordinates.		Differences of deviation.	Whole deviation.	Whole deviation reduced to 60° prism.	Probable error. Cols. VI, VII, VIII.	$n = \frac{\sin \frac{1}{2}(\theta+a)}{\sin \frac{1}{2}a}$		Probable error of n.	Wave-length.
	F. E. F.	C. G. A.	Mean of reduced.	Average deviation from mean.								
I.	II.	III.	IV.	V.	VI.	VII.	VIII.	IX.	X.	XI.	XII.	XIII.
					° ′ ″	° ′ ″	° ′ ″	″	″			μ
388 d	5	6	28.3	4.6		52.9	27.7	16.0	0.18	2.43	267	087	1.7945
389 d	6	6	19.0	4.3	51 2.8	17.8	6.1	.12	2.43	235	087	.8049
390 c	6	6	13.5	3.0		9.6	11.0	33 59.2	.24	2.44	214	087	.8115
391 d	4	6	8.1	1.8		19.9	0.7	48.9	.12	2.43	182	087	.8229
392 d ?	2	} 5	{ 12.1	0.0		23.8	26 56.8	45.0	.00	2.43	170	087	.8262
393 d ?	3		8.6	3.6		28.3	52.3	40.5	.54	2.49	156	089	.8315
394 d	4	5	6.6	2.1		35.0	45.6	33.7	.06	2.43	135	087	.8380
395 d	3	5	5.2	1.7		45.8	34.8	22.9	.42	2.46	101	088	.8500
396 d	5	5	4.9	1.5		55.0	25.6	13.7	.24	2.44	072	087	.8599
397 d	6	6	4.6	1.8	52 0.9	19.7	7.7	.24	2.44	053	087	.8661
398 d	5	6	5.5	2.1		7.6	13.0	1.6	.18	2.43	032	087	.8735
399 d	5	6	5.4	1.7		17.3	3.3	32 51.3	.18	2.43	001	087	.8840
400 d	6 Ω	5	4.2	1.8		28.5	25 52.1	40.0	.30	2.45	1.526066	087	.8962
401 d	6	6	5.1	1.8		35.8	44.8	32.7	.12	2.43	943	087	.9040
402 d	5 a	5	8.0	2.7		43.1	37.5	25.4	.12	2.43	921	087	.9122
403 d ?	2	0	8.7	0.0		44.8	35.8	23.7	.30	2.45	915	087	.9146
404 d	6	6	9.3	2.9		50.8	29.8	17.6	.24	2.44	896	087	.9211
405 c	6	6	12.6	3.8		58.9	21.7	9.5	.12	2.43	871	087	.9299
406 d	6	6	16 6	3.9	53 6.1	14.5	2.3	.12	2.43	848	087	.9388
407 c	6	6	18.2	3.3		15.0	5.6	31 53.3	.06	2.43	820	087	.9500
408 d	6	4	23.2	5.5		20.7	24 59.9	47.6	.12	2.43	802	087	.9546
409 d	6	6	34.2	3.9		28.9	51.7	39.4	.06	2.43	777	087	.9640
410 d	6	6	41.6	3.5		35.6	45.0	32.7	.24	2.44	756	087	.9720
411 d ω₁	6	6	43.3	3.4		44.9	35.7	23.3	.30	2.45	726	087	.9825
412 b	6	6	20.8	2.4		57.7	22.9	10.5	.18	2.44	686	087	.9980
413 c b	6	6	23.8	2.8	54 5.7	14.9	2.4	.18	2.44	661	087	2.0070
414 d	6	6	47.2	3.4		20.7	23 59.9	30 47.4	.36	2.46	614	088	.0250
415 d	6	6	48.6	3.9		30.3	50.3	37.7	.30	2.45	583	087	.0364
416 b ω₂	6	0	37.8	3.0		40.7	39.9	27.3	.18	2.44	551	087	.0490
417 c b	6	6	39.2	3.8		49.9	30.7	18.1	.12	2.43	522	087	.0604
418 d	5	6	43.6	4.0	0 54 56.8	39 23 23.8	39 30 11.1	0.18	2.44	1.526500	.0000087	2.0686
412 b	7	7	14.2	3.6								
413 d ω₁	5	6	18.7	5.5								
414 d	4	0										
415 d	5	5	39.8	6.9								
416 b ω₂	7	7	29.8	5.4								
417 c b	6	7	31.3	5.5								
418 d	0	4	26.1	7.0								2.0686
419 d	6	6	41.4	6.9	0 55 7.2	39 23 13.4	39 30 0.7	0.42	2.47	1.526467	.0000088	.0818
420 d	5	7	41.8	6.3		14.1	6.5	29 53.8	.48	2.48	446	089	.0900
421 d ?		6 {	43.0	6.2		22.4	22 58.2	45.4	.48	2.48	420	089	.1006
422 d ?	}	6	40.0	5.5		29.3	51.3	38.5	.30	2.45	398	087	.1090
423 c	4	7	42.3	6.4		34.1	46.5	33.7	.18	2.44	383	087	.1150
424 d	3	6	42.5	7.1		41.8	38.8	26.0	.24	2.44	350	087	.1246
425 d	6	6	40.5	7.2	56 3.8		16.8	3.9	.18	2.44	289	087	.1519
426 c	6	7	38.4	5.8		13.9	6.7	28 53.7	.18	2.44	258	087	.1645
427 d	5	6	36.8	5.2		25.2	21 55.4	42.4	.24	2.44	222	087	.1790
428 d	5	6	38.1	7.0		30.0	50.6	37.6	.18	2.44	207	087	.1855
429 d	4	5	35.7	5.8		40 1	40.5	27.4	.12	2.44	175	087	.1974
430 d	6	3	34.6	8.2		43.3	37.3	24.2	.24	2.45	166	087	.2018
431 d ?	4	4	39.1	4.2		54.0	26.6	13.5	.30	2.45	132	087	.2151
432 d ?	4	5	38.7	5.3	57 8.3		12.3	27 59.1	.36	2.46	087	088	.2338
433 d	5	4	32.5	7.0		35.9	20 44.7	31.4	.30	2.45	000	087	.2595
434 d	5	6	32.3	6.0		43.9	36.7	23.4	.36	2.46	1.525975	068	.2796

TABLE 21.—*Absorption lines in the infra-red solar spectrum, from bolographs of the rock-salt prismatic spectrum*—Cont'd.

Designation.	Number of observations by—		Ordinates.		Differences of deviation.	Whole deviation.	Whole deviation reduced to 60° prism.	Probable error. Cols. VI, VII, VIII.	$n = \frac{\sin \frac{1}{2}(\theta + a)}{\sin \frac{1}{2} a}$	Probable error of n.	Wavelength.
	F. E. F.	C. G. A.	Mean of reduced.	Average deviation from mean.							
I.	II.	III.	IV.	V.	VI.	VII.	VIII.	IX.	X.	XI.	XII.	XIII.
					° ′ ″	° ′ ″	° ′ ″	″	″			μ
435 d?	4	3	29.4	6.7		52.0	28.6	15.2	0.30	2.45	949	087	2.2905
436 d?	3	3	27.9	0.6		55.6	25.0	11.6	.06	2.44	938	087	.2950
437 d	4	6	30.3	5.5		58.3	21.3	7.9	.12	2.44	926	087	.2999
438 c	6	7	29.3	6.4	58 13.0	7.0	26 53.6	.18	2.44	881	087	.3180
439 d	4	5	30.9	7.1		20.2	0.4	46.9	.30	2.45	860	087	.3268
440 d	6	7	25.9	5.5		29.5	19 51.1	37.6	.30	2.45	831	087	.3390
441 d₁	5	5	26.3	5.4		36.7	43.0	30.4	.12	2.44	808	087	.3488
442 d₂	4	4	22.5	4.2		39.8	40.8	27.3	.12	2.44	799	087	.3520
443 d	6	7	21.9	5.4		52.6	28.0	14.4	.24	2.45	759	087	.3690
444 d₁	5	} 7	20.9	5.5	{	59.6	21.0	7.4	.54	2.50	736	089	.3780
445 d₂	6					59 4.4	16.2	2.6	.54	2.50	721	089	.3850
446 d	5	6	19.4	5.5		18.4	2.2	25 48.5	.18	2.44	678	087	.4030
447 c	6	7	16.5	5.3		24.8	18 55.8	42.1	.18	2.44	657	087	.4115
448 d	5	6	16.8	5.0		39.1	41.5	27.7	.30	2.46	612	088	.4309
449 c	7	7	12.7	4.4		49.2	31.4	17.6	.12	2.44	581	087	.4446
450 d?	0	3	14.0	4.2	1 0 1.4	19.2	5.3	.54	2.50	542	089	.4610
451 d?	6	0				4.1	16.5	2.6	.54	2.50	534	089	.4650
452 d?	0	6	10.4	4.4		7.4	13.2	24 59.3	.30	2.46	523	088	.4690
453 d?	5	0				9.5	11.1	57.2	.60	2.51	516	090	.4718
454 d	3	6	7.5	3.7		14.5	6.1	52.2	.24	2.45	501	087	.4782
455 c	6	7	5.9	2.8		19.1	1.5	47.6	.18	2.44	486	087	.4850
456 d	4	6	5.3	2.6		30.5	17 50.1	36.1	.12	2.44	451	087	.5000
457 d	4	5	4.6	3.2		38.0	42.6	28.6	.12	2.44	427	087	.5100
458 d?	4	0				42.5	38.1	24.1	.72	2.54	413	091	.5162
459 a X	{ 6	7	1.1	1.2		44.8	35.8	21.8	1.32	2.78	405	099	.5200
	6	7	0.5	0.2	4 39.4	13 41.2	20 26.2	1.26	2.72	1.524667	097	.8412
460 d	0	4	0.9	0.5		35.8	44.8	29.9	0.12	2.45	678	087	.8390
461 d	5	6	1.2	0.6		49.7	30.9	15.9	.18	2.45	634	087	.8585
462 d	4	7	1.8	1.0		59.5	21.1	6.1	.18	2.45	603	087	.8716
463 c	6	7	2.2	1.0	5 14.3	6.3	19 51.2	.24	2.46	557	088	.8900
464 d	6	7	3.0	1.4		24.9	12 55.7	40.6	.18	2.45	523	087	.9075
465 d	4	2	4.5	1.5		30.3	50.3	35.1	.24	2.46	506	088	.9158
466 c	7	7	3.4	1.5		44.0	36.6	21.4	.24	2.46	463	088	.9342
467 d	6	6	3.0	1.6		50.0	30.6	15.4	.30	2.46	445	088	.9430
468 d?	2	3	2.9	1.5	6 0.9	19.7	4.4	.36	2.47	410	088	.9586
469 b	7	7	2.8	1.1		5.0	15.6	0.3	.18	2.45	397	087	.9640
470 d	3	3	3.1	0.7		11.0	9.6	18 54.3	.12	2.45	.378	087	.9728
471 d?	2	4	5.4	1.8		20.7	11 59.9	44.5	.30	2.40	348	088	.9861
472 c	6	7	4.6	1.7		26.3	54.3	38.9	.30	2.40	330	088	.9942
473 d	3	0				30.0	50.6	35.2	.00	2.45	319	087	3.0000
474 c	6	7	4.2	1.5		38.0	42.6	27.2	.24	2.46	293	088	.0108
475 c	6	7	3.2	1.3		52.7	27.9	12.4	.42	2.48	247	089	.0316
476 c x₁	7	7	2.0	1.1	7 0.8	19.8	4.3	.24	2.46	228	088	.0432
477 c a	7	7	3.4	1.3		12.9	7.7	17 52.1	.24	2.46	183	088	.0605
478 d	6	4	3.6	1.4		26.6	10 54.0	38.4	.30	2.47	140	088	.0798
479 c	3	7	2.8	1.2		30.0	50.6	35.0	.12	2.45	129	087	.0856
480 d?	5	2	2.2	0.2		35.4	45.2	29.6	.24	2.46	112	088	.0920
481 c	6	7	2.9	1.1		40.4	40.2	24.5	.24	2.46	096	088	.0998
482 c	6	7	4.6	1.6		53.9	26.7	11.0	.24	2.46	054	088	.1190
483 c	7	7	4.1	1.7	8 8.4	12.2	16 56.4	.24	2.46	008	088	.1391
484 d	6	5	5.6	1.2		16.3	4.3	48.5	.30	2.47	1.523983	088	.1500
485 d	3	} 6	7.4	2.4	{	27.3	9 53.3	37.4	.96	2.63	948	094	.1659
486 d	6					31.2	49.4	23.5	.42	2.48	936	089	.1710
487 d	4	4	5.1	1.5		37.1	43.5	27.6	.42	2.48	921	089	.1796
488 d	5	5	2.6	1.0		49.1	31.5	15.6	.36	2.48	880	089	.1955

TABLE 21.—*Absorption lines in the infra-red solar spectrum, from bolographs of the rock-salt prismatic spectrum*—Cont'd.

Designation.	Number of observations by—		Ordinates.		Differences of deviation.	Whole deviation.	Whole deviation reduced to 60° prism.	Probable error. Cols. VI, VII, VIII.	$n=\frac{\sin \frac{1}{2}(\theta+a)}{\sin \frac{1}{2}a}$	Probable error of n.	Wavelength.
	F. E. F.	C. G. A.	Mean of reduced.	Average deviation from mean.							
I.	II.	III.	IV.	V.	VI.	VII.	VIII.	IX.	X.	XI.	XII.	XIII.
					° ′ ″	″	° ′ ″	″				μ
489 c	5	5	2.3	0.8	54.3	26.3	10.2	6.12	2.46	963	088	3.2032
490 d	4	0			58.4	22.2	6.2	.90	2.48	851	089	.2092
491 d	6	6	2.3	0.8	9 4.9	15.7	15 59.7	.24	2.46	830	088	.2180
492 d ?	0	3	3.0	1.1	8.2	12.4	56.4	.00	2.45	820	087	.2228
493 c X₂	7	7	3.5	1.1	21.6	8 59.0	42.9	.24	2.46	778	088	.2410
494 b	7	7	3.2	1.1	36.9	43.7	27.0	.30	2.47	729	088	.2620
495 d a	5	5	3.9	1.5	42.6	38.0	21.9	.66	2.54	711	091	.2702
496 d ?	2	7	3.1	1.0	10 0.2	20.4	4.2	.06	2.45	656	087	.2942
497 c	6				3.4	17.2	1.0	.42	2.49	646	089	.2990
498 b	7	7	2.7	0.8	15.1	5.5	14 49.2	.24	2.46	609	088	.3150
499 c	6	7	4.6	1.4	36.8	7 43 8	27.4	.42	2.49	540	083	.3436
500 c	6	7	4.5	1.2	46.2	34.4′	18.0	.36	2.48	511	080	.3550
501 d	5	3	5.1	1.6	56.6	24.0	7.6	.24	2.46	478	088	.3699
502 d	5	6	6.1	1.7	11 3.3	17.3	0.8	.12	2.46	457	088	.3795
503 b	7	7	6.2	1.7	22.0	6 57.7	13 41.2	.24	2.46	395	088	.4051
504 c	2	2	7.8	1.8	45.2	35.4	18.8	.60	2.53	325	090	.4350
505 c	2	3	9.3	2.0	59.0	21.6	4.9	.60	2.53	281	090	.4538
506 d		3	9.6	2.2	12 13.4	7.2	12 50.5	.66	2.54	236	091	.4724
507 d	5	2	8.4	1.8	21.7	5 48.9	32.1	.12	2.46	178	088	.4960
508 d	2	3	8.5	1.9	13 1.5	19.1	2.2	.18	2.46	084	088	.5346
509 c	5	6	7.7	1.5	6.5	14.1	11 57.2	.36	2.48	068	089	.5406
501 d ?	2	3	6.1	1.3	15.5	5.1	48.1	.84	2.60	040	099	.5512
511 c	2	5	6.6	2.0	33.0	4 47.6	30.5	.42	2.49	1.522985	089	.5704
512 c	2	3	8.0	1.8	42.6	38.0	20.9	.36	2.48	953	089	.5865
513 c	3	5	6.8	2.3	56.7	21.9	4.7	.30	2.48	904	089	.6072
514 c	4	3	8.4	1.9	14 15.9	4.7	10 47.5	.60	2.53	849	090	.6300
515 d ?	3	0			28.1	3 52.5	35.2	.36	2.48	811	089	.6450
516 d ?	2	0			39.1	41.5	24.2	.48	2.50	776	089	.6590
517 b	7	7	5.8	1.6	49.2	31.4	14.0	.36	2.48	744	089	.6716
518 d ?	3	6	6.5	2.0	15 1.8	18.8	1.4	.36	2.48	705	089	.6880
519 d c	3	6	6.9	1.4	21.0	2 59.6	9 42.1	.18	2.47	644	088	.7138
520 d ?	3	3	7.6	1.6	25.0	55.6	38.1	.12	2.46	632	088	.7178
521 c	6	6	6.2	2.1	37.9	42.7	25.2	.48	2.51	591	090	.7336
522 c	6	6	5.7	1.8	59.0	21.6	4.0	.36	2.49	524	089	.7596
523 d ?	2	2	8.2	0.9	16 7.9	12.7	8 55.0	.36	2.49	496	089	.7708
524 c	4	3	6.7	1.5	22.2	1 58.4	40.7	.24	2.47	451	088	.7886
525 c	5	5	5.9	1.4	41.4	39.2	21.4	.18	2.47	396	088	.8123
526 d	5	6	5.3	1.6	59.5	21.1	3.2	.30	2.48	333	080	.8345
527 d ?	2	0			17 9.2	11.4	7 58.5	.30	2.48	303	089	.8485
528 d	2	5	4.8	1.8	16.6	4.0	46.1	.36	2.49	279	088	.8555
529 d ?	2	0			23.2	0 57.4	39.4	.00	2.45	258	088	.8642
530 d ?	2	0			30.0	50.6	32.6	.30	2.48	237	089	.8722
531 c	7	6	4.0	0.9	34.2	46.4	28.4	.48	2.51	224	090	.8771
532 d ?	2	0			41.3	39.3	21.3	.06	2.46	201	088	.8858
533 d ?	2	0			45.3	35.3	17.3	.30	2.48	189	089	.8909
534 c	7	7	4.1	1.1	18 11.4	9.2	6 51.1	.66	2.55	106	091	.9222
535 d	4	5	4.0	1.5	19 32.7	38 58 47.9	5 29.4	6.24	6.71	1.521850	240	4.0211
536 d	4	3	4.0	0.8	20 21.6	57 59.0	4 40.3	7.98	8.35	695	298	.0610
537 a Y	6	6	0.3	0.0	21 42.8	56 37.8	3 18.8	2.58	3.57	439	127	.1790
	6	6	0.3	0.0	27 2.4	51 18.2	38 57 58.0	9.00	9.23	1.520429	333	.4988
538 c	6	6	0.5	0.2	26 57.4	23.2	58 3.0	1.44	2.87	444	102	.4939
539 c	0	5	1.0	0.2	27 38.6	50 42.0	57 21.6	1.08	2.71	314	097	.5368
540 d	4	6	1.4	0.4	28 24.3	49 56.3	56 35.7	0.54	2.54	170	091	.5845
541 d	4	3	1.9	0.1	53.6	27.0	6 3	1.26	2.73	077	099	.6156
542 c	7	7	1.3	0.4	29 17.2	3.4	55 42.6	0.36	2.51	002	090	.6402

TABLE 21.—*Absorption lines in the infra-red solar spectrum, from bolographs of the rock-salt prismatic spectrum*—Cont'd.

Designation.	Number of observations by—		Ordinates.		Differences of deviation.	Whole deviation.	Whole deviation reduced to 60° prism.	Probable error. Cols. VI, VII, VIII.	$n=\frac{\sin\frac{1}{2}(\theta+a)}{\sin\frac{1}{2}a}$	Probable error of n.	Wavelength.
	F. E. F.	C. G. A.	Mean of reduced.	Average deviation from mean.							
I.	II.	III.	IV.	V.	VI.	VII.	VIII.	IX.	X.	XII.	XIII.
					° ′ ″	° ′ ″	° ′ ″	″			μ
543 c	5	4	1.0	0.4	49.1	48 31.5	10.6	1.02	2.69	1.519901	096	4.6729
544 d		5	1.0	0.4	57.7	22.9	2.0	0.72	2.59	874	092	.6829
545 d c	5	3	0.9	0.4	30 10.3	10.3	54 49.3	.72	2.59	834	092	.6960
546 d	0	4	1.0	0.4	17.2	3.4	42.4	.30	2.50	813	089	.7026
547 d	4	6	1.1	0.4	22.8	47 57.6	36.8	.54	2.54	795	091	.7087
548 d	0	3	1.4	0.2	41.5	39.1	18.0	1.50	2.90	736	104	.7284
549 c	7	7	0.9	0.4	50.2	21.4	0.2	0.66	2.57	645	092	.7582
550 c	7	7	0.8	0.3	31 27.1	46 53.5	53 32.2	.60	2.56	591	091	.7759
551 d	0	5	1.1	0.1	37.4	43.2	21.9	.48	2.53	558	090	.7866
552 d c	3	6	0.7	0.2	54.2	26.4	5.6	.48	2.53	506	090	.8029
553 d c	3	3	1.0	0.0	32 3.3	17.3	52 55.9	1.14	2.74	477	098	.8129
554 d ?	0	4	1.0	0.4	12.9	7.7	46.2	.48	2.53	447	090	.8228
555 b	7	6	0.7	0.3	46.9	45 33.7	12.1	0.72	2.59	339	092	.8590
556 d	7	5	0.7	0.3	33 24.1	44 56.5	51 34.8	.48	2.54	321	091	.8678
557 b	7	6	0.6	0.3	44.2	36.4	14.6	.48	2.54	158	091	.9185
558 d ?	0	6	0.6	0.3	48.9	31.7	9.9	.36	2.52	142	090	.9239
559 d ?	0	6	0.8	0.3	59.3	21.3	50 59.4	.36	2.53	110	090	.9345
560 c	7	6	0.7	0.3	34 8.6	12.0	50.1	.42	2.53	080	090	.9440
561 d	0	6	0.7	0.3	18.9	6.7	44.8	.54	2.55	063	091	.9495
562 d ?	0	6	0.8	0.2	29.4	43 51.2	29.2	.36	2.52	014	090	.9654
563 c	5	5	0.5	0.2	35.0	45.6	23.6	1.02	2.69	1.518997	096	.9714
564 d ?		6	0.4	0.2	47.2	33.4	11.3	1.36	2.52	953	090	.9836
565 d	6	6	0.4	0.2	52.7	27.9	5.2	.42	2.53	941	090	.9888
566 d b		6	0.5	0.2	58.6	21.8	49 59.7	.18	2.50	921	089	.9955
567 d		4	0.7	0.2	35 9.3	11.3	49.1	.48	2.54	888	091	5.0063
568 d ?	0	5	0.3	0.2	22.4	42 58.2	36.0	.42	2.53	847	090	.0195
569 d	5	3	0.3	0.2	29.9	50.7	28.5	1.08	2.72	823	097	.0263
570 d ?	0	4	0.6	0.2	45.5	35.1	12.8	0.30	2.51	774	090	.0422
571 c	5	6	0.5	0.2	56.3	24.2	2.0	.48	2.54	739	091	.0537
572 d	4	5	0.4	0.2	36 21.9	41 58.7	48 36.3	.66	2.56	658	092	.0784
573 c	4	6	0.2	0.1	56.0	24.6	2.0	.30	2.51	550	090	.1000
574 d	4	2	0.3	0.0	37 23.8	40 56.8	47 34.1	.30	2.52	462	090	.1321
575 d	4	4	0.1	0.1	59.8	20.8	46 58.0	.66	2.58	348	092	.1725
576 d	4				38 34.5	39 46.1	23.1	.78	2.62	238	094	.2050
577 c	4	4	0.2	0.1	39 27.5	38 53.1	45 29.9	.66	2.59	070	092	.2537
578 d	4	3	0.0	0.0	40 29.0	37 51.6	44 28.2	1.74	3.05	1.517875	109	.3084
579 d	0	6	0.1	0.1	1 41 2.7	36 37 17.9	36 43 54.4	0.30	2.52	1.517768	.0000090	5.3286

S. Doc. 20——16

ANNALS OF ASTROPHYSICAL OBSERVATORY.

TABLE 22.—*Absorption lines in the infra-red solar spectrum, from bolographs of the glass prismatic spectrum.*

Designation. (Explained on p. 132.)	Number of observations.		Ordinates, May 16, 1898. Bolograph II.		Differences of deviation.	Whole deviation.	Probable error of VI.	$n = \dfrac{\sin \frac{1}{2}(\theta + a)}{\sin \frac{1}{2} a}$	Wave-length.
I.	II.	III.	IV.	V.	VI.	VII.	VIII.	IX.	
	F. E. F.	C. G. A.			° ′ ″	° ′ ″	″		μ
1 d	4	2	....	....	−0 0 4.2	46 46 15.2	0.30	1.605255	0.7601
2 c	4	4	3.3	28.6	±0 0 0.0	11.0	.24	243	.7604
3 d	3	0	....	....	+0 0 3.4	7.6	.12	233	.7606
4 d	4	3	3.4	29.5	5.8	5.2	.24	226	.7607
5 d b	4	0	....	....	8.9	2.1	.18	217	.7610
6 d	3	3	3.9	33.8	12.5	45 58.5	.30	206	.7612
7 d	3	0	....	....	14.9	56.1	.24	200	.7614
8 d	2	0	....	....	18.5	52.5	.00	189	.7616
9 d	3	2	4.2	36.4	30.2	40.8	.12	155	.7625
10 d	4	4	4.2	36.4	34.0	37.0	.06	144	.7628
11 d	3	2	....	....	37.0	34.0	.12	136	.7630
12 d	4	4	4.4	38.1	41.8	29.2	.12	122	.7634
13 d	4	4	4.6	39.9	46.3	24.7	.18	109	.7636
14 c	4	4	4.9	42.5	52.6	18.4	.06	091	.7641
15 c	4	4	5.1	44.2	59.3	11.7	.06	071	.7646
16 c b	4	4	5.3	46.0	1 5.5	5.5	.18	054	.7650
17 c	4	4	5.5	47.7	13.0	44 58.0	.18	032	.7656
18 c	4	4	5.8	50.3	19.1	51.9	.06	014	.7661
19 c	4	4	6.0	52.0	26.8	44.2	.18	1.604992	.7666
20 d	2	0	....	....	32.4	38.6	.00	976	.7670
21 c	2	4	6.3	54.6	40.9	30.1	.18	951	.7677
22 c	4	4	6.4	55.5	43.2	27.8	.24	944	.7679
23 c	4	4	6.4	55.5	50.5	20.5	.12	923	.7684
24 c	4	3	6.5	56.4	58.1	12.9	.06	901	.7690
25 d	4	3	7.0	60.7	3 13.9	42 57.1	.54	682	.7748
26 d ?	0	4	7.8	67.6	6 1.1	40 9.9	.30	198	.7880
27 d ?	0	3	8.0	69.4	50.0	39 21.0	.24	057	.7919
28 d ?	0	4	8.0	69.4	7 4.1	6.9	.30	016	.7931
29 d ?	0	4	8.1	70.2	22.2	38 48.8	.36	1.603963	.7945
30 c	3	3	8.1	70.2	40.7	30.3	.36	910	.7960
31 c	0	4	8.3	72.0	8 27.8	37 43.2	.24	773	.8000
32 d ?	0	4	8.3	72.0	36.9	34 1	.54	747	.8008
33 d ?	0	3	8.2	71.1	9 1.5	9.5	.72	676	.8028
34 d	0	4	8.4	72.8	10.1	0.9	.30	651	.8036
35 c	4	4	8.4	72.8	21.9	36 49 1	.30	617	.8046
36 d ?	0	3	8.4	72.8	10 32.1	35 38.9	.30	413	.8106
37 d ?	0	3	8.2	71.1	37.1	33.9	.36	399	.8110
38 d	0	4	8.1	70.2	41.7	29.3	.84	386	.8115
39 d	0	4	7.9	68.5	51.7	19.3	.42	357	.8123
40 d ??	0	2	....	....	59.7	11.3	.00	334	.8130
41 c	4	4	7.6	65.9	11 4.7	6.3	.18	319	.8135
42 c	4	4	7.6	65.9	13.2	34 57.8	.24	294	.8141
43 c	4	4	7.3	63.3	21.3	49.7	.24	271	.8150
44 c	4	4	7.4	64.2	27.6	43.4	.12	253	.8155
45 c b	4	4	7.0	60.7	36.9	34.1	.18	226	.8162
46 c	4	4	7.4	64.2	45.0	26.0	.06	202	.8170
47 c	4	4	7.2	62.4	54.3	16.7	.12	175	.8178
48 d	4	0	....	....	12 0.2	10.8	.24	158	.8184
49 d	4	3	....	....	6.6	4.4	.30	140	.8189
50 d	3	2	....	....	12.0	33 59.0	.06	124	.8194
51 d	4	3	....	....	16.6	54.4	.12	111	.8198
52 d	3	0	....	....	20.6	50.4	.12	099	.8201
53 d	2	2	....	....	26.4	44.6	.30	082	.8206
54 d ?	2	0	....	....	28.7	42.3	.72	076	.8209
55 d b	4	3	....	....	33.7	37.3	.18	061	.8213
56 d	3	3	....	....	42.2	28.8	.06	037	.8221
57 d ?	3	0	....	....	46.7	24.3	.30	024	.8225
58 b	4	4	7.0	60.7	50.6	20.4	.06	012	.8228

TABLE 22.—*Absorption lines in the infra-red solar spectrum, from bolographs of the glass prismatic spectrum*—Cont'd.

Designation.	Number of observations.		Ordinates. May 16, 1898. Bolograph II.		Differences of deviation.	Whole deviation.	Probable error of VI.	$n = \dfrac{\sin \frac{1}{2}(\theta + a)}{\sin \frac{1}{2} a}$	Wavelength.
I.	II.	III.	IV.	V.	VI.	VII.	VIII.	IX.	
	F.E.F.	C.G.A.			° ′ ″	° ′ ″	″		μ
59 d	4	0	....	....	59.4	11.6	0.06	1.602987	0.8236
60 c	4	4	8.5	73.7	13 6.5	4.5	.06	966	.8243
61 d	4	3	8.6	74.6	15.9	32 55.1	.42	939	.8251
62 c	4	4	8.4	72.8	21.6	49.4	.12	922	.8256
63 c	4	4	8.8	76.3	29.1	41.9	.12	900	.8263
64 d } c	3	} 4	8.3	72.0	{ 39.6	31.4	.00	870	.8273
65 d	4				42.5	28.5	.36	862	.8275
66 d	4	4	8.5	73.7	47.9	23.1	.18	846	.8280
67 c	4	4	8.5	73.7	56.0	15.0	.06	823	.8288
68 c	4	4	8.7	75.4	14 2.7	8.3	.12	803	.8294
69 d } d	3	} 4	8.8	76.3	{ 9.3	1.7	.06	784	.8300
70 d	4				18.5	31 57.5	.12	772	.8304
71 d f	2	3	9.0	78.0	20.6	50.4	.18	751	.8310
72 c	4	4	8.9	77.2	29.7	41.3	.12	725	.8319
73 d	2	4	9.1	78.9	40.2	30.8	.18	694	.8329
74 c	4	4	9.1	78.9	44.5	26.5	.24	682	.8333
75 d f	0	4	10.6	91.9	17 7.2	29 3.8	.42	268	.8467
76 d f	0	3	10.6	91.9	17.9	28 53.1	.30	237	.8477
77 d } b	4	2	10.6	91.9	28.6	42.4	.42	206	.8488
78 b	4	4	10.1	87.6	36.0	35.0	.06	185	.8495
79 c	4	4	10.6	91.9	52.3	18.7	.18	137	.8511
80 d f	3	2	10.8	93.6	18 0.4	10.6	.24	114	.8519
81 d	4	4	10.6	91.9	6.6	4.4	.30	096	.8525
82 d } b	4	0	....	....	12.7	27 58.3	.24	078	.8531
83 b	4	4	9.4	81.5	20.1	50.9	.12	056	.8538
84 d f	4	0	....	....	28.8	42.2	.60	031	.8546
85 d	4	3	10.8	93.6	34.4	36.6	.18	015	.8552
86 d f	0	3	11.1	96.2	49.9	21.1	.36	1.601970	.8567
87 d f	0	4	11.2	97.1	19 0.7	10.3	.48	939	.8578
88 d f	0	3	11.2	97.1	15.4	26 55.6	.72	896	.8592
89 d f	0	4	11.4	98.8	51.6	19.4	.36	791	.8630
90 d	4	4	11.3	98.0	20 7.2	3.8	.18	746	.8645
91 d f } b	3	0	....	....	12.6	25 58.4	.00	730	.8650
92 b	4	4	10.5	91.0	19.0	52.0	.06	711	.8657
93 d	3	0	....	....	25.9	45.1	.00	691	.8664
94 d	4	4	11.1	96.2	31.0	40.0	.18	676	.8670
95 d	3	3	11.8	102.3	36.8	34.2	.06	659	.8676
96 d	4	4	11.8	102.3	44.1	26.9	.18	638	.8683
97 d f f	0	3	12.0	104.0	21 8.1	2.9	.54	569	.8708
98 d	4	4	12.0	104.0	45.3	24 25.7	.24	461	.8748
99 c	4	4	12.2	105.8	56.6	14.4	.30	428	.8760
100 d	4	4	12.3	106.6	22 5.4	5.6	.12	403	.8769
101 d f	0	3	12.3	106.6	15.1	23 55.9	.48	375	.8780
102 d	4	4	12.3	106.6	23.2	47.8	.12	351	.8788
103 d f	3	3	....	....	31.9	39.1	.66	325	.8798
104 c	4	4	12.1	104.9	38.0	33.0	.12	308	.8804
105 d f	2	2	12.5	108.4	45.8	25.2	.60	285	.8812
106 d f	3	4	12.4	107.5	52.9	18.1	.18	265	.8820
107 d	2	3	12.4	107.5	23 4.9	6.1	.06	230	.8833
108 d	3	3	12.5	108.4	12.6	22 58.4	.36	207	.8842
109 c	4	3	12.3	106.6	27.2	43.8	.24	165	.8858
110 c	4	4	12.4	107.5	31.9	39.1	.36	151	.8863
111 d	2	4	12.5	108.4	44.5	26.5	.48	114	.8877
112 d f	2	4	12.5	108.4	24 1.7	9.3	.24	065	.8897
113 d f	3	3	12.3	106.6	12.1	21 58.9	.42	035	.8909
114 d	2	3	12.1	104.9	17.6	53.4	.06	019	.8915
115 d f	2	2	12.0	104.0	22.6	48.4	.30	004	.8920
116 c	4	4	11.6	100.6	29.7	41.3	.18	1.600963	.8982

TABLE 22.—*Absorption lines in the infra-red solar spectrum, from bolographs of the glass prismatic spectrum*—Cont'd.

Designation.	Number of observations.		Ordinates. May 16, 1898. Bolograph II.		Differences of deviation.	Whole deviation.	Probable error of VI.	$n=\dfrac{\sin\frac{1}{2}(\theta+'a)}{\sin\frac{1}{2}a}$	Wavelength.
I.	II.	III.	IV.	V.	VI.	VII.	VIII.	IX.	
	F.E.F.	C.G.A.			° ′ ″	° ′ ″	″		μ
117 d	4	4	11.1	96.2	43.0	28.0	0.36	945	0.8944
118 d	4	3	10.8	93.6	46.6	24.4	.60	934	.8948
119 c ⎱b	4	4	10.6	91.9	51.8	19.2	.66	919	.8954
120 c	4	4	10.1	87.6	25 1.1	9.9	.72	892	.8965
121 c	4	4	10.1	87.6	8.8	2.2	.60	870	.8974
122 d?	4	3			16.5	20 54.5	.54	848	.8982
123 b	4	4	9.1	78.9	23.1	47.9	.48	828	.8990
124 c	4	3	9.7	84.1	33.5	37.5	.18	798	.9003
125 c ⎱b	4	4	10.0	86.7	38.3	32.7	.30	784	.9008
126 c	4	4	10.2	88.4	47.1	23.9	.24	759	.9019
127 c	4	4	10.4	90.2	54.4	16.6	.48	738	.9028
128 d?	4	2	11.0	95.4	59.9	11.1	.54	722	.9035
129 c	4	4	11.3	98.0	26 9.4	1.6	.36	694	.9046
130 c	4	4	11.0	95.4	18.3	19 52.7	.18	668	.9056
131 c	4	4	10.3	89.3	26.7	44.3	.24	644	.9066
132 b	4	4	9.9	85.8	34.4	36.6	.30	621	.9075
133 c ⎱b	4	4	9.8	85.0	42.4	28.6	.36	598	.9085
134 c	3	3	10.0	86.7	48.5	22.5	.18	580	.9092
135 d	3	1	10.2	88.4	53.8	17.7	.06	566	.9098
136 d	4	4	10.1	87.6	59.5	11.5	.72	542	.9105
137 c	4	4	10.0	86.7	27 5.8	5.2	.84	520	.9114
138 b	4	4	10.0	86.7	15.7	18 55.3	.60	501	.9126
139 b	4	4	9.4	81.5	26.9	44.1	.18	468	.9139
140 d?	2	0			29.3	41.7	.54	461	.9141
141 d	4	4	10.4	90.2	38.2	32.8	.18	436	.9152
142 b	4	4	9.6	83.2	45.9	25.1	.48	413	.9162
143 d	4	0			55.1	15.9	.72	387	.9172
144 b ⎱Group.	4	4	10.0	86.7	28 4.0	7.0	.54	361	.9183
145 d	4	4	11.0	95.4	11.6	17 59.4	.48	338	.9192
146 c	4	4	11.1	96.2	19.7	51.3	.24	315	.9201
147 d	4	4	11.0	95.4	30.8	40.2	.30	283	.9215
148 c	4	4	10.7	92.8	37.4	33.6	.18	263	.9223
149 d ⎱Distant group.	4	3	10.6	91.9	46.9	24.1	.60	236	.9235
150 d	4	4	10.3	89.3	55.3	15.7	.24	211	.9244
151 c	4	.4	10.4	90.2	29 2.0	9.0	.18	192	.9253
152 d	4	2			9.8	1.2	.36	169	.9262
153 c	4	4	10.1	87.6	14.5	16 56.5	.36	156	.9268
154 d?	4	0			24.7	46.3	.48	126	.9281
155 d	4	4	8.4	72.8	31.1	39.9	.12	107	.9289
156 d	4	4	8.1	70.2	36.5	34.5	.18	092	.9296
157 d	4	4	7.3	63.3	44.1	26.9	.12	070	.9305
158 d	4	4	7.2	62.4	48.7	22.3	.24	056	.9310
159 d	4	4	6.6	57.2	56.4	14.6	.30	034	.9320
160 d	4	4	5.9	51.2	30 2.8	8.2	.48	015	.7327
161 c ρ ⎱b	4	4	5.7	49.4	7.9	3.1	.30	000	.9334
162 c	4	4	5.3	46.0	16.4	15 54.6	.18	1.599976	.9345
163 c	4	4	5.6	48.6	21.2	49.8	.30	962	.9351
164 b	4	4	5.7	49.4	31.7	39.3	.06	931	.9363
165 b	4	4	5.7	49.4	42.7	28.3	.18	900	.9376
166 c	4	4	6.3	54.6	50.9	20.1	.36	876	.9387
167 d	4	1			56.6	14.4	.30	859	.9394
168 c	4	4	7.0	60.7	31 9.2	1.8	.24	822	.9410
169 c	4	4	7.0	60.7	16.1	14 54.9	.18	802	.9420
170 d	4	4	6.8	59.0	22.6	48.4	.42	783	.9427
171 b	4	4	6.4	55.5	29.5	41.5	.06	763	.9435
172 b	4	4	5.9	51.2	39.5	31.5	.00	734	.9450
173 b	4	4	6.4	55.5	53.2	17 8	.18	694	.9466
174 d?	2	0			55.4	15.6	.12	688	.9469

TABLE 22.—*Absorption lines in the infra-red solar spectrum, from bolographs of the glass prismatic spectrum*—Cont'd.

Designation.	Number of observations.		Ordinates, May 16, 1896. Bolograph II.		Differences of deviation.	Whole deviation.	Probable error of VI.	$n = \frac{\sin \frac{1}{2}(\theta + a)}{\sin \frac{1}{2} a}$	Wavelength.
I.	II.	III.	IV.	V.	VI.	VII.	VIII.	IX.	
	F.E.F.	C.G.A.			° ′ ″	° ′ ″	″		μ
175 c	4	4	7.1	61.6	32 1.8	9.2	0.24	669	0.9476
176 c	4	4	6.9	59.8	10.7	0.3	.36	643	.9488
177 b	4	4	6.4	55.5	22.3	13 47.7	.18	607	.9504
178 b a	3	3			40.8	30.2	.06	556	.9527
179 d?	3	0			47.5	22.5	.18	536	.9535
180 d σ	3	2			55.1	15.9	.30	514	.9546
181 b	4	4	5.4	46.8	33 0.3	10.7	.24	499	.9552
182 b	4	4	5.6	48.6	8.5	2.5	.24	475	.9563
183 b b	4	4	5.8	50.3	16.1	12 54.9	.18	453	.9574
184 c	4	4	6.2	53.8	28.0	43.0	.06	418	.9589
185 b	4	4	5.8	50.3	35.2	35.8	.06	397	.9598
186 d	4	0			41.1	29.9	.12	380	.9606
187 c	4	4	7.1	61.6	47.2	23.8	.06	363	.9614
188 b	4	4	7.2	62.4	59.0	12.0	.18	328	.9630
189 b	4	4	7.6	65.9	34 11.1	11 59.9	.06	293	.9646
190 c	4	4	8.1	70.2	17.0	54.0	.12	276	.9654
191 d	4	3	8.4	72.8	24.8	46.2	.18	253	.9665
192 c	4	4	8.3	72.0	29.2	41.8	.12	240	.9670
193 d?	3	0			34.8	36.2	.24	224	.9678
194 c	3	3	9.8	85.0	43.4	27.6	.00	199	.9690
195 c	3	3	10.2	88.4	52.3	18.7	.12	173	.9701
196 c	3	3	10.2	88.4	58.3	12.7	.06	156	.9709
197 c	4	4	10.2	88.4	35 8.6	2.4	.36	125	.9724
198 c	4	4	9.9	85.8	17.0	10 54.0	.18	101	.9735
199 c	4	4	9.3	80.6	24.8	46.2	.12	078	.9746
200 d b r	4	3	10.0	86.7	31.1	39.9	.30	060	.9755
201 c	4	4	9.7	84.1	39.0	32.0	.18	037	.9765
202 d	4	3	10.2	88.4	43.5	27.5	.06	024	.9771
203 d	4	3	10.8	93.6	53.4	17.6	.24	1.598995	.9784
204 c	4	4	10.6	91.9	58.3	12.7	.18	981	.9791
205 c	4	4	10.7	92.8	36 5.6	5.4	.12	950	.9800
206 c	4	4	11.8	96.0	12.4	9 58.6	.18	940	.9811
207 c	3	4	11.8	102.3	22.8	48.2	.30	910	.9825
208 d?	2	0			24.9	46.1	.00	903	.9827
209 c	4	4	11.8	102.3	29.7	41.3	.12	889	.9834
210 d??	2	0			32.7	38.3	.66	881	.9838
211 d	4	4	12.5	108.4	37.2	33.8	.48	868	.9845
212 c	4	4	12.4	107.5	44.4	26.6	.18	847	.9854
213 d??	3	0			51.9	19.1	.54	825	.9865
214 c	4	4	12.8	111.0	37 3.0	8.0	.24	793	.9880
215 c	3	3	13.1	113.6	17.1	8 53.9	.06	751	.9899
216 d	2	2	13.1	113.6	23.6	47.4	.24	732	.9909
217 c	3	4	13.3	115.3	42.5	28.5	.30	677	.9935
218 d	0	4	13.5	117.0	53.9	17.1	.24	644	.9950
219 d??	0	2	13.5	117.0	38 31.7	7 39.3	.36	534	1.0005
220 d	4	4	13.4	116.2	49.1	21.9	.30	483	1.0030
221 d	4	3	13.3	115.3	39 0.2	10.8	.24	451	1.0046
222 c	4	4	13.1	113.6	8.0	3.0	.06	428	1.0056
223 d	3	2	13.2	114.4	11.4	6 59.6	.24	418	1.0062
224 c	4	4	13.6	117.9	18.6	52.4	.06	397	1.0072
225 d	2	3	13.6	117.9	29.0	42.0	.18	367	1.0087
226 d	0	4	13.7	118.8	38.1	32.9	.84	341	1.0100
227 d?	0	3	13.9	120.5	58.4	12.6	.60	282	1.0130
228 d??	0	3	13.8	119.6	40 14.9	5 56.1	.60	233	1.0155
229 c	0	3	13.9	120.5	46.0	25.0	.24	143	1.0202
230 c	3	3	13.9	120.5	41 2.9	8.1	.12	093	1.0226
231 d??	0	3	13.9	120.5	42 51.5	3 19.5	.00	1.597777	1.0391
232 c	2	3	13.7	118.8	43 39.0	2 32.0	.30	638	1.0465

TABLE 22.—*Absorption lines in the infra-red solar spectrum, from bolographs of the glass prismatic spectrum*—Cont'd

Designation.	Number of observations.		Ordinates, May 16, 1898. Bolograph II.		Differences of deviation.	Whole deviation.	Probable error of VI.	$n=\frac{\sin \frac{1}{2}(\theta+e)}{\sin \frac{1}{2}a.}$	Wave-length.
I.	II.	III.	IV.	V.	VI.	VII.	VIII.	IX.	
	F.E.F.	C.G.A.			° ′ ″	° ′ ″	″		μ
233 d ? ?	0	2	....	....	44 43.0	1 28.0	0.00		451	1.0566
234 c	3	3	13.3	115.3	45 1.0	10.0	.48		399	1.0594
235 c	3	3	13.6	117.9	12.1	0 58.9	.12		367	1.0612
236 d ? ?	0	2	....	....	24.5	46.5	.00		331	1.0632
237 d	2	2	....	....	42.4	28.6	.12		278	1.0661
238 c	3	3	13.4	116.2	49.4	21.6	.06		258	1.0671
239 d	3	3	13.4	116.2	46 0.1	10.9	.42		227	1.0682
240 c	3	3	13.1	113.6	5.9	5.1	.06		210	1.0698
241 d	3	3	13.6	117.9	17.5	59 53.5	.18		176	1.0716
242 c	2	3	13.5	117.0	28.7	42.3	.12		143	1.0734
243 c	3	3	13.5	117.0	42.4	28.6	.48		103	1.0756
244 d	3	2	13.8	118.8	49.3	21.7	.36		083	1.0768
245 c	3	3	13.5	117.0	55.6	15.4	.54		065	1.0778
246 c	3	3	13.5	117.0	47 4.6	6.4	.12		039	1.0792
247 d	3	3	13.4	116.2	13.6	58 57.4	.30		012	1.0807
248 c	3	3	13.3	115.3	20.4	50.6	.12	1.506993	960	1.0818
249 c	3	3	13.2	114.4	31.5	39.5	.24		960	1.0836
250 d	3	1	....	....	36.9	34.1	.30		944	1.0845
251 d	3	3	13.3	115.3	42.5	28.5	.12		929	1.0855
252 d	3	3	12.7	110.1	50.1	20.9	.12		906	1.0866
253 c	3	3	13.0	112.7	55.3	15.7	.06		891	1.0875
254 c	3	3	12.9	111.8	48 5.5	5.5	.12		861	1.0892
255 c	3	3	12.8	111.0	14.8	57 56.2	.18		834	1.0908
256 d ?	3	0	....	....	24.3	46.7	.36		805	1.0924
257 c	3	3	12.0	104.0	28.6	42.4	.06		794	1.0930
258 c	3	3	11.9	103.2	37.0	34.0	.06		769	1.0945
259 c	3	3	11.7	101.4	47.5	23.5	.06		739	1.0963
260 c	4	4	11.3	98.0	49 1.1	9.9	.18		699	1.0986
261 d	2	0	....	....	7.0	4.0	.54		682	1.0996
262 c	4	4	11.0	95.4	9.7	1.3	.36		674	1.1000
263 d	4	4	10.9	94.9	17.2	56 53.8	.60		652	1.1012
264 c	4	4	10.2	88.4	24.6	46.4	.36		631	1.1025
265 d	4	4	10.2	88.4	30.7	40.3	.42		612	1.1036
266 c	4	4	9.6	83.2	38.0	33.0	.24		591	1.1048
267 b	4	4	9.0	78.0	47.6	23.4	.18		563	1.1064
268 c	4	4	8.9	77.2	50.8	11.2	.06		528	1.1085
269 d	4	4	9.0	78.0	50 8.2	2.8	.30		503	1.1099
270 d ?	2	0	....	....	15.7	55 55.3	.00		481	1.1112
271 c	4	4	7.9	68.5	17.2	53.8	.24		477	1.1115
272 b	4	4	6.1	52.9	29.3	41.7	.12		442	1.1135
273 d	4	4	5.2	45.1	41.7	29.3	.18		405	1.1156
274 b	4	4	4.4	38.1	48.1	22.9	.06		386	1.1168
275 c	4	4	4.3	37.3	51 2.6	8.2	.30		343	1.1193
276 d	4	4	4.4	38.1	8.2	2.8	.18		328	1.1203
277 b } b	.4	4	3.8	32.9	19.8	54 51.2	.24		294	1.1223
278 c	4	4	4.4	38.1	28.6	42.4	.18		268	1.1238
279 b	4	4	3.9	33.8	39.1	31.9	.24		238	1.1256
280 c	4	4	4.0	34.7	51.3	19.7	.18		202	1.1276
281 c	4	4	4.2	36.4	59.4	11.6	.24		178	1.1290
282 d ? φ	3	0	....	....	52 1.6	9.4	.12		172	1.1294
283 d ?	3	0	....	....	12.6	53 58.4	.06		140	1.1314
284 d ?	3	0	....	....	17.4	53.6	.36		126	1.1322
285 d ?	3	0	....	....	25.7	45.3	.30		101	1.1336
286 b	4	4	3.9	33.8	28.8	42.2	.30		092	1.1341
287 d	4	3			36.4	34.6	.24		070	1.1355
288 c } c	3	4	6.6	52.0	48.8	22.2	.30		034	1.1376
289 d }	3	2	6.2	53.8	51.4	19.6	.12		026	1.1381
290 d } a	4	3	5.8	50.3	53 1.3	9.7	.30	1.595097		1.1397

TABLE 22.—*Absorption lines in the infra-red solar spectrum, from bolographs of the glass prismatic spectrum*—Cont'd.

Designation.	Number of observations.		Ordinates. May 16, 1898. Bolograph II.		Differences of deviation.	Whole deviation.	Probable error of VI.	$n=\frac{\sin\frac{1}{2}(\theta+a)}{\sin\frac{1}{2}a}$	Wavelength.
I.	II.	III.	IV.	V.	VI.	VII.	VIII.	IX.	
	F.E.F.	C.G.A.			° ′ ″	° ′ ″	″		μ
291 c	4	4	5.6	48.6	6.3	4.7	0.42	983	1.1406
292 c	3	4	4.5	39.0	23.8	52 47.2	.24	932	1.1436
293 d ? b	4	0			26.1	44.9	.12	925	1.1440
294 d	4	4	4.7	40.7	30.3	40.7	.06	913	1.1448
295 c	4	4	4.8	41.6	37.2	33.8	.12	893	1.1460
296 d	4	4	6.0	52.0	50.4	20.6	.30	854	1.1483
297 b	4	4	5.8	50.3	54.3	16.7	.24	843	1.1490
298 d	4	4	6.5	56.4	54 1.8	9.2	.24	822	1.1503
299 b	3	4	6.3	54.6	8.1	2.9	.18	802	1.1514
300 b	4	4	6.6	57.2	16.9	51 54.1	.12	777	1.1530
301 c	4	4	7.8	67.6	25.6	45.4	.18	751	1.1545
302 c	4	4	8.4	72.8	36.9	34.1	.24	718	1.1565
303 c	4	3	8.7	75.4	45.0	26.0	.30	695	1.1580
304 c	4	4	8.8	76.3	49.8	21.2	.18	681	1.1587
305 d	2	3	9.8	85.0	57.4	13.6	.30	658	1.1601
306 d ?	3	0			59.3	11.7	.12	653	1.1604
307 d	4	3	10.2	88.4	55 1.0	10.0	.30	648	1.1607
308 c	4	4	10.5	91.0	10.9	0.1	.54	619	1.1625
309 b	4	4	10.1	87.6	19.1	50 51.9	.42	595	1.1639
310 d	4	4	10.8	93.6	26.6	44.4	.18	573	1.1651
311 d	3	4	11.2	97.1	34.9	36.1	.12	549	1.1666
312 d ?	3	2	11.5	99.7	41.4	29.6	.30	529	1.1678
313 c	4	4	11.5	99.7	46.9	24.1	.30	514	1.1688
314 d	4	4	11.7	101.4	54.6	16.4	.42	499	1.1702
315 d	4	3	11.3	98.0	56 0.0	11.0	.30	475	1.1712
316 b	4	4	11.7	101.4	4.7	6.3	.24	461	1.1720
317 c	4	4	11.7	101.4	12.7	49 58.3	.12	438	1.1736
318 b	4	4	11.1	96.2	25.1	45.9	.12	402	1.1758
319 c	4	4	11.8	102.3	36.2	34.8	.24	369	1.1778
320 b	4	4	10.7	92.8	48.7	22.3	.12	333	1.1801
321 d ?	4	0			53.9	17.1	.54	318	1.1810
322 c	4	4	11.4	98.8	57 0.4	10.6	.18	290	1.1822
323 d	2	3	11.9	103.2	12.0	48 59.0	.12	265	1.1844
324 d ?	2	0			17.6	53.4	.48	249	1.1853
325 b	2	3	11.1	96.2	0 57 25.4	45 48 45.6	1.02	1.595 226	1.1868

Designation.	Number of observations.		Ordinates. May 17, 1898. Bolograph V.		Differences of deviation.	Whole deviation.	Probable error.	$n=\frac{\sin\frac{1}{2}(\theta+a)}{\sin\frac{1}{2}a}$	Wavelengths.
I.	II.	III.	IV.	V.	VI.	VII.	VIII.	IX.	
	F.E.F.	C.G.A.			° ′ ″	° ′ ″	″		μ
318 b	3	3	13.0	108.3					
319 c	4	3	13.5	112.5					
320 b	4	2	12.6	105.0					
321	0	0							
322 c	4	3	13.0	108.3					
323 d ?	0	2	13.4	111.6					
324	0	0							
325 b	4	3	12.9	107.5	0 57 25.4	45 48 45.6	0.42	1.595226	1.1868
326 d	4	3	13.9	115.8	59.1	31.9	.12	186	1.1894
327 c	4	3	13.6	113.3	55.4	15.6	.36	138	1.1924
328 c	4	3	13.5	112.5	58 4.0	7.0	.18	112	1.1940
329 d ?	2	1	13.6	113.3	9.7	1.3	.06	096	1.1950
330 d	4	3	13.2	110.0	16.3	47 54.7	.06	077	1.1963

TABLE 22.—*Absorption lines in the infra-red solar spectrum, from bolographs of the glass prismatic spectrum*—Cont'd.

Designation.	Number of observations.		Ordinates, May 17, 1898. Bolograph V.		Differences of deviation.	Whole deviation.	Probable error.	$n = \frac{\sin \frac{1}{2}(\theta + a)}{\sin \frac{1}{2} a}$	Wavelengths.
I.	II.	III.	IV.	V.	VI.	VII.	VIII.	IX.	
	F.E.F.	C.G.A.			° ′ ″	° ′ ″	″		μ
331 b	4	3	12.9	107.5	22.7	48.3	0.24	058	1.1975
332 d	0	2	13.5	112.5	32.0	39.0	.24	031	1.1992
333 c	4	3	13.3	110.8	37.3	33.7	.18	015	1.2002
334 b	4	3	12.5	104.1	47.4	23.6	.18	1.594986	1.2021
335 d	4	4	13.3	110.8	54.5	16.5	.24	965	1.2036
336 c	4	3	13.5	112.5	59 5.4	5.6	.48	933	1.2055
337 c	4	4	13.4	111.6	13.1	46 57.9	.06	910	1.2070
338 c	4	4	13.5	112.5	26.6	44.4	.18	871	1.2095
339 d ?	0	2	13.7	114.1	32.6	38.4	.72	853	1.2106
340 c	4	4	13.8	115.0	41.4	29.6	.24	828	1.2124
341 d	4	1	....	....	55.0	16.0	.24	788	1.2150
342 d	3	1	....	....	1 0 4.4	6.6	.24	761	1.2167
343 c	4	3	14.3	119.1	17.9	45 53.1	.24	721	1.2194
344 c ?	2	0	....	....	27.7	43.3	.00	692	1.2212
345 c	4	4	14.3	119.1	46.6	24.4	.30	637	1.2249
346 d	4	4	14.6	121.6	57.9	13.1	.18	604	1.2271
347 d	3	3	14.5	120.8	1 10.1	0.9	.06	568	1.2294
348 d		2	14.6	121.6	16.3	44 54.7	.84	551	1.2306
349 d ?	3	2	13.9	115.6	3 25.6	42 45.4	.06	173	1.2565
350 d ?	3	0	....	....	33.1	37.9	.78	150	1.2581
351 d ?	3	0	....	....	36.0	35.0	.84	142	1.2587
352 d	4	3	13.4	111.6	41.0	30.0	.36	127	1.2596
353 d	4	3	13.1	109.1	49.8	21.2	.30	101	1.2615
354 d	4	3	13.0	108.3	57.0	14.0	.54	080	1.2630
355 d ?	2	0	....	....	4 6.0	5.0	.00	054	1.2650
356 b	4	4	12.1	100.8	10.9	0.1	.12	040	1.2658
357 d	4	0	....	....	18.8	41 52.2	.36	016	1.2675
358 d ?	2	0	....	....	30.8	40.2	.24	1.593961	1.2700
359 c	2	2	14.2	118.3	39.1	31.9	.30	957	1.2717
360 d	2	2	14.3	119.1	47.3	23.7	.24	933	1.2734
361 d	2	2	14.3	119.1	54.4	16.6	.84	912	1.2750
362 d	3	2	14.1	117.5	5 5.3	5.7	.00	880	1.2771
363 d	3	2	14.0	116.6	9.8	1.2	.54	867	1.2782
364 b	4	4	13.7	114.1	15.3	40 55.7	.24	851	1.2794
365 d	3	4	14.6	121.6	26.6	44.4	.24	813	1.2816
366 d	2	4	14.7	122.5	52.1	18.9	.30	743	1.2870
367 d	1	3	14.6	121.6	6 14.6	39 56.4	.42	677	1.2918
368 d	1	3	14.4	120.0	22.4	48.6	.24	654	1.2935
369 c	3	4	14.1	117.5	28.6	42.4	.48	636	1.2948
370 d	4	0	....	....	43.8	27.2	.54	592	1.2980
371 c	3	3	....	....	48.0	23.0	.66	579	1.2990
372 d	3	0	....	....	57.3	13.7	.36	552	1.3010
373 b	4	4	13.3	110.8	7 1.1	9.9	.30	541	1.3018
374 c	4	4	13.5	112.5	12.9	38 58.1	.24	506	1.3044
375 d	3	0	....	....	18.4	52.6	.42	490	1.3055
376 c	4	4	13.2	110.0	27.8	43.2	.12	463	1.3076
377 d	3	2	12.8	106.6	41.6	29.4	.60	426	1.3105
378 c	4	4	12.7	105.8	45.2	25.8	.48	412	1.3115
379 d	4	4	12.9	107.5	51.3	19.7	.24	394	1.3126
380 d	4	3	12.8	106.6	8 0.6	10.4	.24	367	1.3147
381 d	4	4	12.4	103.3	8.1	2.9	.24	344	1.3164
382 c	4	4	12.2	101.6	13.5	37 57.5	.12	329	1.3175
383 d	4	0	....	....	20.9	50.1	.48	307	1.3192
384 b	4	4	11.9	99.1	28.7	42.3	.30	284	1.3209
385 d	3	1	....	....	34.5	36.5	.24	267	1.3222
386 d	4	4	12.0	100.0	45.2	25.8	.30	236	1.3246
387 d	3	3	11.3	94.1	51.0	20.0	.48	219	1.3260

TABLE 22.—*Absorption lines in the infra-red solar spectrum, from bolographs of the glass prismatic spectrum*—Cont'd.

Designation.	Number of observations.		Ordinates. May 17, 1896. Bolograph V.		Differences of deviation.	Whole deviation.	Probable error.	$n = \dfrac{\sin \frac{1}{2}(\theta + a)}{\sin \frac{1}{2} a}$	Wavelengths.
I.	II.	III.	IV.	V.	VI.	VII.	VIII.	IX.	
	F.E.F.	C.G.A.			° ′ ″	° ′ ″	″		μ
388 c	4	4	10.5	87.5	57.4	13.6	0.30	200	1.3274
389 b	4	4	9.8	81.6	9 8.1	2.9	.18	169	1.3299
390 d } b	3	0			12.5	36 58.5	.24	156	1.3308
391 c	4	4	10.4	86.6	17.6	53.4	.12	141	1.3320
392 c	4	4	10.0	83.3	27.8	43.2	.18	111	1.3342
393 c	4	4	9.6	80.0	34.7	36.3	.06	091	1.3357
394 d	4	0			42.0	29.0	.24	070	1.3374
395 c	4	4	8.8	73.3	47.7	23.3	.12	053	1.3385
396 c	4	4	6.3	52.5	10 5.3	5.7	.06	001	1.3425
397 d	4	0			14.7	35 56.3	.18	1.592974	1.3445
398 c	4	4	3.8	31.7	23.1	47.9	.18	949	1.3464
399 c	1	3	2.9	24.2	36.1	34.9	.18	911	1.3493
400 d †	1	2	1.9	15.8	56.7	14.3	.06	851	1.3538
401 c †	1	1	2.0	16.7	12 31.3	33 39.7	.00	574	1.3750
402 c †	2	2	2.0	16.7	13 44.0	32 27.0	.60	360	1.3915
403 d †	1	2	2.0	16.7	53.4	17.6	1.26	333	1.3938
404 c	4	3	2.1	17.5	14 15.9	31 55.1	0.18	267	1.3990
405 c	4	3	2.3	19.2	36.0	35.0	.30	208	1.4036
406 c	2	3	2.5	20.8	46.7	24.3	.30	176	1.4060
407 c	3	3	2.4	20.0	58.4	12.6	.42	142	1.4087
408 d	2	3	3.0	25.0	15 4.5	6.5	.48	124	1.4101
409 d	4	0			7.0	4.0	.36	117	1.4107
410 b	4	4	2.8	23.3	19.0	30 52.0	.06	082	1.4135
411 c	4	4	3.5	29.2	33.2	37.8	.06	040	1.4167
412 c	4	4	3.1	25.8	40.5	30.5	.06	018	1.4184
413 d	4	0			46.9	24.1	.48	000	1.4199
414 b	4	4	4.0	33.3	58.1	12.9	.12	1.591967	1.4225
415 c	4	4	4.5	37.5	16 8.0	3.0	.18	938	1.4248
416 b	4	4	4.1	34.2	17.3	29 53.2	.06	909	1.4270
417 d	3	2	5.5	45.8	26.2	44.3	.18	884	1.4289
418 b	4	4	4.1	34.2	36.4	34.6	.42	855	1.4313
419 c	4	4	4.8	40.0	46.3	24.7	.18	825	1.4335
420 d †	2	3	4.8	40.0	55.8	15.2	.30	797	1.4356
421 c	4	4	4.7	39.2	17 0.1	10.9	.12	785	1.4367
422 c	4	4	5.4	45.0	9.6	1.4	.12	757	1.4388
423 c	4	4	5.5	45.8	17.5	28 53.5	.12	734	1.4406
424 d	4	4	5.8	48.3	22.6	48.4	.72	719	1.4418
425 b	4	4	5.4	45.0	35.2	35.8	.12	682	1.4447
426 d	3	0			49.6	21.4	.00	640	1.4459
427 b	4	4	5.4	45.0	56.4	14.6	.00	620	1.4495
428 d	4	3	7.1	59.1	18 3.0	8.0	.36	600	1.4510
429 b	4	4	7.5	62.5	16.8	27 54.2	.06	560	1.4542
430 d	4	4	8.3	69.1	26.6	44.4	.12	531	1.4564
431 d †	3	0			32.2	38.8	.24	515	1.4576
432 c } b	4	4	7.4	61.6	40.4	30.6	.12	490	1.4596
433 c	4	4			51.4	19.6	.12	458	1.4621
434 b	4	4	6.5	54.1	19 5.4	5.6	.42	415	1.4653
435 d † } b	4	0			22.5	26 48.5	.48	367	1.4692
436 b	4	4	6.1	50.8	24.4	46.6	.24	361	1.4696
437 d	4	3	7.0	58.3	31.2	39.8	.24	341	1.4712
438 d †	3	0			36.1	34.9	.36	327	1.4724
439 c	4	4	7.3	60.8	48.1	22.9	.06	292	1.4751
440 c } b	4	4	7.3	60.8	52.9	18.1	.12	278	1.4762
441 b	4	4	7.5	62.5	20 4.4	6.6	.06	244	1.4789
442 d	4	4	8.4	70.0	11.2	25 59.8	.06	223	1.4804
443 c } b	4	4	8.7	72.5	24.3	46.7	.12	185	1.4835
444 b	4	4	8.4	70.0	33.4	37.6	.06	159	1.4855
445 d	4	3	9.9	82.5	44.7	26.3	.18	125	1.4880
446 b	4	4	10.0	83.3	54.6	16.4	.12	096	1.4904

TABLE 22.—*Absorption lines in the infra-red solar spectrum, from bolographs of the glass prismatic spectrum*—Cont'd.

Designation.	Number of observations.		Ordinates. May 17, 1896. Bolograph V.		Differences of deviation.	Whole deviation.	Probable error.	$n = \dfrac{\sin \frac{1}{2}(\theta + a)}{\sin \frac{1}{2} a}$	Wave-lengths.
I.	II.	III.	IV.	V.	VI.	VII.	VIII.	IX.	
	F.E.F.	C.G.A.			° ′ ″	° ′ ″	″		μ
447 d?	3	1	....	....	21 0.4	10.6	0.24	079	1.4916
448 c	4	4	10.5	87.5	9.4	1.6	.06	053	1.4938
449 c	4	4	10.9	90.8	21.6	24 49.4	.18	017	1.4966
450 d?	4	0	....	....	27.8	43.2	.48	1.590999	1.4980
451 d	4	3	10.7	89.1	38.0	33.0	.12	969	1.5003
452 b	4	4	10.2	85.0	45.6	25.4	.12	947	1.5021
453 d?	0	2	11.4	95.0	22 3.1	7.9	.42	895	1.5061
454 b	4	4	11.2	93.3	7.1	3.9	.06	883	1.5070
455 c	3	3	11.6	96.6	27.2	23 43.6	.18	824	1.5117
456 c	4	4	11.6	96.6	42.0	29.0	.24	781	1.5151
457 d	4	4	11.7	97.5	50.4	20.6	.54	756	1.5170
458 d	4	4	11.9	99.1	23 1.6	9.4	.78	723	1.5196
459 c	3	4	11.9	99.1	17.9	22 53.1	.54	675	1.5234
460 c	4	4	11.8	98.3	37.9	33.1	.30	616	1.3279
461 c	4	4	11.8	98.3	57.0	14.0	.24	560	1.5324
462 c?	0	3	11.6	96.6	25 6.2	4.8	.30	523	1.5482
463 d?	0	3	11.6	96.6	22.4	20 48.6	.30	310	1.5519
464 d	4	3	10.9	90.8	26 26.9	19 44.1	.12	120	1.5668
465 d?	4	0	....	....	39.5	31.5	.48	063	1.5696
466 c	4	3	10.7	89.1	48.5	22.5	.18	057	1.5716
467 c	4	3	....	....	58.1	12.9	.18	029	1.5740
468 d	3	3	10.9	90.8	27 21.5	18 49.5	.42	1.588959	1.5793
469 b	4	4	10.6	88.3	50.3	20.7	.18	875	1.5859
470 c	4	4	10.8	90.0	28 20.0	17 51.0	.42	788	1.5928
471 d	3	3	10.8	90.0	31.5	39.5	.42	753	1.5955
472 c	4	4	10.4	86.6	42.8	28.2	.12	721	1.5981
473 d	3	3	10.5	87.5	51.5	19.5	.66	695	1.6001
474 c	4	4	10.4	86.6	29 9.2	1.8	.24	643	1.6042
475 c	4	3	10.4	86.6	18.9	16 52.1	.48	614	1.6065
476 d	4	4	10.5	87.5	30 1.9	9.1	.24	468	1.6161
477 d	4	4	10.2	85.0	51.8	15 19.2	.36	341	1.6277
478 d	4	3	9.9	82.5	31 23.8	14 47.2	.42	247	1.6350
479 d?	0	3	9.9	82.5	30.7	40.3	.42	226	1.6367
480 d?	3	2	9.8	81.6	32 14.4	13 56.6	.24	101	1.6468
481 d?	3	2	9.7	80.8	20.4	50.6	.30	080	1.6482
482 d?	3	2	9.7	80.8	35.6	35.4	.12	036	1.6516
483 c	4	4	9.4	78.3	33 12.3	12 57.7	.12	1.588924	1.6605
484 d?	3	0	....	....	24.8	45.2	.12	891	1.6631
485 d	3	4	9.4	78.3	39.7	31.3	.18	847	1.6665
486 d?	3	0	....	....	56.2	14.8	.36	798	1.6703
487 d?	3	0	....	....	34 14.1	11 56.9	.24	745	1.6744
488 c	4	3	8.8	73.3	35 24.8	10 46.2	.12	537	1.6910
489 d?	3	1	....	....	36.9	34.1	.00	502	1.6937
490 d	4	3	8.7	72.5	42.4	28.6	.12	485	1.6950
491 d	4	3	8.4	70.0	36 15.0	9 56.0	.54	389	1.7025
492 c	4	3	8.3	69.1	24.6	46.4	.06	361	1.7046
493 c	4	3	8.3	69.1	38.9	32.1	.24	319	1.7080
494 d	3	3	8.2	68.3	47.2	23.8	.06	295	1.7099
495 c	4	4	7.9	65.8	37 8.7	2.3	.18	232	1.7151
496 c	4	4	7.9	65.8	19.4	8 51.6	.12	200	1.7176
497 d	4	3	8.0	66.6	31.6	39.4	.18	164	1.7205
498 d	4	4	7.9	65.8	39.2	31.8	.12	142	1.7222
499 d	3	3	7.9	65.8	46.5	24.5	.24	120	1.7239
500 c	4	4	7.5	62.5	54.6	16.4	.18	096	1.7258
501 c	4	4	7.6	63.3	38 11.8	7 59.2	.24	046	1.7299
502 d	4	0	....	....	17.0	54.0	.12	030	1.7310
503 d	4	3	7.7	64.1	27.8	43.2	.60	1.587998	1.7335
504 c	4	4	7.5	62.5	33.3	37.7	.30	982	1.7348

TABLE 22.—*Absorption lines in the infra-red solar spectrum, from bolographs of the glass prismatic spectrum*—Cont'd.

Designation.	Number of observations.		Ordinates, May 17, 1898. Bolograph V.		Differences of deviation.	Whole deviation.	Probable error.	$n=\dfrac{\sin \frac{1}{2}(\theta + a)}{\sin \frac{1}{2} a}$	Wavelengths.
I.	II.	III.	IV.	V.	VI.	VII.	VIII.	IX.	
	F. E. F.	C. G. A.			° ′ ″	° ′ ″	″		μ
505 d	4	4	7.4	61.6	49.2	21.8	0.36	935	1.7385
506 b	4	4	7.1	59.1	56.9	14.1	.18	913	1.7404
507 d	4	4	7.6	63.3	39 8.3	2.7	.12	879	1.7430
508 c	4	4	7.3	60.8	16.7	6 54.3	.12	854	1.7450
509 c	4	4	7.0	58.3	32.4	38.6	.24	808	1.7485
510 c	4	4	7.0	58.3	40.5	30.5	.18	784	1.7505
511 d ?	3	0			57.8	13.3	.30	733	1.7545
512 c	3	2			40 2.8	8.2	.48	718	1.7555
513 b	3	4	4.9	40.8	17.3	5 53.7	.18	675	1.7590
514 d	4	4	5.7	47.5	28.5	42.5	.30	642	1.7615
515 c	4	4	5.4	45.0	35.6	35.4	.36	621	1.7631
516 c	4	4	5.4	45.0	39.2	31.8	.72	611	1.7641
517 c	4	4	5.0	41.6	53.3	17.7	.12	569	1.7674
518 c	4	4	4.8	40.0	41 1.5	9.5	.06	545	1.7698
519 c	4	4	4.5	37.5	12.4	4 58.6	.18	513	1.7718
520 c	4	4	4.4	36.7	23.4	47.6	.12	481	1.7744
521 c	4	4	3.9	32.5	39.5	31.5	.18	433	1.7782
522 d	4	4	4.3	35.8	46.5	24.5	.24	412	1.7798
523 d	4	0			54.6	16.4	.24	389	1.7818
524 d ?	2	3	3.6	30.0	42 5.1	5.9	.12	358	1.7842
525 c	4	4	3.2	26.7	14.9	3 56.1	.12	329	1.7865
526 d ?	3	0			22.2	48.8	.24	307	1.7882
527 c	4	4	3.2	26.7	29.0	42.0	.06	287	1.7897
528 b	4	4	1.9	15.8	46.0	25.0	.30	237	1.7937
529 d	4	0			53.1	17.9	.18	216	1.7951
530 d	4	1			59.1	11.9	.48	198	1.7966
531 c	4	4	1.8	15.0	43 11.4	2 59.6	.36	162	1.7995
532 d ?	4	1			22.8	48.2	.06	138	1.8022
533 d	4	4	1.4	11.7	32.0	39.0	.06	101	1.8044
534 d	4	4	1.4	11.7	43.6	32.4	.12	083	1.8058
535 c	4	4	1.1	9.2	51.6	19.4	.24	044	1.8090
536 c	2	2			44 8.9	2.1	.00	1.586998	1.8130
537 c	2				16.1	1 54.9	.00	971	1.8147
538 c ?	0	3	0.3	2.5	51 22.4	44 54 48.6	.90	1.585713	1.9143
539 c	0	3	0.3	2.5	54.5	16.5	.18	618	1.9219
540 c	0	3	0.3	2.5	52 9.1	1.9	.96	575	1.9252
541 c	0	3	0.6	5.0	22.0	53 49.0	.24	537	1.9283
542 c	3	3	0.3	2.5	35.3	35.7	.12	498	1.9314
543 c		3	0.3	2.5	45.0	26.0	.00	469	1.9336
544 c	3	3	0.5	4.2	53 9.7	1.3	.12	396	1.9395
545 c	3	3	0.7	5.8	25.0	52 46.0	.48	351	1.9430
546 b	3	3	0.5	4.2	59.1	11.9	.36	250	1.9510
547 c	3	3	0.9	7.5	54 31.5	51 39.5	.90	154	1.9586
548 b	0	1			55 10.6	0.4	.00	038	1.9677
549 b	0	1			1 55 47.3	44 50 23.7	.00	1.584990	1.9768

TABLE 23.—*Absorption lines in the infra-red solar spectrum, from bolographs.*

Designation. (Explained on p. 132.)	Wave-lengths. μ	Designation.	Wave-lengths. μ	Designation.	Wave-lengths. μ	Designation.	Wave-lengths. μ	Designation.	Wave-lengths. μ
1 d	0.7601	64 d	0.8273	128 d ?	0.9035	190 c	0.9654	253 c	1.0675
2 c	.7604	65 d ?	.8275	129 c	.9046	191 d	.9665	254 c	1.0692
3 d	.7606	66 d	.8280	130 c	.9056	192 c	.9670	255 c	1.0908
4 d	.7607	67 c	.8288	131 c	.9068	193 d ?	.9678	256 d ?	1.0924
5 d	.7610	68 c	.8294	132 b	.9075	194 c	.9690	257 c	1.0930
6 d	.7612	69 d	.8300	133 c	.9085	195 c	.9701	258 c	1.0945
7 d	.7614	70 d	.8304	134 c	.9092	196 c	.9709	259 c	1.0963
8 d	.7618	71 d ?	.8310	135 d	.9098	197 c	.9724	260 c	1.0986
9 d	.7625	72 c	.8319	136 d	.9105	198 e	.9735	261 d	1.0996
10 d	.7628	73 d	.8329	137 c	.9114	199 c b	.9746	262 c	1.1000
11 d	.7630	74 c	.8333	138 b	.9126	200 d	.9755	263 d	1.1012
12 d	.7634	75 d ?	.8467	139 b	.9139	201 c	.9765	264 c	1.1025
13 d	.7636	76 d ?	.8477	140 d ?	.9141	202 d	.9771	265 d	1.1036
14 c	.7641	77 d	.8488	141 d	.9152	203 d	.9784	266 c	1.1043
15 c	.7646	78 b	.8495	142 b	.9162	204 c	.9791	267 b	1.1064
16 c	.7650	79 c	.8511	143 d	.9172	205 c	.9800	268 c	1.1085
17 c	.7656	80 d ?	.8519	144 b	.9183	206 c	.9811	269 d	1.1090
18 c	.7661	81 d	.8525	145 d	.9192	207 c	.9825	270 d ?	1.1112
19 c	.7666	82 d	.8531	146 c	.9201	208 d ?	.9827	271 c	1.1115
20 d	.7670	83 b	.8538	147 d	.9215	209 c	.9834	272 b	1.1135
21 c	.7677	84 d ?	.8546	148 c	.9223	210 d ? ?	.9838	273 d	1.1156
22 c	.7679	85 d	.8552	149 d	.9235	211 d	.9845	274 b	1.1163
23 c	.7684	86 d ?	.8567	150 d	.9244	212 c	.9854	275 c	1.1193
24 c	.7690	87 d ?	.8578	151 c	.9253	213 d ? ?	.9865	276 d	1.1203
25 d	.7748	88 d ?	.8592	152 d	.9262	214 c	.9880	277 b b	1.1223
26 d ?	.7880	89 d ?	.8680	153 c	.9268	215 c	.9899	278 c	1.1238
27 d ?	.7919	90 d	.8645	154 d ?	.9281	216 d	.9909	279 b	1.1256
28 d ?	.7931	91 d ?	.8650	155 d	.9289	217 c	.9935	280 c	1.1276
29 d ?	.7945	92 b	.8657	156 d	.9296	218 d	.9950	281 c	1.1290
30 c	.7960	93 d	.8664	157 d	.9305	219 d ? ?	1.0005	282 d ?	1.1294
31 c ·	.8000	94 d	.8670	158 d	.9310	220 d	1.0030	283 d ?	1.1314
32 d ?	.8008	95 d	.8676	159 d	.9320	221 d	1.0046	284 d ?	1.1322
33 d ?	.8028	96 d	.8683	180 d ?	.9327	222 c	1.0056	285 d ?	1.1336
34 d	.8036	97 d ? ?	.8708	161 c	.9334	223 d	1.0062	286 b	1.1341
35 c	.8046	98 d	.8743	162 c	.9345	224 c	1.0072	287 d	1.1355
36 d ?	.8106	99 c	.8760	163 c	.9351	225 d	1.0087	288 c	1.1376
37 d ?	.8110	100 d	.8769	164 b	.9363	2.6 d	1.0100	289 d ? ?	1.1381
38 d	.8115	101 d ?	.8780	165 b	.9376	227 d ?	1.0130	290 d	1.1397
39 d	.8123	102 d	.8788	166 c	.9387	228 d ? ?	1.0155	291 c	1.1406
40 d ? ?	.8130	103 d ?	.8798	167 d	.9394	229 c	1.0202	292 c	1.1436
41 c	.8135	104 c	.8804	168 c	.9410	230 c	1.0226	293 d ?	1.1440
42 c	.8141	105 d ?	.8812	169 c	.9420	231 d ? ?	1.0391	294 d	1.1448
43 c	.8150	106 d ?	.8820	170 d	.9427	232 c	1.0465	295 c	1.1480
44 c	.8155	107 d	.8833	171 b	.9485	233 d ? ?	1.0566	296 d	1.1483
45 c	.8162	108 d	.8842	172 b	.9450	234 c	1.0594	297 b	1.1490
46 c	.8170	109 c	.8858	173 b	.9466	235 c	1.0612	298 d	1.1503
47 c	.8178	110 c	.8863	174 d ?	.9469	236 d ? ?	1.0632	299 b	1.1514
48 d	.8184	111 d	.8877	175 c	.9476	237 d	1.0661	300 b	1.1530
49 d	.8189	112 d ?	.8897	176 c	.9488	238 c	1.0671	301 c	1.1545
50 d	.8194	113 d ?	.8909	177 b	.9504	239 d	1.0688	302 c	1.1565
51 d	.8198	114 d	.8915	178 b	.9527	240 c	1.0698	303 c	1.1580
52 d	.8201	115 d ?	.8920	179 d ?	.9535	241 d	1.0716	304 c	1.1587
53 d	.8205	116 c	.8923	180 d	.9546	212 c	1.0734	305 d	1.1601
54 d ?	.8209	117 d	.8944	181 b	.9552	243 c	1.0756	306 d ?	1.1604
55 d	.8213	118 d	.8948	182 b	.9563	244 d	1.0768	307 d	1.1607
56 d	.8221	119 c b	.8954	183 d	.9574	245 c	1.0778	308 c	1.1625
57 d ?	.8225	120 c	.8965	184 c	.9589	246 c	1.0792	309 b	1.1639
58 b	.8228	121 c	.8974	185 b	.9598	247 d	1.0807	310 d	1.1651
59 d	.8236	122 d ?	.8982	186 d	.9606	248 c	1.0818	311 d	1.1656
60 c	.8243	123 b	.8990	187 c	.9614	249 d	1.0836	312 d ?	1.1678
61 d	.8251	124 c	.9003	188 b	.9630	250 d	1.1845	313 c	1.1688
62 c	.8256	125 c b	.9008	189 b	.9646	251 d	1.0895	314 d	1.1702
63 c	.8263	126 c	.9019			252 d	1.0696	315 d	1.1712
		127 c	.9028						

ANNALS OF ASTROPHYSICAL OBSERVATORY.

TABLE 23.—*Absorption lines in the infra-red solar spectrum, from bolographs*—Continued.

Designation.	Wave-lengths. μ	Designation.	Wave-lengths. μ	Designation.	Wave-lengths. μ	Designation.	Wave-lengths. μ	Designation.	Wave-lengths. μ	Designation.	Wave-lengths. μ
316 b	1.1720	379 d	1.3126	442 c	1.4596	505 c	1.7151	568 b	1.9510
317 c	1.1736	380 d	1.3147	443 c }b	1.4621	506 c	1.7176	569 c	1.9586
318 b	1.1756	381 d	1.3164	444 b	1.4653	507 d	1.7205	570 b	1.9677
319 c	1.1778	382 c	1.3175	445 d? }	1.4692	508 d	1.7222	571 b	1.9763
320 b	1.1801	383 d	1.3192	446 b }b	1.4696	509 d	1.7239	572 d)	1.9825
321 d?	1.1810	384 b	1.3209	447 d	1.4712	510 c	1.7258	573 b }ω?	1.9980
322 c	1.1822	385 d	1.3222	448 d ?	1.4724	511 c	1.7299	574 c }b	2.0070
323 d	1.1844	386 d	1.3246	449 c	1.4751	512 d	1.7310	575 d)	2.0250
324 d?	1.1853	387 d	1.3260	450 c }b	1.4762	513 d	1.7335	576 d)	2.8364
325 b	1.1868	388 c	1.3274	451 b	1.4789	514 c	1.7348	577 b }	2.0490
326 d	1.1894	389 b }b	1.3290	452 d)	1.4804	515 d	1.7385	578 c }b	2.0604
327 c	1.1924	390 d	1.3306	453 c }b	1.4835	516 b	1.7404	579 d)	2.0686
328 c	1.1940	391 c	1.3320	454 b }	1.4855	517 d	1.7430	580 d	2.0818
329 d?	1.1950	392 c	1.3342	455 d	1.4880	518 c	1.7450	581 d	2.0900
330 d	1.1963	393 c	1.3357	456 b	1.4904	519 c	1.7485	582 d?	2.1006
331 b	1.1975	394 d	1.3374	457 d?	1.4916	520 c	1.7505	583 d?	2.1090
332 d	1.1992	395 c	1.3385	458 c	1.4928	521 d?	1.7545	584 c	2.1150
333 c	1.2002	396 c	1.3425	459 c	1.4966	522 c	1.7555	585 d	2.1240
334 b	1.2021	397 d	1.3445	460 d?	1.4980	523 b	1.7590	586 d	2.1519
335 d	1.2036	398 c	1.3464	461 d	1.5003	524 d	1.7615	587 c	2.1645
336 c	1.2055	399 d	1.3499	462 b	1.5021	525 c	1.7631	588 d	2.1790
337 c	1.2070	400 d	1.3510	463 d?	1.5061	526 c	1.7641	589 d	1.1853
338 c	1.2095	401 d?	1.3538	464 b	1.5070	527 c	1.7674	590 d	2.1974
339 d?	1.2106	402 d	1.3590	465 c	1.5117	528 c	1.7693	591 d	2.2018
340 c	1.2124	403 d	1.3618	466 c	1.5151	529 c	1.7718	592 d?	2.2151
341 d	1.2150	404 d	1.3647	467 d	1.5170	530 c	1.7744	593 d?	2.2338
342 d	1.2167	405 d	1.3679	468 d	1.5196	531 c	1.7782	594 d	2.2695
343 c	1.2194	406 d	1.3691	469 c	1.5234	532 d	1.7796	595 d	2.2796
344 c?	1.2212	407 d	1.3734	470 c	1.5279	533 d	1.7818	596 d?	2.2905
345 c	1.2249	408 c?	1.3750	471 c	1.5324	534 d?	1.7842	597 d?	2.2960
346 d	1.2271	409 d	1.3780	472 c?	1.5482	535 c	1.7865	598 d	2.2999
347 d	1.2294	410 d	1.3815	473 d?	1.5519	536 d?	1.7882	599 c	2.3180
348 d	1.2306	411 d Ψ	1.3863	474 d	1.5668	537 c	1.7897	600 d	2.3268
349 d?,	1.2565	412 d	1.3915	475 d?	1.5696	538 b	1.7937	601 d	2.3390
350 d?	1.2581	413 d?	1.3938	476 c	1.5716	539 d	1.7951	602 d,	2.3488
351 d?	1.2587	414 c	1.3990	477 c	1.5740	540 d	1.7966	603 d? }c	2.3520
352 d	1.2596	415 c	1.4036	478 d	1.5793	541 c	1.7995	604 d	2.3650
353 d }b	1.2615	416 c a	1.4060	479 b	1.5859	542 d?	1.8022	605 d,	2.3780
354 d	1.2630	417 c	1.4087	480 c	1.5928	543 d	1.8044	606 d }c	2.3850
355 d?	1.2650	418 d	1.4101	481 d	1.5955	544 d	1.8058	607 d	2.4030
356 b	1.2658	419 d	1.4107	482 c	1.5981	545 c	1.8090	608 c	2.4115
357 d	1.2675	420 b	1.4135	483 d	1.6001	546 c)	1.8130	609 d	2.4309
358 d?	1.2700	421 c	1.4167	484 c	1.6042	547 c }	1.8147	610 c	2.4446
359 c	1.2717	422 c	1.4184	485 c	1.6085	548 d	1.8220	611 d?	2.4610
360 d	1.2734	423 d	1.4199	486 d	1.6161	549 d?	1.8262	612 d?	2.4650
361 d	1.2750	424 b	1.4225	487 d	1.6277	550 d?	1.8315	613 d?	2.4690
362 d	1.2771	425 c	1.4248	488 d	1.6350	551 d	1.8380	614 d?	2.4718
363 d	1.2782	426 b	1.4270	489 d?	1.6367	552 d	1.8500	615 d .	2.4782
364 b	1.2794	427 d	1.4289	490 d?	1.6468	553 d	1.8599	616 c	2.4850
365 d	1.2816	428 b	1.4313	491 d?	1.6482	554 d	1.8661	617 d	2.5000
366 d	1.2870	429 c	1.4335	492 d?	1.6516	555 d	1.8735	618 d	2.5100
367 d	1.2918	430 d?	1.4356	493 c	1.6605	556 d	1.8840	619 d?	2.5162
368 d	1.2935	431 c	1.4367	494 d?	1.6631	557 d	1.8962	620 a X	2.5200 / 2.8442
369 c	1.2948	432 c	1.4388	495 d	1.6665	558 d	1.9040		
370 d	1.2980	433 c	1.4406	496 d?	1.6703	559 d	1.9122	621 d	2.8390
371 c	1.2990	434 d	1.4418	497 d?	1.6744	560 c?	1.9143	622 d	2.8585
372 d	1.3010	435 b	1.4447	498 c	1.6910	561 c	1.9219	623 d	2.8716
373 b	1.3018	436 d	1.4459	499 d?	1.6937	562 c	1.9252	624 c	2.8930
374 c	1.3044	437 b	1.4495	500 d	1.6950	563 c	1.9288	625 d	2.9075
375 d	1.3055	438 d	1.4510	501 d	1.7025	564 c	1.9314	626 d	2.9158
376 c	1.3076	439 b	1.4542	502 c	1.7046	565 c Ω	1.9336	627 c	2.9342
377 d	1.3105	440 d	1.4564	503 c	1.7080	566 c a	1.9395	628 d	2.9430
378 c	1.3115	441 d?	1.4576	504 d	1.7099	567 c	1.9430	629 d?	2.9586

199

TABLE 23.—*Absorption lines in the infra-red solar spectrum, from bolographs*—Continued.

Designation.	Wave-lengths. μ	Designation.	Wave-lengths. μ	Designation.	Wave-lengths. μ	Designation.	Wave-lengths. μ	Designation.	Wave-lengths. μ
630 b	2.9640	653 d ?	3.2228	676 d ?	3.6450	698 a Y	4.1790	720 d ?	4.9345
631 d	2.9728	654 c	3.2410	677 d ?	3.6590		4.4988	721 c	4.9440
632 d ?	2.9861	655 b x₂	3.2620	678 b	3.6716	699 c	4.4939	722 d	4.9495
633 c	2.9942	656 d	3.2702	679 d ?	3.6880	700 d	4.5368	723 d ?	4.9654
634 d	3.0000	657 d ?	3.2942	680 d	3.7138	701 d	4.5845	724 c	4.9714
635 c	3.0108	658 c	3.2990	681 d }c	3.7178	702 d	4.6156	725 d ?	4.9836
636 c	3.0316	659 b	3.3150	682 c	3.7336	703 c	4.6402	726 d }b	4.9888
637 c	3.0432	660 c	3.3430	683 c	3.7596	704 c	4.6739	727 d	4.9955
638 c	3.0605	661 c	3.3550	684 d ?	3.7708	705 d }c	4.6829	728 d	5.0063
639 d }x₁	3.0798	662 d	3.3690	685 c	3.7886	706 d	4.6960	729 d ?	5.0195
640 c a	3.0856	663 d	3.3795	686 c *	3.8122	707 d	4.7026	730 d	5.2268
641 d ?	3.0920	664 b	3.4051	687 d	3.8345	708 d	4.7087	731 d ?	5.0422
642 c	3.0998	665 c	3.4350	688 d ?	3.8465	709 d	4.7284	732 c	5.0537
643 c	3.1190	666 c	3.4530	689 d	3.8555	710 c	4.7582	733 d	5.0784
644 c	3.1391	667 d	3.4734	690 d ?	3.8642	711 c	4.7750	734 c	5.1000
645 d	3.1500	668 d	3.4960	691 d ?	3.8722	712 d	4.7866	735 d	5.1381
646 d	3.1659	669 d	3.5346	692 c	3.8771	713 d }c	4.8039	736 d	5.1725
647 d	3.1710	670 c	3.5406	693 d ?	3.8858	714 d	4.8139	737 c	5.2050
648 d	3.1796	671 d ?	3.5512	694 d ?	3.8909	715 d ?	4.8238	738 c	5.2537
649 d	3.1955	672 c	3.5704			716 b	4.8590	739 d	5.3084
650 c	3.2032	673 c	3.5865	695 c	3.9222	717 d	4.8978	740 d	5.3386
651 d	3.2092	674 c	3.6072	696 d	4.0211	718 b	4.9185		
652 d	3.2180	675 c	3.6800	697 d	4.0810	719 d ?	4.9239		

The final results of the preceding tables are, as the reader has already been reminded, graphically depicted in Plates XX and XXI.

The work on the infra-red (bolometric) prismatic spectrum, commenced at Allegheny in 1881, was renewed at this observatory in 1892, and now, July, 1900, is published in the present form. It may conveniently be considered in two parts. The first extends from the limits of the visible spectrum to the extreme limits of the "heat" spectrum as known or suspected in 1881, i. e., to Ω, then vaguely known as the end of the spectrum. The second part is the region whose existence was first discovered in that year by the writer, with the bolometer, upon Mount Whitney at an elevation of 12,000 feet. In Pl. XX these divisions may be readily distinguished.

COMPARISON OF THE PRECEDING RESULTS WITH THE SOLAR SPECTRUM MAPS OF HIGGS AND ABNEY AND WITH INFRA-RED METALLIC LINES OF SNOW, LEWIS, AND OTHERS.

Taking the work of Abney and Higgs as subjects of comparison, the observation is that the present method extends to great spectral regions wholly inaccessible to photography. In the very upper portions of the infra-red, e. g., from 0.8 to 1.0, photography doubtless is more convenient and complete than bolometry. There is a region in this upper portion where the two processes are fairly comparable; below this photographic processes almost wholly fail, and the further region belongs at present to the bolometer alone.

The reader may care to compare the results of the bolographic analysis of the solar spectrum with the maps of grating spectra prepared photographically by Higgs

and by Abney. As he can readily do this, either by means of the table or of the normal spectrum map, what is said here will merely call attention to some points where injustice might inadvertently be done to the bolographic results in such a comparison.

In the first place, it need hardly be recalled that more lines have been found in a given space by the photographic than by the bolographic processes, because with their capacity for long exposures with cumulative effects, the spectrum photographed may be far more extended and the slit much narrower than is at present possible with the bolometer. In consequence of this greater richness of detail, some faint lines are found in the photographs where none at all are discovered in the bolographs, and frequently two or more lines of the photographs correspond to one in the bolographs.

In the second place the system of designating intensities employed in the tables is different from that of Abney, and gives at first glance a different impression from the direct photographs of Higgs. Thus in Abney's tables we find several lines in succession to which he has assigned the intensity 4 or 5 (that is very strong), while in the bolographs they would be called d, that is, of very little prominence separately; but a bracket marked b would be placed against the whole group in the table, showing that these separate minute deflections were found in the bottom and sides of one great deflection. Abney and Higgs each used so great dispersion that bright strips show in such places between the dark lines, so that the latter appear very strong. In the bolographs, on the other hand, the width of the bolometer and the relatively slight dispersion prevent the corresponding appearance on the plate. The reader may therefore consider in his comparison that all lines included in a bracket marked b are possibly separately of intensity b, though designated otherwise.

In the third place, the wave-lengths assigned to the lines in the tables are, as has been already said, liable to inaccuracy to the extent of one or two Ångstrom units, on account of the errors of wave-length determination, plotting, and interpolation in the plot. Further, as the single deflections of the bolographs corresponding to two or more lines in the Higgs photographs are often 1 or 2 mm. wide, it is by no means certain that the minimum as determined in the reduction of the bolographs should correspond within one or two Ångstrom units to the mean position of the several lines it represents, especially if these lines are of unequal intensity. Finally, it is universally found by observers that errors of determination may not infrequently amount to two or three times the probable error deduced in the usual manner from the residuals of the separate observations from the mean; so that, in case of the bolographs, errors of position of a half or two-thirds of a second of arc may be occasionally expected, and such would correspond to nearly one Ångstrom unit. Thus, when we take all these factors together, occasional variations from Higgs's map of three or four Ångstrom units are possible.

As an example of what may be expected from the bolographs the following table gives a comparison of the results with the rock-salt prism and Higgs's map as far as the latter extends, which includes the first 49 lines in the rock-salt table. On the whole, the agreement is excellent. The wave-lengths seldom differ by more than two Ångstrom units, and the relative prominence of the lines is about as we should expect. The chief discrepancies which occur are the following: One of the doubles in the A series is absent, and the presence of several lines marked doubtful in the bolographs and one marked d with no question mark, seem hardly substantiated:

TABLE 24.—*Comparison of results of bolographs with the salt prism and Higgs's photographic map.*

Smithsonian Observatory observations.		Higgs's map of the solar spectrum.		Intensity.[1]		Remarks.
Designation.	Wave-length.	Wave-lengths.	Number of lines.	c.	d.	
	μ	μ				
1 d ?	0.7596	0.7594 / 98	6	4	2	
2 d	.7600	.7599				
3 d }b	.7604	....	11	11	......	Relatively open spaces at 7597 to 7600, at 7611.5, and at 7615.
4 d	.7609	.7611				
5 d ?	.7612	.7612 / 14	2	2	......	
6 d	.7614	.7615 / 165	2	2		
7 d	.7621	.7621	1	1		
8 d	.7624	.7623 / 25	2	2		
9 d	.7629	.7627 / 29	2	2		
10 d	.7631	.7631 / 33	2	2		
11 d A	.7634	.7635 / 37	2	2		
12 d	.7638					
13 d	.7641	.76395 / 412	2	2	......	
14 d }b	.7650	.7649 / 51	2	2	......	Double between these omitted.
15 d	.7653	.7654 / 553	2	2		
16 d	.7660	.7659 / 61	2	2		
17 d	.7666	.7665 / 66	3	2	1	
18 d	.7668	.76708 / 719	2	......	2	
19 d	.7676	.76768 / 779	2	......	2	
20 d	.7680	.76805 / 84	3	......	3	17 faint lines between these.
21 d ?	.7772	.7772 / 76	3	......	3	

[1] The scale of intensities annexed roughly corresponds with that adopted for the bolographs. Thus, corresponding with line 40 c, we find in Higg's map 4 lines of intensity c and 1 of intensity d; in all, 5 lines.

TABLE 24.—*Comparison of results of bolographs with the salt prism, etc.*—Continued.

Smithsonian Observatory observations.		Higg's map of the solar spectrum.		Intensity.		Remarks.
Designation.	Wave-length.	Wave-lengths.	Number of lines.	c.	d.	
	μ	μ				
22 d	.7826	.7832	1	......	1	5 faint lines between these.
23 d?	.7987	.7999(?)	3	......	3	
24 d?	.8039	.8046(?)	3	......	3	Sparcely filled with faint lines; coincidences doubtful.
25 d?	.8119	.8115(?)	2	......	2	
26 d	.8126	....	......	......	......	
27 d	.8136	.8134 / 36	2	1	1	
28 d	.8143	.8141 / 424	2	......	2	
29 d	.8150	.8147 / 50	5	......	5	Last two strongest.
30 c	.8160	.8155 / 585	2	2	......	First stronger.
31 d, b	.8164	.8162 / 65	3	3	......	First two close.
32 d?	.8173	.8169 / 706	3	1	2	Last strongest.
33 c	.8178	.8177 / 795	4	1	3	First strongest.
34 d?	.8181	.8182 / 838	2	1	1	Do.
35 d	.8183					
36 d	.8190	.81896	1	1		
37 c	.8196	.81934 / 98	3	2	1	Middle weakest.
38 d	.8211	.8212 / 14	2	......	2	
39 d	.8221	.82185 / 22	3	1	2	First strongest.
40 c	.8228	.8227 / 32 } "Z"	5	4	1	Strongest at 8228.
41 d	.8240	.82438	1	1		
42 d	.8257	.8257	1	1		
43 d	.8274	.8272 / 77	3	2	1	First weakest.
44 d	.8280	.8280 / 824	2	1	1	First weaker.
45 d	.8290	.8288 / 90	2	1	1	First stronger.
46 d	.8308	.8305	2	1	1	First weaker.
47 d	.8321	.8317 / 22	3	2	1	Do.
48 d?	.8331	.8327 / 36		......	5	Strongest at 8330.
49 d?	0.8340	0.83395	1	1		

S. Doc. 20——17

INFRA-RED METALLIC LINES.

Lists containing in all about eighty metallic lines, either observed or predicted to occur in the infra-red, have been cited in Chapter I, and very probably there may be other observations or predictions of this kind which have been overlooked, as there was no attempt at an exhaustive search of the literature of the subject.

A partial comparison made here of these results with the infra-red solar spectrum has led to the conclusion that it is not yet the time to announce coincidences between metallic and infra-red solar lines.

Chapter VII.

THE VARIATIONS OF ABSORPTION IN THE INFRA-RED SOLAR SPECTRUM.

The most casual observer of such solar energy curves as are shown in Pls. XIX to XXIII can not fail to notice the steady increase in the importance of the great absorption bands as he looks through the visible spectrum down to the red and infrared. Another most prominent feature is seen to be, that beginning with α there is a striking similarity between the larger bands. Each is abrupt on the short wave-length side, but gradually fades away toward the longer waves. There appears also a marked "head" and "tail". Such a similarity of appearance suggests a similarity of cause, and this can be indeed at least partially affirmed.[1] For the bands α B and A have long been known to be mainly of terrestrial origin, and to be partly caused by the absorption of the aqueous vapor in the earth's atmosphere, and partly by the absorption of its oxygen. Within recent years it has been shown that the bands called here Ψ at 1.4 μ, Ω at 1.8 μ, and X at 2.6 μ are coincident with regions of strong water vapor absorption, and hence are presumably of terrestrial origin. Carbon dioxide vapor is similarly active in absorption at 4.4 μ, so that the band Y is also to be regarded as probably terrestrial. The general appearance of the bands $\rho\sigma\tau$ and Φ is so similar to that of these terrestrial ones just named as to lead one to suspect that they, too, are telluric, and this has been proved by Abney.[2] Thus the nine great bands, α, B, A, $\rho\sigma\tau$, Φ, Ψ, Ω, X and Y are doubtless largely of terrestrial origin.

Since the absorption of the earth's atmosphere thus plays so important a part in decreasing our supply of radiation from the sun, it is interesting to study its effect more thoroughly. The present investigations, while not primarily designed for this purpose, have furnished much material for such study. There have been taken since 1893 hundreds of bolographs, or solar energy curves, extending from A to X, a lesser number taken since 1895 extending as far as 5μ, and within the last years some of fast speed covering the whole solar spectrum from 0.4 μ to 6 μ. Besides these, there

---

[1] See the author's remarks in Am. Jour. Sci., third series Vol. XXXVI, p. 397, 1888.
[2] Royal Society Proceedings, vol. 35, p. 80, 1883.

have been taken a number of fast-speed bolographs with terrestrial energy sources extending from A to near 8μ.

An examination of these many curves has been made for the purpose of detecting variations of their form, and of following out such variations to see if they were periodic or irregular, gradual or sudden, and whether they are connected with meteorological conditions or not. It is not, of course, certain that all changes in the form of the energy curve of the solar radiation are due to terrestrial causes, for we are well aware that there are changes in the sun itself, and these may, for all that we know, produce analagous though probably smaller effects. Yet the presumption is strongly in favor of a telluric origin of many of the infra-red bands, especially when we consider that the earth's atmosphere is subject to such enormous local changes in humidity, temperature, and composition, and that the mass of it to be penetrated by the solar rays changes during each day and with each season.

Unfortunately for the value and completeness of the bolographic records just alluded to, they have these several shortcomings. First, they must necessarily be lacking in cloudy and rainy weather like any other solar records. Second, there have been a number of intervals of one or more months when the apparatus was in use for other investigations or was undergoing improvements so radical that no bolographs were attempted. Third, many of the bolographs, especially of the earlier years, are so inferior to later ones as hardly to be proper for comparison. Fourth, mirrors may have tarnished more or less, and salt prisms "fogged" slightly; the slit and bolometer strip may have subtended more or less, the adjustments may have been more or less perfect, and the minor disturbances due to magnetic, mechanical and electrical causes may have been little or much—all these things exerting their influence on the form of the curve. Thus it can not be claimed that the results to be given are absolute, but it is believed that they contain facts of interest.

TYPICAL ENERGY CURVES.

Superposed curves[1] are shown in Pls. XXX A and XXX B, which will serve to give preliminary ideas relative to the seasonal changes to be observed. In the first group we have typical spring, summer, autumn, and winter bolographs of the spectrum from A to near X. For convenience in examination they have been placed on a background of parallel lines of equal intensity of radiation, with a scale of ordinates such that the height of the maximum between Ψ and Ω is 100 on the scale for each good curve, i. e., they are all reduced to the same vertical scale for this place. The reader will note that these bolographs have been selected for their general form rather than for accuracy of position or richness of detail.

---

[1] It has been thought best to take some of these curves from earlier observations when, however, the progressive "drift" was still evident. This progressive movement of the needle "drift" causes a regular lifting of the curve as a whole without disturbing its parts, and its results are indicated here by the inclination of the base lines of the curves in these two plates.

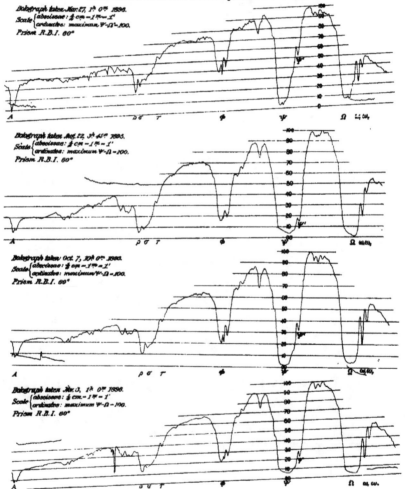

BOLOGRAPHS OF THE INFRA-RED SOLAR SPECTRUM OF A 60° ROCK-SALT PRISM, TYPIFYING THE YEARLY VARIATIONS IN ABSORPTION IN SPRING, SUMMER, AUTUMN AND WINTER.

OBSERVATIONS OF S. P. LANGLEY WITH THE ASSISTANCE OF C. G. ABBOT.

BOLOGRAPHS OF THE INFRA-RED SOLAR SPECTRUM OF A 60° ROCK-SALT PRISM, TYPIFYING THE YEARLY VARIATIONS IN ABSORPTION IN SPRING, SUMMER, AUTUMN AND WINTER.

OBSERVATIONS OF S. P. LANGLEY WITH THE ASSISTANCE OF C. G. ABBOT.

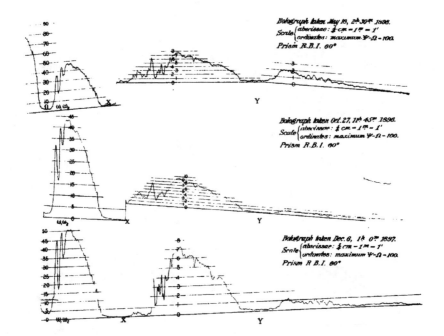

A more general similarity will be found between numbers I and III, the spring and autumn representatives, than between any other pair. The general direction of the changes in form is somewhat as follows:

Spring: Radiation relatively very great from A to Φ, Φ of medium depth, Ψ narrow, showing absorption moderate on its long wave-length side, and thus ψ' high up on the curve, Ω narrow, absorption moderate on long-wave side, ω_1 ω_2 moderately deep, and with the sharp maxima just preceding them very high, thus indicating both fairly great absorption in ω_1 and ω_2 and slight absorption immediately preceding each of them.

Summer: The relative energy in the region from A to Φ is much reduced, while Ψ and Ω have become very broad, owing to a great increase of absorption on their long wave-length sides. ω_1 and ω_2 have become less prominent partly because of this increase of absorption below Ω and partly because of actual increase in the height of their minima.

Autumn: Beginning usually in September, the relative amount of energy from A to Φ gradually grows and attains its maximum, as already noted, in the spring months. Ψ and Ω become much narrower usually in the latter part of September and remain so until some time in November. ω_1 and ω_2 increase steadily in depth, beginning in August, and their prominence reaches its maximum in October and retains it little altered until the next summer comes on.

Winter: It is usually but not universally the case, in the months of December and January, that the bands Ψ and Ω have the broad, flat bottom shown in the illustration. In other respects the curve is very like those of autumn and spring.

The second group of curves contains typical examples of spring, autumn, and winter bolographs of the region from Ω to beyond Y. A representative of summer is absent because bolographs of this region have not been taken in summer in sufficient number to lead to any general idea of the conditions which then prevail in this region. Scales of ordinates have been annexed on which, as in the A–Ω illustrations, 100 is the height of the maximum between Ψ and Ω. In the present illustrations the scales change suddenly at several points owing to alterations of the slit width. It will be seen that in this region we have on the whole less seasonal variation than in the region from A–Ω.

From what has been said it will be seen that the principal changes to be observed between different seasons are in the relative amounts of energy from A–Φ, and in the changes of absorption in Ψ and Ω.

IRREGULAR VARIATIONS OF ABSORPTION.

We turn now to changes more irregular and rapid in character. The variation of Ψ and Ω has been included among those of a seasonal character, yet while Ψ and Ω are never seen broad and flat in spring, still, in the transition from winter to spring, from spring to summer, from summer to autumn, and even in midwinter, their character may change from narrow to broad and back again several times in a week. Furthermore, an evidence of the telluric character of the absorption which produces them is found in the fact that the absorption increases with the declining sun. Examples of these changes are furnished by curves of the following dates, and many others could be cited:

December 18, 1893, several curves, Ψ and Ω very narrow; height of ψ', 50

December 20, 1^h 40^m p. m., broad; height of ψ', 10 to 20.

December 21, 10^h a. m., narrow again; height of ψ', 50.

December .2, 9^h 40^m a. m., very broad; height of ψ', 20.

December 23, 10^h 30^m a. m., still more broad.

December 27, curves bad, but indicate decidedly less absorption.

February 9, 1898. Three curves taken at about noon, at 2^h 30^m p. m., and at 3^h 40^m p. m., respectively, show a very striking change from moderate absorption to great absorption in the regions Ψ and Ω.

Another region where the energy is greatly decreased with declining sun, as long since noticed by the writer, is that between X and x_2. A marked example of this is furnished by the bolometric curves of December 10, 1896. One which was begun at 3 o'clock reaching X near 3^h 20^m hardly indicates any energy at all between X and x_2, and only half as much between x_1 and x_2 as a curve started a half hour after noon of the same date, which was normal at these points, and taken under precisely the same conditions except as regards the position of the sun.

Our atmospheres absorption in the infra-red is on the whole much less than in the visible spectrum, as was shown by the writer nearly twenty years ago. Indeed, low sun, though accompanied by the local effects above mentioned, produces little general diminution of energy at wave-lengths greater than 1μ, while its effect through the visible spectrum and up to a wave-length of 0.9μ is very marked.

In some bolographs the absorption band Φ contains three very deep indentations in its bottom, while in others the third, and especially the second, are much less prominent. This is not the effect of poor definition, for the change in character has repeatedly been observed in a single day, with no intermediate alteration of the spectroscope. Moreover, some bolographs taken with wide bolometers and wide slits show deeper deflections in Φ than other curves where the apparatus was arranged for greater detail. No particular time of the year is attended by these variations, nor are they associated with any of the changes heretofore noted.

Just after the curve has risen from the long-wave side of φ we come upon a strip covering about $2'$ of (rocksalt prism) spectrum which varies considerably in character. A good idea of the change in appearance is given by saying that some curves here approximately follow the base and perpendicular side of a right triangle, while others follow its hypotenuse. Of course we find intermediate characters between these two extremes. There appears to be no association between the variations in this region and those heretofore noted.

Near the middle of the maximum between Φ and Ψ will be seen a deep deflection, quite unique in its appearance. The depth of this depression varies considerably, and generally at the same time with ω_1 and ω_2. Its changes are not, however, so marked as those of ω_1 and ω_2, and sometimes take place partially independently. Nevertheless, the behavior of the three deflections is on the whole so similar that one is inclined to believe them all due to the same cause, acting, however, perhaps in different degree.

The bottoms of the great absorption bands Ψ and Ω are usually either raggedly broken into minor deflections, or broad and flat, or slightly rounded, according as the condition of little or great absorption in these bands is present. Occasionally, however, three nicely rounded deflections appear in the bottom of Ω, and either three or five at the same time in Ψ. At one such occasion, March 24, 1896, these deflections were 5 to 8 mm. in depth, and were accompanied by a most unusual narrowness of Ψ and Ω. This condition remained, with slight modifications, until March 28, when the absorption was far greater, though still very moderate, and the deflections in question had disappeared. April 8 brought a partial return to the condition of March 24, but on April 13 the absorption became very great, and all deflections in the bottoms of Ψ and Ω disappeared. The absorption, however, had decreased again before April 22, but it is not known how long it remained slight.

Several bolographs of March 28, 1899, seemed to indicate sharp maxima in Y, but as there was considerable disturbance of the galvanometer at that time not too implicit trust should be given these observations.

Many minor changes have been noted in certain of the smaller deflections, such as two deflections appearing where there were three, variations in depth in different bolographs and the like; but these minor variations have not been so often observed as to make it certain that they are not due to accidental disturbances of the apparatus rather than to changes in absorption, so that they will not here be enumerated. Enough has been said, however, to indicate the great modifications which take place in the infra-red from time to time. For a more elaborate record, though less extended in point of time than that given above, the following table of relative ordinates of bolographs of 1896–97 is inserted, and the reader may, if he chooses, compare it with Table 20 of Part I, containing similar results for 1897–98. The ordinates given are at the minima of certain absorption deflections. The depth of these deflections somewhat exceeds 5 mm., 1 mm., or 0.1 mm., according as they are marked b, c, or d. The numbers designating the deflections agree with those in Tables 18 and 21. Should it appear desirable the many bolographs of former years may at some future time be measured in ordinates to make the record more complete. Indeed, a rough set of measures already made have been compared with meteorological data for the same period by means of plots. However, nothing certain could be determined through this comparison, and it is not here given.

TABLE 31.—*Relative ordinates, bolographs of 1896.*

Designation.	Oct. 26. Bolograph IX.	Oct. 31. Bolograph II.	Oct. 31. Bolograph VII.	Nov. 2. Bolograph II.	Dec. 10. Bolograph III.	Dec. 10. Bolograph IX.
	$12\frac{1}{4}^h$–$1\frac{1}{4}^h$.	10^h–11^h.	1^h–2^h.	10^h–11^h.	11^h–12^h.	$1\frac{1}{4}^h$–$2\frac{1}{4}^h$.
3 b } A	12	10	6	6		
9 o } a	16	13	....	9	....	....
23 o	40	29	23	24	34	26
26 d	....	26	22	25	33	....
28 d	36	25	20	24	30	22
30 d } b	34	24	20	23	30	20
33 d	35	....	20	....	....	20
36 d	36	25	....	....	....	19
38 d	38	....	23	....	33	20
40 b	34	24	21	22	29	17
41 d	....	....	....	....	....	....
42 d	39	29	25	27	34	21
44 c	38	29	24	26	34	20
45 d	38	30	24	27	....	20
47–8 d	39	32	26	28	36	22
—— d	45	35	29	32	49	29
53 d	46	36	30	32	49	31
55 o	44	36	29	32	36	29
56–7 d	46	36	29	32	38	30
59 b	41	34	28	30	34	28
64–5 d	48	38	32	34	....	35
69 b	45	38	29	32	42	36
70–1 d	46	....	31	34	....	38
80 d	50	42	33	37	48	44
84 d	51	42	35	37	46	43
89 d	50	41	33	36	48	42
90–1 d	47	....	30	....	47	40
92–3 c } b	44	36	28	32	43	38
94 c	41	34	26	29	42	36
95–6 o	41	34	27	....	43	37
97 d	42	35	29	30	45	38
100 d	46	39	31	34	49	44
103 d	43	36	28	30	43	41
104 d	42	35	26	29	43	39
105–6 d	43	35	27	30	43	40
109 c	43	35	26	30	43	38
111 c	43	35	25	30	45	38
112 c	43	36	27	30	47	38
116 d	....	....	....	....	....	....
118–9 d	48	39	30	34	52	44
120 c	48	38	29	33	51	43

ANNALS OF ASTROPHYSICAL OBSERVATORY. 211

TABLE 31.—*Relative ordinates, bolographs of 1896*—Continued.

Designation.	Oct. 26. Bolograph IX.	Oct. 31. Bolograph II.	Oct. 31. Bolograph VII.	Nov. 2. Bolograph II.	Dec. 10. Bolograph III.	Dec. 10. Bolograph IX.
	$12\frac{1}{2}^h$–$1\frac{1}{2}^h$.	10^h–11^h.	1^h–2^h.	10^h–11^h.	11^h–12^h.	$1\frac{1}{2}^h$–$2\frac{1}{2}^h$.
130 c	20	17	7	11	23	19
131-2 c } b }ρ	21	17	7	12	25	19
134-5 d	29	24	13	17	33	26
137-8 c	25	22	11	15	29	22
139 c	26	22	12	16	30	23
140-1 d	28	24	....	....	....	24
142 c	28	24	13	18	33	24
143 c	29	25	15	18	33	24
144 c } b σ	31	26	17	20	35	26
145-6 c a	32	....	17	21	35	26
148 c	34	29	19	22	38	26
149-50 d	37	33	23	26	43	31
151 d	39	36	25	28	46	34
152-3 d	43	38	27	30	49	35
154-5 d	55	47	35	32	57	45
158 d } b r	53	45	34	37	56	45
159 d	53	44	35	37	56	45
161-2 c } b	57	48	38	39	57	48
163-4 d	61	53	42	45	62	53
165 d	63	57	46	49	....	59
169 d	69	59	49	....	68	70
170 d	69	60	49	....	....	69
172 c	68	61	49	53	67	69
173 d	70	62	51	....	70	72
186 c	72	68	58	63	75	78
191-2 c	71	69	57	62	75	76
197 d	74	68	56	60	....	76
198 c	72	67	56	59	76	76
201 c	69	63	54	55	73	70
205 d	64	57	47	52	69	64
206-7 d	62	53	40	50	67	60
208-9 d	....	....	....	....	....	....
213-4 d	28	24	19	18	35	23
215 c } b	22	18	14	13	28	19
217 c	20	18	14	12	27	18
219 b	21	19	16	13	24	20
221 d φ	35	30	27	26	37	31
222-3 b a	26	23	20	19	27	23
224 d	33	30	27	26	34	30
225 d	36	....	29	....	39	32
227-8 d	50	43	41	41	48	41
229 c	62	57	53	53	65	61
235 c	68	64	60	59	72	68
236 c	69	66	61	....	74	71
237 c	69	65	61	59	73	68
239 c	71	66	63	....	75	71
242-3 d	69	72	....	....	80	76

TABLE 31.—*Relative ordinates, bolographs of 1896*—Continued.

Designation.	Oct. 26. Bolograph IX. $12\frac{1}{2}^h-1\frac{1}{2}^h$.	Oct. 31. Bolograph II. 10^h-11^h.	Oct. 31. Bolograph VII. 1^h-2^h.	Nov. 2. Bolograph II. 10^h-11^h.	Dec. 10. Bolograph III. 11^h-12^h.	Dec. 10. Bolograph IX. $1\frac{1}{2}^h-2\frac{1}{2}^h$.
244 c	73	69	67	....	77	74
247 c	73	69	68	....	76	75
———	76	73	71	....	80	77
248 d	77	74	72	....	82	78
249 d	78	77	73	....	83	81
250-1 d	81	80	79	....	86	84
254 d	83	81	81	....	89	86
259 d	88	86	85	80	93	89
268 d	75	76	72	....	77	71
270 b	71	72	71	65	74	67
271 d	80	....	79	76	82	83
273 c	83	83	84	79	88	83
276 d	88	86	86	83	92	92
279 d	....	83	82	81	90	86
281 d	....	77	....	75	84	82
285 c	....	71	75	64	76	75
288 d	60	63	64	59	70	68
290 c	46	45	47	52	52	47
311-12 d Ψ	7	3	9	....	7	6
315 c	15	9	13	8	13	9
319 d a	24	....	21	15	22	16
323 b	29	24	26	21	27	20
324 c	38	31	34	28	35	27
326 d	46	....	....	36	45	36
329 c	75	71	72	69	73	72
331 d	....	....	84	86	89	88
333	96	94	94	94	....	96
339 d	100	100	100	100	100	100
345 b	93	94	93	93	92	92
351 b	94	96	93	94	88	87
354 d	97	100	96	98	91	98
359 c	94	98	92	97	90	95
361 c	. 94	98	93	98	90	95
363-4 c	93	97	92	98	89	94
371 d	87	88	85	90	83	88
373 d	83	84	81	84	79	82
376 d Ω	78	78	76	78	74	77
379 c	75	76	71	76	72	74
407 d a	13	6	9	7	10	5
410-11 d	40	....	35	....	35	34
412-3 b ω_1	20	20	17	16	17	14
416 b	37	39	35	37	34	34
417 d ω_2	38	40	36	39	35	36

TABLE 31.—*Relative ordinates, bolographs of 1896*—Continued.

Designation.	Oct. 26. Bolograph VI.	Oct. 27. Bolograph V.	Nov. 2. Bolograph VI.	Dec. 10. Bolograph VI.	Dec. 10. Bolograph XI.
	11^h–12^h.	$11\frac{1}{2}^h$–$12\frac{1}{2}^h$.	$11\frac{1}{2}^h$–$12\frac{1}{2}^h$.	$12\frac{1}{2}^h$–$1\frac{1}{2}^h$.	3^h–4^h.
412 b ω_1	18	20	17	16	14
416 b } ω_2	37	39	37	34	34
417 d	40	....	....	35	37
420–1 d	50	52	51	48	67
423 c	50	52	51	49	69
425 d	48	50	50	47	69
426 c	46	49	47	45	59
428 c	45	47	46	44	54
430 c	45	47	45	43	50
432 d	43	45	....	42	45
433–4 d	40	43	41	39	39
437 c	37	39	38	35	34
438 c	34	36	36	32	31
441 c	29	31	29	27	25
443 d	27	....	....	25	21
444–5 c	25	25	24	22	19
447 c	20	20	18	17	15
448 d	....	....	....	17	14
449 c	15	11	14	13	10
455 c	5	5	4	5	5
459 a X	1.2	1.4	1.2	2.5	....
463 d	1	....	1.1	1.2	1.5
464 d	1.3	....	1.4	1.5	1.8
466 d	2.5	....	....	2.2	1.8
469 c	2.1	2.6	2	2.2	1.8
472 c	3	3	2.7	3	3
474 d x_1	3	3	....	3	3
476 c a	2.3	2.2	1.9	1.8	1.8
479 c	2	2	1.6	1.5	1.5
482–3 c	4	4	3.4	3	3
489–0 b	1.3	1.7	1	0.9	1.5
492 d	....	2.2	....	2.4	2.7
494 b	3	3	2.7	2.4	3
498 b x_2	2.3	3	2	2	2.4
499 c a	4	2.4	4.1	2.4	4
500 d	4	....	....	4	5
502 d	5	6	....	5	7
503 c	6	7	6	6	8
505 d	....	8	7	7	11
506 d	7	8	8	7	12
509 c	6	7	7	6	10
511–12 c	6	7	7	6	10
513 d	....	7	7	6	10
514 d	....	....	6	7	11
517 b	5	8	6	5	8
519 c	5	....	....	6	10
521 c	....	....	....	6	10

TABLE 31.—*Relative ordinates, bolographs of 1896*—Continued.

Designation.	Oct. 26. Bolograph VI. 11^h-12^h.	Oct. 27. Bolograph V. $11\frac{3}{4}^h$-$12\frac{3}{4}^h$.	Nov. 2. Bolograph VI. $11\frac{1}{2}^h$-$12\frac{1}{2}^h$.	Dec. 10. Bolograph VI. $12\frac{1}{4}^h$-14^h.	Dec. 10. Bolograph XI. 3^h-4^h.
522 c	....	....	6	5	9
524 c	....	6	....	6	9
525 c	....	....	....	5	9
526 d	5	5	....	5	9
531 c	4	5	5	4	6
534 c	4	4	5	4	6
535 d	3	4	3	2	3
536 d	....	....	....	2	1.8
537 a Y	0.7	0.7	0.5	0.7	0.8
540 d	1.1	....	....	1.4	1.5
542 c	0.8	....	0.3	1.2	1.8
543 d	0.5	....	....	0.9	1.5
545 d	....	....	0.0	0.9	1.2
549 c	0.3	0.0	....	0.8	1.2
550 c	0.1	0.0	0.1	0.6	1.2
552 c	0.1	0.0	0.3	0.4	1.2
555 c	0.3	0.0	0.5	0.2	1.5
557 c	0.5	0.0	0.3	0.4	0.9
560 c	....	....	....	0.4	1.2
565 c	0.3	0.0	0.1	0.8	0.6
571 c	0.1	0.2	0.1	0.3	0.6
573 d	....	0.0	0.1	0.4	0.3
576 d	....	0.2	....	0.6	0.3

Reasoning from our knowledge of the visible solar spectrum, it appears probable that the smaller deflections in the bolographs are generally due to metallic vapors in the sun, and if such will be usually invariable in magnitude. The broad, deep deflections will, on the other hand, be due to atmospheric absorption, and will be subject to variations with air mass and changes in the constitution of the atmosphere. In accordance with this, it is well known that A is due partially to oxygen and partially to water vapor in the earth's atmosphere, and Paschen and others have shown that there are strong water vapor bands at Ψ, Ω, and X, and a strong band due to carbon dioxide at Y. As already noted Abney attributes the bands $\rho\sigma\tau$ and Φ to water or water vapor.

From the observations which have been made here and described above it may be assumed as probable that the lines numbered 219, 222, 223, 270, 412, 416, and 519 to 579 in Tables 20 and 31, and which are graphically indicated on Plate XX, are atmospheric, and that many others are so is not improbable.

EMISSION AND ABSORPTION EXPERIMENTS.

In several recent experiments at this observatory some additional light is thrown on the subject. Certain emission experiments, which consisted of taking energy curves of the spectrum of the incandescent mantle of the Welsbach light (composed of impure thorium oxide), and of several other mantles, among them pure thorium oxide and iron oxide, are more fully discussed at a subsequent page. In this connection it is sufficient to call attention to three results: First, the two mantles, composed largely or wholly of thorium oxide, gave very strong maxima at 86' below A, corresponding to a wave-length of 4.6 μ, which is near the middle of Y, and no such maxima were found with the iron mantles. Hence it appeared probable at first glance that thorium has a line at 4.6 μ; but as this occurs in the great carbon dioxide band Y, it seemed possible that this strong maximum referred to was due to the emission of carbon dioxide vapor in the lamp, which failed for some reason to be completely absorbed away or "reversed" with the thorium oxide mantles, but was so "reversed" with iron oxide mantles. Second, the curves all showed moderate absorption bands at Ψ and Ω and strong ones at X and Y. Third, a long series of at least fifty absorption bands, beginning just beyond Y, extended down about 10' of arc, or, in other words, between wave-lengths of about 4.5 μ and 7.5 μ. Lines 549, 550, 552, 553, 555, 557, 564–567, 571, 573, 575, 577, and 578 of Table 21 could be identified with some of these, and, very probably, had the purity of the lamp spectrum equaled that of the sun, still other of the solar lines would have been found. All the absorption bands mentioned occurred with all the mantles. The mantles were heated by the flame of the Kitson lamp, which burns vaporized petroleum oil, with strong air draft. It is probable, therefore, that the bands in question were due to gases liberated in burning the lamp, and such would consist chiefly of water vapor and carbon dioxide. This is rendered the more certain by the fact that the spectrum of an incandescent platinum strip showed none of them with certainty.

The absorptive effect of liquid water upon the changeable portion of the spectrum, including Φ, Ψ, and Ω, was investigated On April 1, 1899, the absorption in Ψ and Ω being unusually small, a bolograph of this region beginning just before Φ was taken, with width of slit 0.50 mm. A dry, empty cell, having two thin glass sides, was then interposed in the beam before it had reached the slit, and a second bolograph was taken. The width of slit required to give about the same ordinates as in the first bolograph was 0.70 mm. Several different thicknesses of water were now used in the cell with the following absorption effects.

Thickness of water film, 0.025 mm.: Absorption zero from start (about three minutes above Φ) to finish beyond Ω.

Thickness 1.0 mm.: Absorption zero from start to just beyond line 219 (see Table 21, Part I) in Φ; 0.2 on the maximum between Φ and Ψ; total in the bottom of Ψ up to the deflection ψ'; 0.4 between this point and Ω; total to beyond ω_2; 0.8 from ω_2 to X.

Thickness 6 mm.: Absorption zero to beyond line 219, 0.5 from this point to Ψ; total beyond.

Thickness 13 mm.: Absorption zero to beyond line 219. 0.7 from Φ-Ψ; total beyond.

As the result of these absorption experiments, as well as of the emission experiments, it appears that the band Φ is little influenced by thin layers of water. The absorption of water becomes evident at the long-wave side of Φ, is moderate up to Ψ, very great for a strip about 2' wide on the long-wave side of Ψ, moderate between Ψ and Ω, but still greater than between Φ and Ψ, very great for a little distance below Ω, and very considerable from here on. It was remarkable in these absorption experiments that Ψ and Ω were broadened out on their long wave-lengths sides, but not at all on their short wave-length sides, and in general that the strongest absorption is on the short wave-length side of the great bands, just as has been observed and noted above in examination of the bolographs.

SUMMARY OF CHAPTER VII.

The infra-red is the seat of great terrestrial atmospheric absorption. Certain bands, notably Φ, Ψ, and Ω, are found to vary greatly between different bolographs of the solar spectrum, and the relative intensities of energy at different parts of the spectrum, notably A–Φ and X–Y, change very appreciably. Some of these changes are found to be annual, others irregular. Of the former class the most conspicuous are the decrease in the relative intensity from A to Φ, and the broadening out on the long wave-length sides of Ψ and Ω in summer, and the increased prominence of absorption lines between X and Y, with frequent broadening of Ψ and Ω, in winter. Of the latter class the most conspicuous are the alterations in prominence of certain indentations in Φ and in the amount of absorption in a narrow region on the long-wave side of Φ. The regions of spectrum from A–Φ and that from ω_2 to X remain generally constant in appearance. if we except from consideration the relatively greater intensity of energy general from A–Φ during the winter and spring as compared with summer.

PART II.

SUBSIDIARY RESEARCHES.

Chapter I.

THE DISPERSION OF ROCK SALT AND FLUORITE.

In the appendix will be found a résumé of determinations by various observers of the dispersion of rock salt. These results (I speak of them more freely because I am myself one of the observers) are necessarily based in most cases on studies made with apparatus of less delicacy and precision than that now used, and may lead to error of considerable magnitude when it is attempted to interpolate between them for the wave-lengths corresponding to the infra-red absorption lines included in the tables here published, and inasmuch as it is the wave-lengths of these lines which have perhaps the greatest interest, and in consideration of the special facilities of this observatory, it has been thought best to redetermine by a direct method the dispersion of rock salt, and also to make a comparison between rock salt and fluorite—a substance frequently used for infra-red work. This investigation was carried through in 1899.

THE RELATIVE DISPERSION OF ROCK SALT AND FLUORITE.

The latter part of this investigation is the easier, and was first undertaken. The method employed consisted simply in taking a number of bolographs with a fluorite prism, comparing these with bolographs taken with rock salt, picking out common absorption lines, measuring the position of these lines in the fluorite bolographs, and, after reduction of these measures, comparing their results with those obtained for the salt.

A fluorite prism with faces about 30 mm. wide and 38 mm. high, and with a refracting angle of approximately 60°, was the only one available. As its resolution was only about one-fourth as great as that of the great salt prism R. B., in consequence of its small aperture, and as in further consequence of its small size the amount of radiation passing through it was reduced to about one-tenth that transmitted by the salt prism, which necessitated the use of wider slits and a wider bolometer, the detail in the fluorite bolographs fell far short of that obtained in 1897–98 with salt. It was therefore thought more accurate to compare these with salt prism bolographs of 1896–97, which had been measured and reduced, as has been said, but which were less full of detail than the later ones.

Fifteen bolographs were taken with this fluorite prism, of which eight covered the region of spectrum 0.76μ to 2.4μ, and seven the region from 1.8μ to 4.4μ.

As in the salt bolographs, some were taken in the direction from short wavelengths to long and others in the reverse. All precautions observed in taking the salt bolographs and described in Part I were observed here, except that the temperature was somewhat more variable, as the work was done in June and July of 1898, when the cooling apparatus was inadequate to preserve a constant temperature at $20°$. However, as the fluorite prism was so small, it is not probable that its angle varied appreciably with the temperature, and as the deviation of fluorite is, relatively to rock salt, far less influenced by the temperature, it is probable that the results are not much inferior to those obtained with rock salt in respect to accuracy of positions.

About fifty deflections common to the two kinds of bolographs were measured upon the fluorite bolographs. These were fairly uniformly distributed between wavelengths 0.76μ and 4μ. Beyond the latter point the fluorite dispersion becomes so great and the solar radiation so small that further progress was barred. The results appear in the following table. In reducing to absolute deviation, there arose occasion for knowing the angle of the fluorite prism and the minimum deviation of the A line. These constants were determined with all the precautions described on page 173 of Part I in connection with the salt-prism work. Without including details it may be said simply that as a result of five determinations the angle of the fluorite prism was found to be $59° 58' 0.4'' \pm 0.7''$, and as the result of eight determinations the deviation of A for this prism was found to be $31° 20' 39.4'' \pm 0.2''$. Temperature, $20°$ C.; barometer, 76 cm.

Corresponding constants for the salt prism R. B. I. as used in the comparison were

$$59° 55' 9.0'' \pm 1.6''$$
$$\text{and } 40° 20' 7.2'' \pm 1.6''$$
Temperature, $20°$ C.; barometer, 76 cm.

In the following table appear the results of the comparison. The three columns succeeding that giving the number of the lines contain the difference of deviation from A, absolute deviations and indices of refraction of the lines in the spectrum of the rock-salt prism. As the probable error for these quantities does not vary much between the different lines, the average probable error in each column is placed at its foot, and not given for each line separately. The next three columns give the similar values for the fluorite prism, and, as there is here considerably more variation in accuracy of determination, the probable errors are annexed to each column.

A plot is given in Pl. XXVI A in which the indices of refraction for rock salt are the ordinates and the indices for fluorite are the abscissæ. The original scale of plotting was so large that the sixth decimal place in n could be retained. It will be seen how closely the points lie to the curve, especially in the upper portion, where the absorption lines are sharp and narrow.

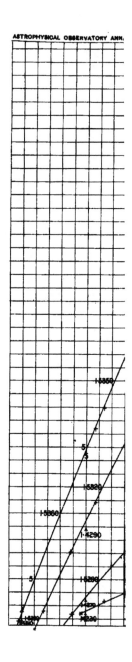

In the last column of the table are given the wave-lengths of the various lines as obtained from researches at this observatory on the dispersion of rock salt described upon a later page.

TABLE 25.—*Comparative dispersion of rock salt and fluorite.*

Number.	Rock salt.			Fluorite.					Wavelength. μ
	Difference of deviation.	Whole deviation.	Index of refraction.	Difference of deviation. $\Delta \theta$.	Probable error of $\Delta \theta$.	Whole deviation. θ.	Probable error of θ.	Index of refraction. n.	Probable error of n.	
*1	0 0 0.0	40 20 7.2	1.536818	0 0 0.00	0.00	31 20 39.4	0.2	1.431020	0.000003	0.7604
2	0 22 9	19 43.3	1.536747	0 7.26	.66	20 32.1	.7	1.430996	3	.7627
3	7 46.3	12 13.9	1.535360	2 55.80	.30	17 43.6	.4	1.430425	2	.8162
4	8 38.1	11 29.1	1.535206	3 15.60	1.32	17 23.8	1.3	1.430357	5	.8227
5	12 16.0	7 51.2	1.534526	4 39.54	0.66	15 59.9	0.7	1.430072	3	.8538
6	13 32.4	6 34.8	1.534287	5 11.82	.54	15 27.6	.6	1.429963	3	.8658
7	16 46.5	3 18.7	1.533676	6 38.96	.60	14 02.4	.7	1.429674	3	.8992
8	17 38.3	2 28.9	1.533520	6 53.44	.90	13 41.0	.4	1.429601	2	.9090
9	18 18.2	1 49.0	1.533427	7 11.88	.54	13 27.5	.6	1.429555	3	.9135
10	18 57.2	1 10.0	1.533274	7 28.20	.60	13 11.2	.7	1.429500	3	.9232
11	20 0.8	0 6.4	1.533075	7 58.32	.42	12 41.1	.5	1.429398	3	.9357
12	20 44.3	39 59 23.0	1.532940	8 18.90	.36	12 20.5	.4	1.429328	2	.9443
13	23 7.0	57 0.2	1.532494	9 26.52	.42	11 18.9	.5	1.429119	3	.9759
14	25 7.8	54 59.4	1.532116	10 18.60	.42	10 20.8	.5	1.428922	3	1.0041
15	29 11.7	50 55.5	1.531319	12 18.00	1.20	8 21.4	1.2	1.428516	5	1.0728
16	32 9.8	47 57.4	1.530796	13 58.08	0.66	6 41.3	0.7	1.428175	3	1.1240
17	32 56.7	47 10.5	1.530649	14 9.00	1.20	6 30.4	1.2	1.428138	5	1 1399
18	33 20.7	46 46.5	1.530574	14 31.20	0.42	6 8.2	0 5	1.428063	3	1.1481
19	34 55.8	45 11.4	1.530277	15 28.08	.60	5 11.3	.7	1.427870	3	1.1833
20	35 43.2	44 25.0	1.530132	15 57.90	1.38	4 41.5	1.4	1.427768	5	1.2016
21	38 10.7	41 56.5	1.529666	17 38.04	0.36	3 1.4	.4	1.427428	2	1.2658
22	40 20.1	39 47.1	1.529261	19 12.48	.60	1 26.9	.7	1.427107	3	1.3306
23	43 16.2	36 51.0	1.528709	21 36.84	1.32	30 59 2.5	1.3	1.426616	5	1.4332
24	44 11.4	35 55.8	1.528536	22 22.92	0.66	58 16.5	0.7	1.426459	3	1.4692
25	46 38.7	33 28.5	1.528074	24 51.60	.96	55 47.8	1.0	1.425953	4	1.5734
26	47 17.5	32 49.7	1.527953	25 33.24	1.26	55 6.2	1.3	1.425812	5	1.6017
27	48 1.2	32 6.0	1.527815	26 24.78	1.26	54 14.6	1.3	1.425637	5	1.6362
28	54 6.8	26 0.4	1.526667	35 3.72	0.78	45 35.7	0.8	1.423869	3	2.0000
29	54 49.9	25 17.3	1.526532	36 22.32	.42	44 17.1	.5	1.423601	3	2.0556
30	59 28.8	20 38.4	1.525656	45 54.48	.90	34 44.9	.9	1.421649	4	2.4160
37	59 53.4	20 13.8	1.525578	46 57.72	1.62	33 41.7	1.6	1.421433	5	2.4496
28	0 54 6.8	39 26 0.4	1.526668	0 35 3.72	0.78	30 45 35.7	0.8	1.423869	3	2.0000
29	54 48.9	25 18.3	1.526535	36 24.12	.78	44 15.3	0.8	1.423595	3	2.0536
30	55 39.7	24 27.5	1.526376	37 54.42	.84	42 45.0	.9	1.423287	4	2.1172
31	56 18.1	23 49.1	1.526255	39 8.28	1.20	41 31.1	1.2	1.423085	5	2.1658
32	56 46.3	23 20.9	1.526166	40 2.04	0.96	40 37.4	1.0	1.422851	4	2.2016
33	58 1.3	22 5.9	1.525930	42 35.52	1.02	38 3.9	1.0	1.422328	4	2.2900
34	58 40.2	21 27.0	1.525808	44 8.82	0.84	36 30.6	0.9	1.422009	4	2.3508
35	59 5.7	21 1,5	1.525729	45 4.44	.72	35 35.0	.7	1.421819	3	2.3846
36	59 28.8	20 38.4	1.525656	46 1.80	1.38	34 37.6	1.4	1.421625	5	2.4160
37	59 53.4	20 13.8	1.525578	47 1.44	1.20	33 37.6	1.2	1.421421	5	2.4496
38	1 0 3.8	14 3.4	1.524413	1 2 49.00	2.40	17 50.0	2.4	1.418182	10	2.9575
39	7 3.2	13 4.0	1.524226	5 42.00	3.00	14 52.0	3.0	1.417569	13	3.0430
40	7 37.7	12 29.5	1.524118	7 34.00	5.00	13 0.0	5.0	1.417205	20	3.0900
41	8 7.9	11 59.3	1.524022	9 24.00	2.70	11 15.0	2.7	1.416829	11	3.1340
42	9 1.6	11 5.6	1.523853	11 42.00	2.70	8 57.4	2.7	1.416356	11	3.2100
43	9 37.8	10 29.4	1.523739	13 45.00	2.50	6 54.0	2.5	1.415935	10	3.2610
44	10 14.2	9 53.0	1.523625	15 16.00	5.00	5 23.0	5.0	1.415621	20	3.3000
45	1 11 25.8	39 8 41.4	1.523399	1 19 8.00	3.00	30 1 32.0	3.0	1.414827	13	3.4090
Mean probable error	00.3	01.7	0.0000088		01.1		01.2		0.000005	

*A line.

Pl. XXVI B gives the dispersion curve for fluorite as thus determined. A small plot in the lower part of the plate shows the general form of the curve, while the full accuracy of the data is retained in the other curves plotted on a large scale.

Table 26 gives a comparison between the results of Paschen[1] and those here obtained on the dispersion of fluorite.

TABLE 26.—*Comparison with results of F. Paschen on the dispersion of fluorite.*

Wave-length, Paschen.	Refractive index, Paschen.	Refractive index, A. P. O.	Differences in refractive index.	Corresponding differences in wave-lengths.
μ				μ
0.76040	*1.43101	1.431021	−0.000011	0.0010
0.8840	1.42982	1.429805	+ .000015	− .0018
1.1786	1.42787	1.427892	− .000022	+ .0037
1.3756	1.42690	1.426892	+ .000008	− .0016
1.4733	1.42641	1.426421	− .000011	+ .0022
1.5715	1.42596	1.425948	+ .000012	− .0024
1.7680	1.42507	1.425020	+ .000050	− .0100
1.9153	1.42437	1.424307	+ .000063	− .0129
1.9644	1.42413	1.424066	+ .000064	− .0130
2.0626	1.42359	1.423558	+ .000032	− .0064
2.1608	1.42308	1.423054	+ .000026	− .0053
2.3573	1.42199	1.421975	+ .000015	− .0028
2.5537	1.42088	1.420861	+ .000079	− .0116
2.6519	1.42016	1.420182	− .000022	+ .0034
2.9466	1.41826	1.418224	+ .000036	− .0050
3.2413	1.41612	1.416099	+ .000021	− .0029

*Sarasin's value of n.

It will be understood, in comparing the work by Paschen with the subsequent work at Washington, that these have proceeded on somewhat different lines. Paschen has taken a fluorite prism, and by means of the bolometer compared it directly with a grating. In Washington a similar fluorite prism has been compared with a rock-salt prism of very much greater size, the determinations of whose wave-length in comparison with a large Rowland grating had been made with a possibility of very much higher degree of precision, owing, if to no other reason, to the far greater size of the apparatus employed, where size is a most important element of accuracy.

It is also to be remarked that the comparison between the fluorite and the salt in Washington was made by direct identification of absorption lines in bolographs of the solar spectrum.

THE DISPERSION OF ROCK SALT.

In the determination of the relation between refractive indices and wave-lengths for the rock-salt prismatic spectrum the general method first employed by me in 1885 has been made use of. This method, which was only adopted after the trial and rejection of a large number of others, consists in the form first employed by me[2] in

---
[1] Annalen der Physik und Chemie, 56, 765, 1895.
[2] American Journal of Science, third series, Vol. XXXII, p. 83, 1886.

mounting a large spectrobolometer upon a massive turntable. Its revolutions include a circle in whose circumference was the first slit, through which passed the rays of the electric arc or other source of heat, onto the concave grating, placed also in the circumference, from which grating they were diffracted to the second slit, which was that of the spectrobolometer, and which, though a large instrument, was here bodily movable about the common center. By the well-known law of the grating, its superimposed spectra fell upon this second slit, which, like the first, was parallel both to the lines of the grating and to the refracting angle of the prism.

This prism finally disentangled, so to speak, these (mostly invisible) superimposed spectra and distributed them at different angles, according to their wave-lengths, to separate positions, where they were finally recognized by the linear bolometer.

These were the early methods employed by the writer (1885) but to the reader of this volume it need hardly be said that such progress has been made in the use of the bolometer in the last fifteen years, not only in its delicacy, but in its preciseness of measurement; and that the sensitiveness, position, size, and scale of use in the present apparatus have all so grown since the early one that it has seemed advisable to repeat these important determinations, particularly as the prism can now be kept at a constant and well-known temperature, and any serious "drift" is no longer to be apprehended at the galvanometer.

Moreover, by the adoption of what appears to be an improved method (due to Mr. Abbot's suggestion), the number of circle readings described in the early memoir has been greatly reduced, none being necessary in the improved method except for the determination of the refracting angle of the prism. As already given, this was $59°\ 56'\ 14.9'' \pm 0.7''$. The index of refraction of "A" is 1.536818 ± 0.0000087. All other angular distances were determined with an accuracy which is believed to exceed that of any ordinary method of circle reading, by the employment of the system of special clockwork already described as in use in this bolographic work. The modified procedure, briefly outlined and as used here in 1899, is as follows:

The beam (here a solar beam) is reflected through the slit of the spectrobolometer, and a short bolograph of the A line or some other well-known region of the infra-red is taken. Next, the beam is directed upon the slit of the grating spectroscope; the grating is adjusted so that some readily identifiable visible solar line (like b, of the third spectrum, for instance) falls upon the spectrobolometer slit. Now, upon letting the clock run till the prism spectrum has passed through about forty minutes from where it was stopped after taking the bolograph of A, sharp maxima will be found corresponding to wave-lengths of $\frac{3}{2} \times 0.5184\mu$ and $3 \times 0.5184\mu$, and will appear upon the photographic plate upon which the galvanometer trace is recorded. The prismatic deviation of these maxima can be accurately determined with respect to the A line by measuring the plate upon the comparator, just as the bolographs were measured.

To avoid, however, the prejudicial delay incident to running the clockwork all the way from A to the maxima in question a simple device was added to the clock by means of which the shaft marked c in Pl. XIII of Part I could be rigidly clamped, while a cone friction clutch in that marked e was unclamped. The circle could thus be turned by a crank on the shaft b while the clock and the plate in the plate carrier stood absolutely still. By means of a rotation counter geared to the shaft m and a divided circle upon the same shaft reading by estimation of tenths to one six-hundredth of a rotation the prism could be set at any desired angular position without any motion of the plate, and (when due regard was paid to backlash) with no more inaccuracy than there would have been had the clock driven it as usual.

It was found best (excepting in the region between 0.76μ and 0.92μ, where the second order was necessarily employed) to use only the maxima found in the radiations of the first order spectrum; for these maxima were always so much richer in energy as to allow of slit widths five or ten times as narrow as could be used with the higher orders.

The Apparatus for Determining the Dispersion of Rock Salt.

Having given in brief the scheme of operations, a more detailed description of the arrangements employed will now be readily understood.

The source of radiation employed was for the most part the sun; but beyond a wave-length of 5μ it was found that the solar radiations remaining were less energetic than those from terrestrial heated bodies. From this point on the source employed was a Kitson vaporized petroleum lamp.[1] This Kitson lamp was kindly placed at the writer's disposal by the makers, and served very satisfactorily in these as well as in other experiments, which will be described in another chapter. In the description of apparatus which follows the solar beam is treated as the source of radiation. No change was made where the lamp was substituted except to dispense with the condensing mirror.

Pl. XXVII shows diagramatically the disposition of the apparatus. The siderostat mirror a sends a beam of light beneath the grating mounting to the plane mirrors b and c (c not shown), of which b reflects a portion of the beam back to d to be analyzed by the spectroscopic apparatus, and c reflects the remainder upward and by a circuitous path to the galvanometer mirror to form the photographic trace on the recording plate. The grating spectroscope is mounted upon a heavy stand of seasoned wood and cast iron which rests upon the north pier of the observatory.

d, e, f is a right-angled iron track upon which the grating and slit carriages move.

---

[1] This so-called Kitson vaporized petroleum lamp was tested photometrically and gave over 600 candle power, a really remarkable photometric result. It was for n and λ provided with a slotted brass chimney in place of the usual glass one.

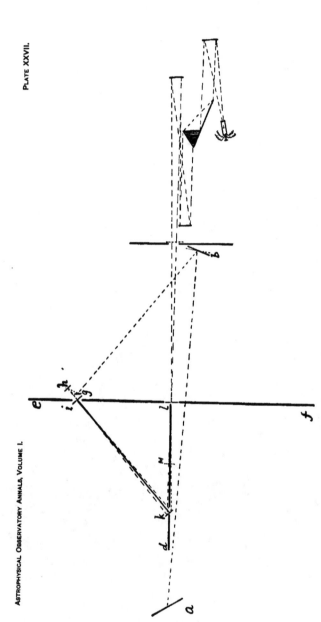

g is a plane mirror which reflects the beam from b nearly vertical upon the spherical concave mirror h, whose solar focus is halfway between the 10 cm. slit i and the concave grating k, and hence at the principal focus of the grating.[1]

The grating is the same employed by the author in his earlier research of this kind, and is that designated as "Grating No. 1" in his publication entitled, "On hitherto unrecognized wave-lengths"[2] and it is one for which he has already acknowledged his obligation to Professor Rowland.

Its constants are as follows:

Radius of curvature...................................millimeters.. 1626
Number of lines to millimeter................................... 142.1
Height of ruled portion...............................millimeters.. 102
Width of ruled portion................................do.... 146
Distance corresponding to 10,000 of Ångstrom's units on the line of slits.........do.... 231

Between the grating k and the second slit l is a plane mirror, m, not in use at the same time as the grating, but which may be interposed when the prism spectroscope alone is desired, as in taking the short bolographs already mentioned for referring the position of the wave-length maxima to known lines in the spectrum. When using the mirror m the mirror b is adjusted so as to send the solar beam directly to m instead of to g. Thence it is reflected through the slit l.

All the arrangements of the spectrobolometer were as described in Part I, except as follows: Two different systems of mirrors were used, both of which were different from those figuring in Part I. The first of these consisted of two collimating mirrors of 102 cm. focus and of 178 cm. focus, respectively, and an image-forming mirror of 230 cm. focus. This arrangement satisfied the energy conditions existing between wave-lengths 0.76 μ and 4.4 μ. Beyond this point collimation was dispensed with and the mirror of 102 cm. focus was used to form the image, so that the arrangement was as illustrated on page 30 of Part I. With this second arrangement no cylindric lens was required in front of the bolometer and a flat salt plate served instead to make the bolometer case air-tight.

The following bolometers have been used:

Number 20. Linear bolometer, 0.08 mm. wide, 10 mm. high. Resistance, 4 ohms.

Number 23. Area bolometer. Exposed area composed of five strips in series. Total width, 1.2 mm.; height, 17 mm. Resistance, 34 ohms.

For the remaining parts of the apparatus the reader is referred to the description in the earlier chapters of this volume.

---

[1] By thus employing the concave mirror the height of the grating spectrum as it comes to a focus at the second slit l is made equal to that of the first slit i whatever be the position of the grating on the arm d l; for the arrangement is such as to render the pencil of rays from any single point on the sun parallel in a vertical plane after leaving the grating, and the divergence of the whole beam is only that caused by the sun's comparatively slight angular aperture. The advantage of this disposition is obvious, for without it the beam would spread out vertically after passing through the solar focus of the grating at the rate of 5 cm. for every 80 cm. traveled, and the major part of the radiations would never reach the bolometer.

[2] American Journal of Science, third series, Vol. XXXII, p. 83, 1886.

PROCEDURE.

All the bolographic apparatus having been adjusted as described in Chapter III of Part I, the solar beam was reflected upon the second slit by means of the mirror m, and a short bolograph was taken either of the A line or of the great band Φ, which has sharp deflections even better suited than A for a reference mark. Readings of the counter and graduated circle of the clock were made by the observer at the galvanometer both before and after this run. This completed, the beam was caused to fall upon the first slit, and by the aid of an eyepiece, temporarily set up, the grating and first slit were so adjusted that a well-known line of the visible spectrum—as, for instance, that of wave-length $0.5616\,\mu$ in the third order—fell centrally in the second slit. This caused radiations of wave-length $1.685\,\mu$ in the first order spectrum to pass through the center of the second slit. The eyepiece was now removed, and an observer at the galvanometer clamped the plate as described on page 224, unclamped the circle, and turned it by the crank on shaft b until the deflection of the galvanometer needle warned him that the prism was at the proper position. The observer noted the magnitude of this deflection, called it out to the one at the grating, who modified slits one and two till the deflection was reduced to about 1 cm. The circle was now set back a couple of minutes of arc and clamped, the reading of its position made, the plate unclamped, the clock started, and the observer retired from the room.

At the end of four minutes the clock was stopped and the position of the circle again read. An observation by the eyepiece at the second slit assured that the grating apparatus had remained in adjustment, and a new line was brought upon the second slit and a new deflection obtained. After the lapse of an hour or so of this harvesting of accurately determined maxima a new short reference bolograph was taken lest there should have occurred some disturbance in the adjustment of the spectrobolometer.

To more clearly show the course of the work, a short extract from the original recorded observations is here inserted:

        Station, Astrophysical Observatory, Washington, D. C.
        Date, January 11, 1899.
        Source of heat, sun.
        Height of slit, 100 mm.
        Prism, RB_1.
        Grating, No. 1.
        Galvanometer, No. 3.
        Time single vibration, $3\frac{1}{4}$ seconds.
        Bolometer, No. 20, 0.08 mm., subtend $7.0''$.
        Observer at grating, C. G. A.
        Observer at galvanometer, F. E. F.
        Remarks: Some disturbance from the movements of workmen employed near the Observatory.
        Immediate object of observation: Determination of relation of n to λ for certain heat maxima.

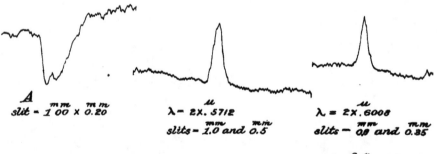

THE DISPERSION OF A 60° ROCK-SALT PRISM.
OBSERVATIONS OF S. P. LANGLEY WITH THE ASSISTANCE OF C. G. ABBOT.

Spectrum speed, 1' in 1 minute; plate speed, 1 cm. in 1 minute. 1'=1 cm. = 1 minute.
Spectrum thrown east.
Temperature in prism, 18.0°; temperature in air, 17.7° (at 9ʰ 30ᵐ).

| Time. | Circle. | Slits. | | Region. | Grade. |
		1	2		
h. m. s.	′ ″	mm.	mm.		
Start 9 56 45	1199 0.1	....	0.20	}A line......	a–b
End 10 2 30	1204 33.9	....	....		
Start 30 30	31 0.1	1.0	0.65	}2×0.5710μ ..	a–b
End 37 30	37 57.2	....	....		
Start 56 15	35 0.0	0.8	0.35	}2×0.6008μ ..	a
End 11 1 0	39 39.9	....	....		
Start 11 42 0	1199 2.1	....	0.20	}A line......	a–b
End 47 0	1204 6.4	....	....		

The complement of these observations is of course the comparator measurements upon the plate. These are as follows:

[Observer F. E. F.]

Start.	Middle.	Maximum.	End.	Height.	Width.	Region.
				cm.	cm.	
6.595 ⎫ 6.595 6.595 ⎭	8.800 ⎫ 8.800 8.799 ⎭	....	12.159 ⎫ 12.158 12.157 ⎭	....	....	A
12.159 ⎫ 12.158 .157 ⎭	15.430 ⎫ 15.430 .429 ⎭	15.446 ⎫ 15.447 .448 ⎭	19.111 ⎫ 19.111 .111 ⎭	2.3	0.8	2×0.5710 μ
19.111 ⎫ 19.111 .111 ⎭	21.068 ⎫ 21.066 .065 ⎭	21.037 ⎫ 21.037 .037 ⎭	23.780 ⎫ 23.780 .779 ⎭	2.0	0.7	2×0.6008 μ
34.979 ⎫ 34.978 .978 ⎭	37.230 ⎫ 37.231 .232 ⎭	....	40.055 ⎫ 40.054 .053 ⎭	....	....	A

Pl. XXVIII shows the first three of these observations reproduced on the original scale 1 cm. = 1' = 1 minute. The reader will remark the narrowness and sharpness of the deflections. If he will compare them with those recorded in the paper "On hitherto unrecognized wave-lengths," already referred to, he will see what advance has been made since that date in the application of the bolometer.

In the reduction of these observations a choice is possible as to which readings shall be taken. Where no choice was indicated by the appearance of the plate the readings of circle and plate at the *end* of each run were made use of. As some of the maxima were slightly unsymmetrical (see differences of 0.017 cm. and 0.030 cm. above, between "middle" and "maximum") it was made a rule to use the "middle" as indicating the position of the deflection rather than the point of "maximum," hoping thus to avoid error from nonsymmetry. As there were several reference points on each plate a question might arise which should be used. In general it was found that the distances of maxima from A as reduced separately from these several reference bolographs would vary slightly but progressively in the same direction, owing very probably to the use of wood in the mounting of the grating and slits. Hence, it was made the rule to determine the shift of the two successive reference lines by their circle readings, and, treating this shift as a linear function of the time, to correct each distance of a maximum from the reference mark preceding it by the proper proportion of the total shift between the two reference marks preceding and following the maximum, taking into account the relative times elapsed between observations.

As an example the reduction of the preceding observations is here given. In the reduction 1cm.=1.000' of arc:

A line taken at $9^h 56^m$.		A line taken at $11^h 46^m$.	
Reading at end of run....	12.158^{cm}	Reading at end of run....	40.054^{cm}
Reading at bottom of A ..	8.800	Reading at bottom of A...	37.231
Difference..........	3.358	Difference..........	2.823
Circle at end.............	1204.565'	Circle at end.............	1204.107'
Position of A.............	1201.207'	Position of A.............	1201.284'
Position difference = 0.077'.		Shift for 10^m = 0.007'.	
Time difference = 110^m.			

	Maximum, $2 \times 0.5710 \mu$ at $10^h 36^m$.	Maximum, $2 \times 0.6008 \mu$ at $10^h 56^m$.
Plate reading at end of this run	19.111	23.780
Plate reading at middle of deflection........................	15.430	21.066
Difference	3.681	2.714
Circle at end of this run..................................	1237.953'	1239.665
Circle at end of A run	4.565	4.565
Difference..	33.388	35.100
Distance A to end of its run	3.358	3.358
Sum	36.746	38.458
Subtracting...	3.681	2.714
Distance A to middle of deflection	33.065	35.744
Correction for shift during elapsed time	—0.028	—0.042
Corrected distance from A................................	33.037^{cm}	35.702^{cm}

The number of different wave-lengths at which the prismatic deviation was thus determined was 38. Nearly all were four times observed. That is, two check observations, usually on different days, were made with spectrum thrown east, and two with spectrum thrown west. These four determinations were thus wholly independent.

RESULTS.

In the tables which follow the results of the investigation are given. Considerable data, which are not indispensable but may be useful, are included in Table 27,

ANNALS OF ASTROPHYSICAL OBSERVATORY. 229

such as the slit widths, heights and widths of max.ma, position of the maximum point of the deflection, and the line set on in the visible spectrum. What immediately concerns us, however, is the column of wave-lengths, the column "Spectrum thrown east or west," and the columns of positions of middle of the deflection, which in Table 27 is given in centimeters from line 219 of Table 18, Part I.

TABLE 27.—*Observations on the dispersion of rock salt, made with the diffraction grating.*

Orders of spectra.	Line set on.	Bolograph and date.	Spectrum thrown.	Slit widths.			Bolometer subtends.	Deflections.				Wave-length.
				Grating.	Prism.	Prism.		Height.	Width.	Maximum at—	Middle at—	
	μ			mm.	"	"		cm.	cm.	cm.	cm.	μ
3/2	0.5328	II, Jan. 3, 1899	E.	2.5	1.25	31.2	7	0.3	1.5	27.019	27.048	0.799
		I, Jan. 20, 1899		3.5	2.0	50.0		.4	2.0	....	26.55	
		I, Jan. 20, 1899		4.0	3.0	75.0		.5	3.0	....	26.25	
		II, Mar. 16, 1899	W.	10.0	10.0	250.0		.7	2.5	....	26.63	
		II, Mar. 16, 1899		10.0	10.0	250.0		.6	2.0	....	26.48	
3/2	.5616	II, Jan. 3, 1899	E.	2.2	1.25	31.2	7	.5	1.5	21.769	21.731	
		I, Jan. 25, 1899	W.	4.0	2.5	62.5		1.5	2.5	....	21.605	.842
		II, Mar. 16, 1899		7.0	5.0	125.0		1.5	2.5	....	21.98	
		II, Mar. 16, 1899		7.0	5.0	125.0		1.5	2.5	....	21.50	
3/2	.5890	II, Jan. 3, 1899	E.	2.2	1.0	25.0	7	1.1	1.3	17.268	17.246	.883
		II, Jan. 3, 1899		2.5	1.0	25.0		.6	1.3	17.359	17.279	
		I, Jan. 20, 1899		3.0	1.5	37.5		1.8	2.3	....	17.157	
		I, Jan. 25, 1899	W.	3.0	1.5	37.5		1.2	1.8	....	17.354	
		II, Mar. 16, 1899		5.0	2.5	62.5		1.2	2.0	....	17.264	
		II, Mar. 16, 1899		5.0	2.5	62.5		1.3	2.0	....	17.39	
3/2	.6022	II, Jan. 3, 1899	E.	2.5	1.0	25.0	7	.6	1.3	15.539	15.440	.903
		II, Jan. 3, 1899		2.5	1.0	25.0		.6	1.3	15.421	15.458	
		I, Jan. 25, 1899	W.	2.5	1.5	37.5		.7	1.0	....	15.589	
		II, Mar. 16, 1899		3.0	2.0	50.0		1.5	1.5	....	15.419	
2/1	.4862	I, Jan. 7, 1899	E.	1.0	.45	11.2	7	2.8	1.0	9.709	9.691	.972
		II, Jan. 7, 1899		1.0	.45	11.2		1.3	.8	9.744	9.712	
		II, Jan. 7, 1899		1.0	.45	11.2		3.0	1.0	9.709	9.689	
		I, Jan. 23, 1899	W.	.8	.4	10.0		1.2	.6	9.724	9.692	
		I, Mar. 11, 1899		1.2	.75	18.7		1.5	.6	9.701	9.706	
2/1	.4958	I, Jan. 7, 1899	E.	1.0	.45	11.2	7	2.3	.8	....	8.316	.992
		II, Jan. 7, 1899		1.0	.45	11.2		2.0	.6	8.408	8.386	
		I, Jan. 23, 1899	W.	.8	.4	10.0		1.6	.8	8.464	8.429	
		I, Mar. 11, 1899		1.25	.80	20.0		2.4	.6	8.264	8.251	
2/1	.5042	I, Jan. 7, 1899	E.	.8	.45	11.2	7	3.0	.8	7.211	7.188	1.008
		II, Jan. 7, 1899		1.0	.45	11.2		2.2	.8	7.201	7.170	
		I, Jan. 23, 1899	W.	.8	.4	10.0		1.7	.7	7.196	7.187	
		I, Mar. 13, 1899		1.0	.5	12.5		1.5	.7	7.118	7.109	
2/1	.5184	I, Dec. 30, 1898	E.	1.5	.75	18.7	7	2.7	1.0	5.193	5.179	1.037
		I, Jan. 3, 1899		1.5	.75	18.7		3.5	1.3	5.226	5.216	
		I, Jan. 23, 1899	W.	2.8	.4	10.0		1.7	.6	5.352	5.337	
		I, Mar. 11, 1899		1.25	.70	17.5		2.0	.7	5.242	5.194	
2,1	.5270	I, Jan. 7, 1899	E.	1.0	.45	11.2	7	2.8	.8	4.298	4.304	1.054
		II, Jan. 7, 1899		1.0	.45	11.2		3.2	.7	4.392	4.370	
		I, Jan. 23, 1899	W.	2.6	.3	7.5		1.5	.6	4.326	4.329	
		I Mar. 11, 1899		1.0	.5	10.0		1.6	.7	4.261	4.217	
2,1	.5405	I, Jan. 7, 1899	E.	1.0	.45	11.2	7	2.5	.8	2.774	2.744	1.081
		II, Jan. 7, 1899		1.0	.45	11.2		2.5	.6	2.794	2.768	
		I, Jan. 23, 1899	W.	.6	.3	7.5		.9	.6	2.912	2.935	
		I, Mar. 11, 1899		1.0	..5	10.0		1.8	.7	2.652	2.647	
2,1	.5529	I, Jan. 7, 1899	E.	1.2	.5	10.0	7	3.3	.9	1.494	1.464	1.106
		II, Jan. 7, 1899		1.0	.45	11.2		3.4	.8	1.462	1.408	
		I, Jan. 23, 1899	W.	.9	.3	7.5		1.0	.6	1.443	1.431	
		I, Mar. 11, 1899		1.0	.5	10.0		.8	.5	1.331	1.313	
		I, Mar. 13, 1899		1.0	.5	10.0		1.5	.6	1.436	1.427	
2/1	.5710	I, Jan. 11, 1899	E.	1.0	.65	16.2	7	2.3	.8	.428	.411	1.142
		I, Jan. 19, 1899		1.0	.65	16.2		1.7	1.0	.512	.539	
		I, Mar. 13, 1899	W.	1.3	.8	20.0		1.2	.6	.371	.410	

TABLE 27.—*Observations on the dispersion of rock salt, made with the diffraction grating*—Continued.

Orders of spectra.	Line set on.	Bolograph and date.	Spectrum thrown.	Slit widths.			Bolometer subtends.	Deflections.				Wavelength.
				Grating.	Prism.	Prism.		Height.	Width.	Maximum at—	Middle at—	
	μ			mm.		''	''	cm.	cm.	cm.	cm.	μ
		I, Jan. 23, 1899		0.6	0.3	7.5		0.5	0.7	0.355	0.368	
2/1	0.5890	I, Dec. 29, 1898	E.	1.50	.75	18.7	7	2.5	1.2	2.174	2.197	1.178
		II, Dec. 29, 1898		1.50	.75	18.7		2.8	1.0	2.158	2.186	
		I, Jan. 19, 1899		1.0	.5	12.5		1.9	.8	2.147	2.215	
		II, Jan. 21, 1899	W.	1.0	.5	12.5		1.4	.5	2.088	2.095	
		I, Mar. 11, 1899		1.0	.6	15.0		1.4	.6	2.066	2.087	
2/1	.6008	I, Jan. 11, 1899	E.	.8	.35	8.7	7	2.0	.7	3.047	3.076	1.202
		I, Jan. 19, 1899		1.0	.5	12.5		2.4	.7	3.084	3.173	
		II, Jan. 21, 1899	W.	.6	.3	7.5		1.0	.5	....	2.948††	
		I, Mar. 11, 1899		.9	.4	10.0		1.0	.4	3.038	3.045	
2/1	.6302	I, Jan. 11, 1899	E.	.8	.35	8.7	7	1.5	.6	5.337	5.383	1.260
		I, Jan. 19, 1899		1.0	.5	12.5		2.5	.8	5.377	5.441	
		II, Jan. 21, 1899	W.	.6	.3	7.5		1.2	.5	5.392	5.424	
		I, Mar. 11, 1899		1.0	.5	12.5		1.5	.6	5.294	5.368	
2/1	.6563	I, Jan. 11, 1899	E.	.8	.35	8.7	7	2.0	.6	7.118	7.148	1.313
		I, Jan. 19, 1899		1.0	.5	12.5		2.5	.8	7.216	7.300	
		I, Jan. 25, 1899	W.	.55	.25	6.2		.5	.5†	....	6.978	
		I, Jan. 23, 1899		.6	.3	7.5		.8	.5	7.128	7.139	
		I, Mar. 11, 1899		1.0	.5	12.5		1.0	.5	7.195	7.237	
3/1	.4958	I, Jan. 19, 1899	E.	.8	.45	11.2	7	1.3	....	12.077	12.113	1.487
		II, Jan. 19, 1899		1.0	.5	12.5		1.8	.6	11.933	12.006	
		I, Mar. 13, 1899	W.	1.3	.8	20.0		1.2	.7	11.998	12.011	
		I, Jan. 23, 1899		1.1	.6	15.0		1.4	.7	11.968	11.978	
		II, Jan. 25, 1899		1.5	.75	18.7		2.4	.8	12.027	12.087	
3/1	.5184	I, Dec. 30, 1898	E.	1.5	.75	18.7		2.7	.7	13.718	13.738	1.555
		I, Jan. 3, 1899		1.5	.75	18.7		3.7	.9	13.746	13.758	
		I, Mar. 11, 1899	W.	1.0	.5	12.5		1.3	.5	13.599	13.633	
		I, Jan. 23, 1899		.6	.3	7.5		.7	.5	13.522	13.530	
		II, Jan. 25, 1899		.9	.4	10.0		1.2	.5	13.608	13.678	
3/1	.5456	II, Jan. 20, 1899	E.	1.0	.5	12.5	7	1.1	1.2	15.466	15.490	1.637
		II, Jan. 20, 1899		1.0	.5	12.5		.7	.7	15.524	15.545	
		I, Jan. 25, 1899	W.	.6	.3	7.5		.8	.8	15.298	15.332	
		I, Jan. 23, 1899		.7	.4	10.0		.9	.5	15.358	15.378	
		II, Jan. 25, 1899		1.0	.5	12.5		1.8	.6	15.320	15.394	
3/1	.5616	I, Jan. 19, 1899	E.	.8	.45	11.2	7	3.0	.7	16.319	16.357	1.685
		I, Jan. 19, 1899		1.0	.5	12.5		1.9	.7	16.322	16.392	
		II, Jan. 19, 1899		1.0	.5	12.5		1.0	.5	16.342	16.364	
3/1	.5890	I, Dec. 29, 1898	W.	1.5	.75	18.7	7	8.0	1.2	17.771	17.893	1.767
		II, Dec. 29, 1898		1.5	.75	18.7		4.6	1.0	17.641	17.663	
		I, Jan. 25, 1899	E.	.7	.3	7.5		.8	.5	17.694	17.760	
		I, Jan. 25, 1899		.7	.3	7.5		.7	.5	17.682	17.708	
4/1	.5184	I, Dec. 30, 1898	W.	1.5	.75	18.7	7	.7	.6	....	22.544	2.074
		I, Jan. 3, 1899		1.5	.75	18.7		3.0	.8	22.498	22.497	
		I, Jan. 25, 1899	E.	1.0	.4	10.0		.8	.6	22.411	22.434	
		II, Jan. 25, 1899		1.2	.5	12.5		1.1	.6	22.421	22.456	
4/1	.5456	I, Jan. 19, 1899	E.	1.0	.5	12.5	7	1.6	.7	23.913	23.962	2.182
		II, Jan. 19, 1899		1.0	.5	12.5		2.0	.5	23.931	23.948	
		I, Jan. 25, 1899	W.	1.0	.4	10.0		1.0	.5	23.829	23.888	
		II, Jan. 25, 1899		1.2	.5	12.5		1.3	.6	23.894	23.961	
4/1	.5616	I, Jan. 19, 1899	E.	1.0	.5	12.5	7	2.5	.8	24.798	24.845	2.246
		II, Jan. 19, 1899		1.0	.5	12.5		2.0	.5	24.734	24.744	
		I, Jan. 23, 1899	W.	1.0	.5	12.5		1.2	.5	24.690	24.726	
		II, Jan. 25, 1899		1.3	.6	15.0		1.2	.7	24.704	24.741	
4/1	.5890	I, Dec. 29, 1898	E.	1.5	.75	18.7	7	1.8	1.0	26.135	26.219	2.356
		II, Dec. 29, 1898		2.0	.75	18.7		1.8	1+	26.202	26.353	
		I, Jan. 25, 1899	W.	1.0	.5	12.5		.8	.5	26.143	26.134	
		II, Jan. 25, 1899		1.5	.6	15.0		1.3	.5	26.133	26.146	
8/2	.5890	I, Feb. 28, 1899		1.0	4.0	120.0	210	2.5	9.2	*26.92	*26.84	
		II, Mar. 1, 1899		....	....	....		2.5	10.0	....	*27.01	
9/2	.5890	I, Feb. 28, 1899	W.	10.0	8.0	240.0	210	3.0	12.0	30.24	30.92	2.650

*Omit these from mean.

TABLE 27.—*Observations on the dispersion of rock salt, made with the diffraction grating*—Continued.

Orders of spectra.	Line set on.	Bolograph and date.	Spectrum thrown.	Slit widths.			Bolometer subtends.	Deflections.				Wavelength.
				Grating.	Prism.	Prism.		Height.	Width.	Maximum at—	Middle at—	
	μ			*mm.*		''	''	*cm.*	*cm.*	*cm.*	*cm.*	μ
5/1	0.5616	II, Jan. 20, 1899	E.	10.0	5.0	125	7	0.5	2.2	....	32.597	2.808
		II, Jan. 20, 1899		10.0	5.0	125		.5	1.5		32.755†	
5/1	.5890	II, Dec. 29, 1898	E.	5.0	5.0	125	7	.2	2.5	....	33.00†	2.945
10/2	.5890	I, Feb. 28, 1899	W.	10.0	2.5	75	210	2.8	10.0	34.46	34.54	
6/1	.5184	I, Jan. 3, 1899	E.	8.0	5.0	125	7	.3	1.5	36.41	35.84	3.110
		I, Jan. 3, 1899		8.0	5.0	125		.3	.8		36.27	
		I, Jan. 25, 1899	W.	7.0	4.0	100		1.0	2.5	....	35.138	
		I, Jan. 25, 1899		7.0	4.0	100		1.0	2.0		35.282	
6/1	.5456	II, Jan. 19, 1899	E.	4.0	2.0	50	7	.5	.7	....	37.165	3.274
		I, Jan. 20, 1899		4.0	2.0	50		2.0	.5		37.36	
		I, Jan. 25, 1899	W.	7.0	4.0	100		1.0	2.5		37.092	
		I, Jan. 26, 1899		7.0	4.0	100		1.1	2.0	37.476	37.184	
6/1	.5616	II, Jan. 19, 1899	E.	4.0	2.0	50	7	.7	1.5	38.532	38.508	3.370
		I, Jan. 20, 1899		4.0	2.5	50		1.0	2.0		38.485	
		II, Jan. 25, 1899	W.	5.5	3.5	87		1.0	2.3	38.566	38.292	
		I, Jan. 26, 1899		5.5	3.5	87		1.0	1.7	38.384	38.401	
7/1	.5184	I, Jan. 3, 1899	E.	3.0	1.5	37	7	1.5	1.0		41.715	3.629
		I, Jan. 3, 1899		3.0	1.5	37		1.4	1.0	....	41.74	
		II, Jan. 25, 1899	W.	3.0	1.5	37		1.2	.8	41.730	41.729	
		I, Jan. 26, 1899		3.0	1.5	37		1.3	.9	41.743	41.741	
7/1	.5456	II, Jan. 20, 1899	E.	4.0	2.0	50	7	1.2	1.5	44.380	44.353	3.819
		II, Jan. 25, 1899	W.	3.5	2.0	50		1.9	1.5	44.234	44.176	
		I, Jan. 26, 1899		3.5	2.0	50		1.5	1.4	44.421	44.289	
7/1	.5890	I, Dec. 29, 1898	E.	4.0	3.0	75	7	.4	2.0		48.32	4.123
		II, Jan. 20, 1899		5.0	3.0	75		.5	2.5	49.07	48.68	
		II, Jan. 20, 1899		8.0	5.0	125		.6	3.0	....	48.51	
		I, Mar. 6, 1879		6.0	4.0	120	210	1.0	10.0	48.64	48.74	
8/1	.5890	I, Dec. 29, 1898	E.	4.0	3.0	75	7	.4	2.0	....	57.2	4.712
		I, Dec. 30, 1898		10±	10±	250		.3	5.2		57.52	
		I, Mar. 6, 1899		10.0	8.0	240	210	.8	8.0	....	57.68	
		I, Mar. 8, 1899		10.0	10.0	300		3.0	12.0		58.02	
		I, Feb. 28, 1899	W.	10.0	8.0	240		2.0	8.0	58.72	58.76	
		II, Mar. 1, 1899		6.0	3.0†	90†		3.0	9.2	58.21	57.73	
9/1	.5890	I, Mar. 6, 1899	E.	10.0	12.0	360	210	.5	14.0	....	*62.88	5.801
		I, Mar. 6, 1899		10.0	12.0	360		.3	10.8		66.81	
		I, Mar. 8, 1899		10.0	10.0	300		1.2	12.0	66.78	66.86	
		I, Mar. 8, 1899		10.0	10.0	300		.8	10.0	67.82	68.02	
		I, Feb. 28, 1899	W.	10.0	8.0	240		2.3	10.0	68.84	68.84	
10/1	.5890	I, Feb. 28, 1899	W.	10.0	10.0	300	210	.8	10.0	....	80.00	5.890
		I, Mar. 1, 1899		10.0	12.0	360		.7	12.0		80.65	
		I, Mar. 1, 1899		10.0	12.0	360		.7	12.0	....	81.29	
		II, Mar. 1, 1899		10.0	12.0	360		1.0	16.0		79.97	
		II, Mar. 1, 1899		10.0	12.0	360		.8	16.0		80.82	
11/1	.5890	Mar. 1, 1899	W.	(†)	(†)	(†)		Eye observation.		....	91.49	6.479

*Omit these from mean.

Further reduction was effected in the following manner: Corresponding to each wave-length the observed linear deviations with spectrum thrown east and spectrum thrown west were averaged separately, and the mean of these two separate averages was taken. A number of observations gave the mean distance from A to line 219, above mentioned, as 32.626 cm., and the mean was thus reduced to the A line as reference point. A correction to this mean, obtained as described at page 175 of Part I, was then applied to make the distances from A (determined for a prism angle of 59° 56′ 20.7″) correspond to a prism angle of 59° 53′ 30″, the angle used in reduction of the bolographs of 1897–98.

It will be noted that the results of these experiments as thus far reduced are quite sufficient to give the wave-lengths of the infra-red absorption lines given in Part I. For we have the distances of various wave-lengths from A in terms of the deviation of a salt prism of known angle and the distances of the absorption lines measured in the same way. All that is necessary, therefore, in order to get the wave-lengths of the absorption line, is merely to plot differences of deviation and wave-lengths on a sufficient scale and to interpolate upon this plot for the differences of deviation found for the absorption bands in Table 18 of Part I.

This has accordingly been done, and the wave-lengths of Table 21 of Part I are the result. They are independent of the absolute deviations of A, and are subject to scarcely greater errors than correspond to the probable errors of the differences of deviation from A.

It should be noted that in addition to the points determined by Table 28 five others were used between 0.7604 μ and 0.8424 μ in the above-mentioned plot, which were obtained by comparing bolographs with Higgs's photographs. Of these five points, two (i. e., those corresponding to the A line and the "Z line") have greater weight than our grating determinations (the second-order spectrum was used here) and the curve has been made to pass exactly through them.

The scale of the plot from A to Ω was such that linear deviations could be read to 0.002 cm. and wave-lengths to 0.0001 μ. From Ω to the end of the region covered by bolographs a somewhat less open scale was used upon which the deviations could be read to 0.005 cm. and the wave lengths to 0.0002 μ. It will indicate the high degree of accuracy attained merely to say that the average deviation in wave-lengths from the curve of the 26 points on the first part of the plot was only 0.0007 μ and that of the 17 points on the second part was 0.0209 μ. If, however, we leave out three points from the first part and four points from the second, which in drawing the curve had to be practically neglected, the average deviations remaining would be only 0.0002 μ and 0.0078 μ,[1] respectively.

The uncorrected mean was further treated by multiplication by the factor 59.998″ to reduce to angular deviation from A. The probable error of the angular deviation from A was obtained as follows: The separate observations with spectrum thrown west were subtracted from their mean, and the same procedure was followed with the observations with spectrum thrown east. These residuals from the separate means were then treated in the usual manner and the result reduced to seconds of arc.

---

[1] The deviations remain almost as small as in the first part till we come to a wave-length of 4.14 μ.

TABLE 28.—*Reduced observations on the dispersion of rock salt.*

Wavelength.	Distance from A line.	Correction to reduce to prism angle of bolographs.	Linear deviation corrected to prism angle of bolographs.	Angular deviation from A line.	Probable error of deviation from A line.	Angular deviation.[*]
μ	cm.	cm.	cm.	° ′ ″	″	° ′ ″
0.7992	6.040	0.010	6.030	0 6 2.4	5.2	40 15 22.0
.8424	10.922	.019	10.903	10 55.3	3.7	10 29.1
.8835	15.344	.026	15.318	15 20.6	1.1	6 3.8
.9033	17.150	.029	17.121	17 9.0	1.4	4 15.4
.9724	22.928	.039	22.889	22 55.6	0.2	39 58 28.8
.9916	24.280	.041	24.239	24 16.8	1.8	57 7.6
1.0084	25.462	.043	25.419	25 27.7	0.7	55 56.7
.0368	27.394	.047	27.347	27 23.6	1.3	54 0.8
.0540	28.321	.048	28.273	28 19.2	1.3	53 5.2
.0810	29.852	.051	29.801	29 51.1	2.3	51 33.3
.1058	31.213	.053	31.160	31 12.7	1.1	50 11.7
.1420	33.058	.056	33.002	33 3.4	1.3	48 21.0
.1780	34.768	.059	34.709	34 46.0	0.3	46 38.4
.2016	35.686	.061	35.625	35 41.1	1.4	45 43.3
.2604	38.030	.064	37.966	38 1.7	0.8	43 22.7
.3126	39.796	.068	39.728	39 47.7	2.2	41 36.7
.4874	44.670	.076	44.594	44 40.1	1.2	36 44.3
.5552	46.307	.079	46.228	46 18.3	1.0	35 6.1
.6368	48.069	.082	47.987	48 4.0	0.7	33 20.4
.6848	48.997	.083	48.914	48 59.7	0.5	32 24.7
.7070	50.382	.086	50.296	50 22.8	2.0	31 1.6
2.0736	55.108	.094	55.014	55 6.4	0.5	26 18.6
.1824	56.566	.097	56.469	56 33.8	0.7	24 50.6
.2461	57.390	.098	57.292	57 23.3	0.8	24 1.1
.3560	58.839	.100	58.739	58 50.2	1.1	22 34.2
.6505	63.546	.108	63.438	1 3 34.6	....	17 51.8
.8080	65.302	.112	65.190	5 18.0	4.0	16 6.4
.9450	66.396	.113	66.283	6 23.6	....	15 0.8
3.1104	68.258	.117	68.141	8 15.3	4.2	13 9.1
.2736	69.826	.119	69.707	9 49.4	2.1	11 35.0
.3696	71.047	.121	70.926	11 2.7	1.0	10 21.7
.6288	74.358	.127	74.231	14 21.3	0.3	7 3.1
.8192	76.918	.134	76.784	16 54.9	1.4	4 29.5
4.1230	81.188	.139	81.049	21 11.1	1.3	0 13.3
.7120	90.564	.155	90.409	30 33.7	7.3	38 50 50.7
5.3009	100.661	.173	100.488	40 39.5	11.6	40 44.9
.8900	113.172	.193	112.979	53 10.1	11.4	28 14.3
6.4790	124.116	0.212	123.904	2 4 6.7	20.0	38 17 17.7

[*] Prism angle =59° 56′ 20.7″; deviation of A =40° 21′ 24.4″; refractive index of A =1.536818.

TABLE 29.—*The dispersion of rock salt. Summary of results.*

Wave length.	Index of refraction.	Probable error of n.
μ	n	
0.4861	1.553339	
.4920	2646	
.4937	2456	
.4983	1940	
.5173	0006	
.5184	.549862	
.5273	9081	
.5372	8175	
.5530	6855	
.5660	5800	
.5710	5546	Average, 0.000009.
.5758	5194	
.5786	5002	
.5860	4505	
.5893	4273	
.6105	3000	
.6400	1405	
.6563	0635	
.6874	.539307	
.7190	8150	
0.7604	1.536818	

The above values are derived from bolographs of the visible spectrum taken March 1, 1898, and compared with Rowland's map.

Wave length.	Index of refraction.	Probable error of n.	Wave length.	Index of refraction.	Probable error of n.
μ	n		μ	n	
0.7992	1.535691	0.000020	1.6848	1.527638	0.000009
.8424	4778	16	.7670	7377	11
.8835	3952	09	2.0736	6487	09
.9083	3613	10	.1824	6213	09
.9724	2332	09	.2464	6038	09
.9916	2278	11	.3560	5783	09
1.0084	2057	09	.6505	4897	....
.0368	1695	10	.8080	4566	17
.0540	1521	10	.9450	4359	....
.0810	1234	12	3.1104	4008	17
.1058	0979	09	.2736	3712	11
.1420	0633	10	.3696	3481	09
.1780	0312	09	.6288	2856	09
.2016	0139	10	.8192	2372	10
.2604	1.529699	09	4.1230	1564	18
.3126	9368	13	.7120	1.519789	27
.4874	8452	10	5.3009	7874	42
.5552	8144	09*	.8900	5497	42
1.6368	1.527813	0.000009	6.4790	1.513413	0.000069

The above values were obtained through the n and λ experiments.

The absolute deviation of the maxima corresponding to the wave-lengths observed was obtained as follows. In Part I it was shown that the index of refraction at A is 1.536818. From this value we obtain the deviation of A for a prism of refracting angle 59° 56′ 20.7″ to be 40° 21′ 24.4″. Subtracting from this the values of angular deviations from A, we have the last column of Table 28.

Indices of refraction have been computed corresponding to the absolute deviations of Table 28, and they appear in Table 29 with their probable errors, which here, of course, include the probable error in determining the index of refraction for A as given in Part I.

In order to extend the table toward the violet end of the spectrum two very fine bolographs of the whole solar spectrum, taken March 1, 1898, have been measured at several points. These bolographs show about one hundred deflections corresponding to lines in the visible spectrum, but were taken at fast speeds for totally other purposes, else they might very likely have shown several times as many. All the spectrum from 0.4 μ to 6.5 μ is, therefore, mapped by the bolographic method, and the table depends upon visual observations for its absolute refractive indices only in so far that the angle of the prism and the minimum deviation of the A line were thus determined.

In Pl. XXIX is plotted the dispersion curve of rock salt on a scale adequate to retain the accuracy of the data.

In view of the great importance of the investigation just made, and in view also (what appears from the immediately following discussion) of its generality of application, I desire to reproduce the tables just given in a slightly different and perhaps more convenient form, in which is obtained by a proper interpolation the wave-lengths corresponding to successive even deviations in every rock-salt prism whatsoever, whose temperature is 20° and whose angle is 60°, and in a second part the deviation of any rock-salt prism corresponding under these conditions to successive even wave-lengths.

The dispersion of a rock-salt prism of 60° 00′ 00″ angle at 20° C. and 76 cm. barometric pressure.

Deviation.	Wave-length. μ	Wave-length. μ	Deviation.
° ′			° ′ ″
41 50	0.4938	0.50	41 45 42
40	.5111	.75	40 27 4
30	.5301	1.00	00 28
20	.5519	.25	39 47 38
10	.5772	.50	40 16
00	.6062	.75	35 00
40 50	.6402	2.00	31 8
40	.6819	.25	27 46
30	.7324	.50	24 40
20	.7979	.75	21 36
10	.8834	3.00	18 38
00	1.0065	.25	15 40
39 50	1.1903	.50	12 32
40	1.5094	.75	9 14
30	2.0800	4.00	5 43
20	2.885	.25	2 8
10	3.692	.50	38 58 12
00	4.385	.75	53 58
38 50	4.996	5.00	49 56
40	5.536	.25	45 36
30	6.041	.50	40 44
20	6.500	.75	35 38
		6.00	30 20
		.25	25 6
		.50	20 0

HAVE ALL ROCK-SALT PRISMS THE SAME INDICES OF REFRACTION?

In what has just been said I have anticipated the results of the statement now made and which it is hoped is of general interest to all spectrographers. Fifteen years ago my attention was drawn to what then seemed a probability that there was material of frequent use in lecture experiments but to which very little capacity of precision was then accorded—I mean to rock salt, which, nevertheless, in the author's view, was capable not only of almost the highest precision in working but which had the precious quality which glass has not of being everywhere the same, and always for a given temperature and refracting angle, giving the same index for the same wave-length, wholly contrary in this to the well-known behavior of the glass prism. If the writer's anticipations were correct, then prismatic determination of wave-length might become quite comparable in constancy and in accuracy to those of the grating itself. He hesitated, however, for years to make this assertion, until the examination by the latter method of numerous specimens of salt obtained from all over the world, but notably from the Royal Salt Mines at Baden and from the mines in Russia, has convinced him that this anticipation is supported by the results.

The question is one of considerable interest to investigators working on the infra-red radiations, for it is almost necessary to use prismatic spectroscopes in this region to economize energy, and it has been shown that rock salt, by reason of its far-reaching transparency and plentiful occurrence in nature, is a substance well suited for the construction of prisms for this work. Could it be shown that all specimens of rock salt have the same indices of refraction throughout the spectrum at any given temperature and barometric pressure, then radiations might be as certainly specified by giving their refractive indices, or their minimum deviations for a 60° prism as by giving their wave-lengths, and with this advantage, that the observations might be repeated by others without the necessity of first consulting a plot or dispersion formula and then computing the prismatic deviation from the value of n thus found.

With a view to determine this matter three rock-salt prisms, two of which came from salt mines in Russia and the third from those of Baden, have been carefully compared at this observatory. The method of comparison was as follows:

Bolographs of the solar spectrum from F at 0.48μ to about 5μ were taken with each of the prisms and compared to see if they indicated equal dispersion. The index of refraction of the A line was measured with all the precautions described at page 173, Part I, for each of the prisms, and the other refractive indices depend upon these values for A. In the following table is given a summary of these results:

TABLE 30.—*Comparison of different rock-salt prisms.*

Designation of prism.	Observed refracting angle.	Observed deviation of A line.	Indices of refraction at A.	Probable error of column 4.	Designation of line.	Refractive indices.		
						Prism R.B.I.	Prism R.B.II.	Prism S.P.L.T.
R.B.I......	° ′ ″ {59 58 36.1 {59 58 14.9	° ′ ″ 40 21 35.1 40 21 20.7	1.536818	0.0000087	D	1.544278	....	1.544289
					C	1.540635	1.540680	1.540639
					A	1.536818	1.536844	1.536812
R.B.II......	59 58 57.5	40 24 22.1	1.536844	0.000005	....	1.529653	1.529694	1.529675
					ω_1	1.526686	1.526691	1.526677
S.P.L.T....	60 1 58.4	40 27 26.7	1.536812	0.000005	....	1.519339	1.519364	1.519361
					....	1.518739	1.518765	1.518749

It will be observed that the agreement is exact as far as the fourth decimal place in n, and that the variations from the mean may in all cases be reasonably accounted for by assuming them to be due to experimental error. We may then conclude with considerable certainty that all salt prisms at the same temperature and air pressure have equal refractive indices.

Chapter II.

MISCELLANEOUS OBSERVATIONS.

1. THE ACCURACY OF THE BOLOMETER.

By the "accuracy of the bolometer" is to be understood in what follows, first, its capacity to repeat observations of the intensity of radiation from a constant source with unvarying results, and second, the exactness of the proportionality between the intensity of radiation and the throw of the galvanometer.[1] The following experiments may be regarded as supplementary to earlier ones given in a contribution of the writer on the Pritchard wedge photometer.[2]

Date, February 12, 1898.

Object.—To test the accuracy of the bolometer in repeating observations with a constant source of radiation.

Method.—The source of radiation was alternately a lamp and a screen filled with water at a constant temperature. The times of separate exposures to the lamp were uniform with each other, and so were these to the water screen, though the two were not equal. With the cold screen the time of exposure was as short as was consistent with the galvanometers attaining a constant deflection, while with the lamp the time of exposure was only ten seconds, which was as short as consistent with the single swing of the needle. An observer stationed at the galvanometer scale read the constant deflection due to the cold screen and the first swing of the needle due to the lamp. If the source of radiation were constant the difference between these readings should be equal.

Apparatus.—Lamp: A common "student lamp," having an Argand burner, used with "305° test" petroleum oil. This lamp was protected from drafts of air by inclosing it in a box with glass window.

Bolometer: Linear bolometer, having two platinum strips, each 1 mm. wide, and blackened with camphor smoke. The resistance of the strips was about 0.5 ohms each. The bolometer balancing coils (of 40. ohms resistance) and slide wire were contained in water jackets similar to those pictured in Pl. XI of Part I. Battery current, 0.2 ampere.

---

[1] See Am. Jour. of Science, third series, vol. 21, No. 123, 1881, also fourth series, vol. 5, p. 242, 1898.
[2] Memoirs of the Academy of Arts and Sciences, vol. 11, pt. 5, No. 6, p. 305, 1887.

Galvanometer: "White" galvanometer. Resistance, 0.5 ohms. Deflections of 1 mm. on scale at 2 meters corresponded at the time of swing used (six seconds) to 4.2×10^{-8} amperes.

Diaphragms: Between the bolometer and lamp was a diaphragm of blackened brass plate, having a circular aperature of 1 cm. diameter. Between this and the lamp, and suspended by a string, was a blackened copper water jacket containing a thickness of water of 2½ cm., maintained at the room temperature. The exposure was effected by pulling aside the water screen.

Distances.—From bolometer to lamp, 35 cm.; from bolometer to first diaphragm, 30 cm.; from bolometer to second diaphragm, 31 cm.

Observations.

Number.	Series I.			Series II.		
	Scale readings.		Deflections.	Scale readings.		Deflections.
	Cold screen.	Lamp.		Cold screen.	Lamp.	
	cm.	cm.	cm.	cm.	cm.	cm.
1	13.20	62.65	49.45	13.33	62.65	49.32
2	13.20	62.73	.53	13.31	62.74	.43
3	13.17	62.55	.38	13.31	62.72	.11
4	13.19	62.63	.44	13.29	62.69	.40
5	13.18	62.62	.44	13.29	62.67	.38
6	13.20	62.68	.48	13.28	62.63	.35
7	13.23	62.73	.50	13.27	62.64	.37
8	13.24	62.70	.46	13.27	62.66	.39
9	13.27	62.77	.50	13.26	62.64	.38
10	13.27	62.71	.44	13.25	62.61	.36
Mean			49.464			49.380
P. E.			.008			.006
p. e			.028			.020

The two series were separated by about ten minutes, and between them there was, as will be noticed, a slight decrease in the intensity of the radiations from the lamp. But if we assume that the lamp remained perfectly constant throughout each series of ten observations, the probable error of a single bolometric observation comes out less than one-twentieth of 1 per cent. It is quite improbable that the lamp was perfectly constant in its radiation, so that a quite indefinite part of this extremely small variation must be attributed to the lamp. The accuracy of the bolometric observations becomes, indeed, notable when we admit this consideration, for it is at any rate a far more accurate measurer of radiation than the best photometer.

Turning now to the question whether the bolometric indications are truly proportional to the intensity of radiations falling upon the bolometer, it may be observed, first of all, that this matter might be pursued a great deal further than it is the present purpose to go. Thus it might be inquired if the proportionality held for different wave-lengths irrespective of the temperature of the source, and the answer to the inquiry would then necessarily show, among other things, how far the bolometer approached the condition of the theoretical "absolutely black body." The observations here presented, however, will merely show that the bolometer, as actually used with the solar radiation, gave at a single wave-length deflections proportional to the intensity of radiation.

The reader will find these observations among a number made for another purpose and given in the table at page 81 of Part I. The spectrometer circle was set so that solar radiations of a wave-length about $1.7\,\mu$ fell upon the bolometer, and the slit width was varied from 0.50 mm. down to zero, while an observer at the galvanometer read the deflection corresponding with each slit width. It was found that diffraction materially affected the results when the slit was narrower than 0.20 mm., so that only those slit widths greater than 0.20 mm. will here be given, with the exception of zero slit width, which, of course, should correspond with zero deflection:

Slit width.	Deflection observed.		Deflection computed.	Differences.
	First series.	Second series.		
mm.	mm.	mm.	mm.	%.
0.000	0.0	0.0	0.0	0.0
.250	230	250	247	7.3–1.2
.300	290	300	296	2.0–1.3
.500	490	....	494	0.8–

The column, "Deflection computed," is filled in on a basis of strict proportionality. The differences given in the last column average, with the exception of that for the first observation at 0.25 mm. slit width, only 1.3 per cent. When we take into account the very probable variation of the sky between these observations it appears that here again the bolometer is as accurate as the observations.

The general result, then, of the observations given here is to show that the bolometer may be depended upon to accurately repeat observations on the intensity of radiations from a constant source with far more accuracy than any photometer, and to give indications accurately proportional to the intensity of radiations of a given wave-length from a single high-temperature source.

2. RADIATION FROM TERRESTIAL SOURCES.

The primary object of these experiments was to determine the form of the energy curve of the Welsbach mantle. The author is indebted to Dr. Shapleigh, chemist of the Welsbach Light Company, for valuable information and material made use of in the investigation.

One of the most remarkable things about heated mantles of these rare oxides is the fact that while a mantle of ordinary impure thorium oxide gives the familiar brilliant light, perfectly pure thorium oxide heated by the same lamp gives far less light. This fact was strikingly illustrated at the observatory by Dr. Shapleigh. A pure thorium mantle was placed over the flame, and its light was but little brighter than that of the flame itself. The impression of it remaining is of a dull reddish-yellow light, certainly not so bright as an equal area of a hard-wood fire. After the mantle had been allowed to cool, a few spots upon it were soaked by a very weak solution of cerium nitrate, and as thus treated it was again placed over the flame. The parts treated now shone out dazzling bright, while the remainder was dull as ever. Dr. Shapleigh stated that a pure cerium oxide mantle was equally as dull as a pure thorium oxide mantle, and that the maximum brightness was to be secured by the mixture of 2 or 3 per cent of one of the oxides with 97 or 98 per cent of the other.

It appeared interesting to try the experiment of shifting from a pure thorium oxide to an ordinary mantle with the bolometer as the observer. Accordingly bolographs of the rock-salt prismatic spectrum were taken as shown in Pl. XXXI A.[1] A is a curve of pure thorium oxide, B that of the ordinary mantle. Lines of zero radiation have been placed beneath each with as much care to eliminate "drift" as possible, and scales of equal radiation have been annexed, i. e., radiation with equal slit widths. This scale has been taken the same as that employed in Part I, i. e., the maximum deflection in the solar spectrum would be 100 upon it, with identical instrumental conditions.

The most striking thing about the curves A and B is the high sharp maxima at 4.6μ and smaller maxima on each side of 2.6μ. These at first appeared to be the energy maxima proper to thorium, for later bolographs with iron oxide mantles and with an electrically heated platinum strip do not show them. But their position is such as to lead to the suspicion that they may be due, on the other hand, to the emission of carbon dioxide and water vapor, and this explanation is accepted here. The phenomenon is in striking analogy to the partial reversal of lines in solar and stellar spectra.

Another interesting feature of the curves is the presence of absorption bands at Ψ, Ω, X, and Y. The three former are due to water vapor, and the last to carbon dioxide, as has already been observed. These gases are the principal products of combustion of the vaporized petroleum, so that their presence is not surprising. The absorption bands in question did not appear in a later curve of the radiation of an electrically heated platinum strip.

A group of about fifty absorption bands, thickly packed together, extending from the curious maximum at 4.6μ, already mentioned, to the end of the region at which

---

[1] These bolographs were taken on rainy days, when it was necessary to depend on an incandescent electric light for the photographic trace. As this trace was very faint in the originals they were blackened for reproduction.

radiations remained distinguishable, may very probably be carbon dioxide bands. I am not aware that their presence has been noted by other observers, though such notice may have escaped my attention.

By transforming the curves from the prismatic to a wave-length scale the ordinary maximum (if we may speak of the one common to all the mantles and to the platinum strip as "ordinary" in contradistinction to the "extraordinary" sharp maximum already mentioned) is found to be at wave-lengths of about 1.6 μ and 2.2 μ with the Welsbach and iron oxide mantles, respectively. This places the temperature of these sources at 1,357° C. and 887° C., according to the results of Paschen,[1] assuming impure thorium oxide to be as nearly a "black body" as iron and copper oxides.

Bolographs were taken with mantles composed of other substances. Curve C of Pl. XXXI A is taken with a mantle made of fine iron-wire gauze. The iron quickly turns to oxide in the fierce flame to which it is subjected, so that we have practically the curve of iron oxide. Its form is seen to be nearly identical with those of the two thorium oxide mantles, but it is richer in energy and its maximum is at a somewhat longer wave-length, indicating a lower temperature. One notable difference is seen in the absence of the "extraordinary" maximum already noted. All the absorption bands find place in this curve as in the others, and some additional ones obscured in the previously given curves by the "extraordinary" maximum are found here. Attention has been called in Part I to the fact that absorption bands occur in the solar spectrum corresponding to some of these.

Curve D is from an incandescent platinum strip. The current from a dynamo was used in heating it, and this caused the very unsteady record. The curve is given here merely to show the absence of the bands Ψ, Ω, X, and Y, at least in any appreciable extent as compared with their appearance in curves A, B, and C.

In E there are superposed portions of curves to show the "reality" of the long series of absorption bands already spoken of.

SUPPLEMENTARY NOTE.

An opportunity offered itself to further investigate the visible spectra of various mantles shortly before this volume went to press, and as results of some interest were obtained it was thought best to add an account of them.

Owing to the small intensity of energy available it was found necessary to use a wider bolometer and slit, each subtending about 15' of arc. Preliminary experiments with apparatus similar to that already described showed that diffuse radiation of long wave-length vitiated the results in the visible spectrum. Hence it was necessary to employ two spectroscopes, one acting as a "sifting train"[2] to exclude the long

---

[1] F. Paschen, Annalen der Physik und Chemie, 60, p. 707, 1897.
[2] S. P. Langley, Memoirs of the National Academy of Sciences, vol. 4, p. 159, 1886.

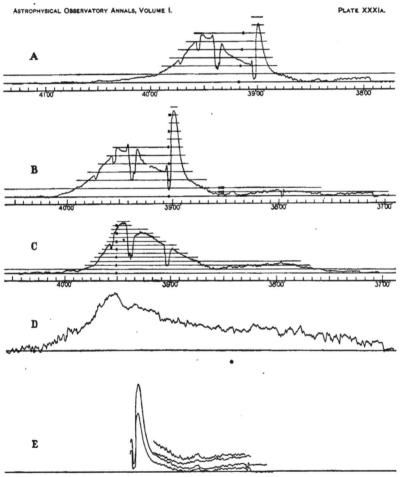

SPECTRUM ENERGY CURVES OF WELSBACH AND OTHER HEATED MANTLES, FROM BOLOGRAPHS WITH A 60° ROCK-SALT PRISM.

Observations of S. P. Langley, with the assistance of C. G. Abbot.

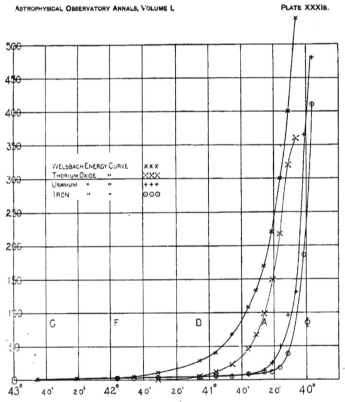

VISIBLE SPECTRUM ENERGY CURVES OF WELSBACH AND OTHER HEATED MANTLES, FROM BOLOMETRIC EXAMINATION WITH 60° ROCK-SALT PRISMS.

Observations of S. P. Langley, with the assistance of C. G. Abbot.

wave-length radiations from the slit of the second. Under these circumstances the curves given in Pl. XXXI B were obtained as the result of eye observations at the bolometer. They are given just as observed, without correction for slit width, and are on the scale of dispersion of a 60° rock-salt prism.

It will be noted how rich, relatively, the Welsbach mantle is in the visible spectrum as compared with the others examined, but how little energy any of them give in the visible spectrum as compared with the infra-red.[1] Thus we see how wasteful the best lamps are as sources of light. With the Welsbach mantle the lamp in question rivaled the electric arc in brilliancy, yet the energy of its visible radiations was almost insensible in comparison with that which it expended in the infra-red.

---

[1] S. P. Langley, American Journal of Science, third series, vol. XL, p. 97, 1890.

Chapter III.

CONSTRUCTION OF A SENSITIVE GALVANOMETER.
PURPOSES FOR WHICH IT WAS CONSTRUCTED.

The writer's own work on the bolometer and on the galvanometer brought the latter instrument, in the year 1892, up to a condition of sensitiveness and precision which it is believed had never before been reached, but, as he has already stated, its subsequent improvement has nevertheless been continuous and has been largely and indeed increasingly due to those gentlemen who have assisted him. This is particularly the case with regard to the matters considered in the present chapter, and it seems proper to say that these recent improvements, and especially the theoretical discussion, are almost entirely due to Mr. Abbot.

The bolographic investigation to discover and determine the position of lines in the infra-red solar spectrum, begun in 1892, had been continued with successive alterations and improvements in instrumental equipment till the close of 1896, when the results obtained in the last three months of that year, and forming the basis of the most complete and accurate map of the infra-red solar spectrum that could then be produced, were collected and reduced for publication. The inflections of the curves then adjudged to be "real," i. e., corresponding to solar or atmospheric absorption lines and bands, numbered 222.

The obstacle to further progress was shown by critical examination to be the impurity of the spectrum, resulting largely from the necessary use of a slit of the prejudicially great angular width, 4″, in the region from A to X, and of four times this width from X to beyond Y. On December 10, 1896, a rearrangement of apparatus had been made for the purpose of economy of energy, and the battery current was increased from 0.075 to 0.095 ampere. These alterations caused such increase in the galvanometer deflections that it became possible to decrease the width of the slit nearly one-half (from 8″ to 4″ in the upper part of the infra-red, and from 27″ to 15″ in the lower), and it was found that a great gain in detail resulted. Further progress in this direction by the means already cited could not be made, and the sensitiveness of the galvanometer was at the maximum value that could be employed without unduly increasing the time of swing.

It thus became necessary, if additional detail was to be sought, that a galvanometer of greater sensitiveness should be procured. With this acquisition it would become possible to reduce the width of the slit, the width of the bolometer, and perhaps also the battery current, with accompanying gains in purity of the spectrum and in freedom from disturbances of the circuit. As the effect of earth tremors would be, it was hoped, no more injurious for the proposed galvanometer than for the old, and as with proper astaticism it was probable that magnetic fluctuations in the earth's field would also be of slight importance, and as improvements in the sheltering of the circuit from drafts and changes of temperature were possible, it seemed probable that the accidental deflections of a galvanometer of considerably greater sensitiveness could be kept within about the old bounds, and that thus the added sensitiveness might be made useful.

In other researches with the bolometer this additional sensitiveness would also be of value. Thus it was proposed to attempt to detect the heat from the brighter stars; to examine the spectrum of the Welsbach lamp; to determine with all possible accuracy the dispersion curve for rock salt, and to examine the reflecting powers of surfaces of various kinds, as, for example, the metals and various paints, in all of which great sensitiveness was to be desired.

CONSIDERATIONS GOVERNING THE CONSTRUCTION.

(a) THE COILS.

From Maxwell's Electricity and Magnetism, Vol. II, paragraph 718, we have for the magnetic force along the axis of a coil of wire at any point on the axis

$$f = \gamma \frac{l \sin \theta}{r^2} \quad \quad \quad \quad \quad \quad \quad \quad (1)$$

where r is the distance from the circumference of the coil in question to the point; l the length of the coil; θ the angle between the axis and a line joining the circumference with the aforesaid point, and γ is the current.

From this it follows that the most favorable form of coil section for a galvanometer is given by the expression

$$r^2 = H^2 \sin \theta \quad \quad \quad \quad \quad \quad \quad \quad (2)$$

in which H is the extreme radius of the coil.

Maxwell proceeds to show in the two next following articles that for a given value of the electromotive force in a circuit containing an external resistance besides that of the galvanometer, the diameter of the wire of which the coils of the galvanometer are composed should increase, proceeding from the center outward, according to one or other of certain laws depending on the insulation of the wire.

This case is not that of the bolometric circuit, however, so that it is necessary to consider the best resistance of the galvanometer as governed by the purpose for which it is designed, and then to decide with regard to the most suitable sizes of wire to be employed in constructing its coils.

(b) Best Resistance for Bolometer Circuit.

It appears that the resistance of the galvanometer will be fixed by that most desirable for the bolometer, and that this latter, for the proposed analysis of the infra-red spectrum, is not far from 25 ohms. In setting this figure it is supposed that the width of the bolometer to be employed will be about 0.02 mm., and for this width a platinum strip of about 25 ohms resistance per centimeter is the best suited. As our bolometer frames take strips about a centimeter long, the resistance is thus fixed.

Let the resistance of the two bolometer arms be a and $a + \delta a$ (where δa is a small fractional increase of a, caused by a given small rise in temperature of one strip, so that $\delta a = \dfrac{a}{m}$, m being a very large number).

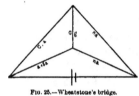

Fig. 25.—Wheatstone's bridge.

Let the resistance of each of the balancing coils be na, n being a multiplier as yet undetermined, and let the current in the bolometer arm be C; that in the galvanometer circuit be g, the resistance of the galvanometer being G. The battery and galvanometer connections are as represented in the figure. Then by the application of Kirchoff's laws (see Maxwell's Electricity and Magnetism, second edition, Vol. I, paragragh 347):

$$g = C \frac{n\, \delta a}{2\, na + G(i+n)} \quad \ldots \ldots \ldots \ldots \ldots (3)$$

But since $\delta a = \dfrac{a}{m}$

$$g = \frac{C}{m} \cdot \frac{1}{2 + \dfrac{G(1+n)}{a\, n}} \quad \ldots \ldots \ldots \ldots \ldots (4)$$

It is required to know what values of G and n will make the galvanometer deflection a maximum for given values of C, m and a. Let $\triangle =$ the deflection of the galvanometer. Then:

$$\triangle = k\, g\, f(G)$$

where k is a constant peculiar to the galvanometer and $f(G)$ is a function of G, which, as we shall later find, is between the square and cube root of G, but nearest to the square root. Let us suppose, then, that

$$\triangle = kg\, \sqrt{G}.$$

Substituting for g the value as given above we find that $\triangle$ is a maximum with respect to G when

$$\frac{G}{a} = 2\, \frac{n}{1+n} \quad \ldots \ldots \ldots \ldots \ldots \ldots \ldots \ldots (5)$$

and with respect to n when n is infinitely great.

As n can not, however, actually be made infinite, it becomes interesting to see how fast the deflection increases with n, providing that the relation $\dfrac{G}{a} = \dfrac{2\, n}{1+n}$ be continually retained. Under these circumstances for the following values of n the deflections are proportional to the numbers below:

n	$\frac{1}{2}$	1	1.5	2	3	4	∞
Δ	44	71	77	82	87	89	100

Thus it appears that about 40 per cent is gained by increasing n from 1 to ∞. Of this nearly one-fourth is gained by making $n = 1.5$, and over one-third is gained when $n = 2$. As the increase of n means proportionally increased heating of the coils, with attendant unsteadiness of the galvanometer, it is doubtful if n can profitably be greater than 2. As a matter of fact a pair of coils of 44 ohms resistance were ready prepared, which would thus go nicely with the bolometer of the proposed resistance 25 ohms, and it was decided to use them.

Corresponding to coils of 44 ohms and a bolometer of 25 ohms, the best resistance of the galvanometer would be 33 ohms. Computations were, however, made for a galvanometer of 25 ohms resistance, as a more round number, so to speak, and of more general usefulness to others who might be able to use the computations In the winding the resistance was brought up to about 30 ohms, and the resistance of the bolometer actually constructed is 28 ohms, so that the actual arrangement nearly fulfills the requirement[1]

$$\frac{G}{a} = \frac{2n}{1+n}$$

Having thus settled upon a suitable resistance for the galvanometer, we have next to inquire whether the resistance of the separate coils should be $\frac{G}{4}$, G, or 4 G, i. e., whether the galvanometer should be connected in one series, in two parallel series, or all in parallel.

In these three cases the current in each coil would be g, $\frac{g}{2}$ and $\frac{g}{4}$ and the deflections due to one coil for a given current being assumed proportional to the square root of the resistance, would be as $\frac{\sqrt{G}}{2} : \sqrt{G} : 2\sqrt{G}$, so that the whole deflection for the four coils would be the same in each case (i. e., $2g\sqrt{G}$), and there is no choice on this score. But the possible combinations of four 25-ohm coils would be those more generally useful, and this would lead to the selection of the second arrangement.[2]

---

[1] If the supposition is made that the deflection is proportional to $g\sqrt[3]{G}$ instead of $g\sqrt{G}$, then Δ is a maximum with respect to G when

$$\frac{G}{a} = \frac{n}{1+n}$$

But as here remarked, the latter assumption is the more nearly true for coils of the form constructed.

[2] If we suppose the deflection proportional to the cube root of G the first arrangement will be superior, the relative efficiency being as 62:50:40. But even in this case the more general usefulness of the galvanometer of 25 ohms over one with 6-ohm coils would lead to its selection.

BEST CONSTRUCTION FOR COILS.

It now remains to decide as to the best sizes of wire to use in constructing coils of about 25 ohms each of the cross section determined by the equation (2) of page 245,

$$r^2 = H^2 \sin \theta$$

except as modified by conditions of practical importance. These conditions are, that space must be left to allow the needle to hang between the coils and to turn through 30° or more, and that coils can not be wound to the centre, even if, by reason of the appreciable length of the magnets, the space at the very center were not rendered ineffective.

It was therefore determined to wind the coils so that the interior radius should be one millimeter, and so that a space 2 millimeters wide would be left between the coils. The accompanying figure shows in full lines the coil section as thus modified, and in dotted lines the section determined by the equation

$$r^2 = H^2 \sin \theta.$$

Let the radius of the vacant space at the center be a and the half width of the space between the coils be b. Let the diameter of the wire of which the coil is composed be $(d + t)$, of which d is the diameter of the bare wire and t the gain by insulation. Let L be the length of this wire, which may be wound within a coil of the proposed form and of exterior radius H, and let F be the force exerted at the center by a given current. Then

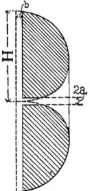

Fig. 26.—Form of galvanometer coil section.

$$L = \frac{2\pi}{(d+t)^2} \left[\int_{\sqrt{a^2+b^2}}^{\sqrt[4]{H^2 a}} \int_{\sin^{-1}\frac{a}{r}}^{\cos^{-1}\frac{b}{r}} r^2 \sin\theta \, dr \, d\theta + \int_{\sqrt[4]{H^2 a}}^{H - \frac{1}{4}\frac{b^2}{H}} \int_{\sin^{-1}\frac{r^2}{H^2}}^{\cos^{-1}\frac{b}{r}} r^2 \sin\theta \, dr \, d\theta \right] \quad \ldots (7)$$

and from (1), page 245,

$$F = \gamma \frac{2\pi}{(d+t)^2} \left[\int_{\sqrt{a^2+b^2}}^{\sqrt[4]{H^2 a}} \int_{\sin^{-1}\frac{a}{r}}^{\cos^{-1}\frac{b}{r}} \sin^2\theta \, dr \, d\theta + \int_{\sqrt[4]{H^2 a}}^{H - \frac{1}{4}\frac{b^2}{H}} \int_{\sin^{-1}\frac{r^2}{H^2}}^{\cos^{-1}\frac{b}{r}} \sin^2\theta \, dr \, d\theta \right] \quad \ldots (8)$$

The expressions for L and F are approximate, but are correct to within one part in 200, where H is not less than $3b$.

Integrating these expressions by means of series, and neglecting powers of $\frac{a}{H}$ and $\frac{b}{H}$ higher than the third, we have:

$$L = \frac{1.53\,H^3}{(d+t)^2}\left[1 - \frac{2b}{H} - 0.2\left(\frac{b}{H}\right)^2 - 1.7\left(\frac{a}{H}\right)^3 + 2\frac{a^2 b}{H^3} + 1.7\left(\frac{b}{H}\right)^3\right] \quad \ldots \ldots (9)$$

$$F = \gamma \frac{H}{(d+t)^2}\left[1.45 - \left\{2\log_\varepsilon \frac{H}{\sqrt{a^2+b^2}} + \frac{\pi}{2}\frac{\sqrt{a^2+b^2}}{b} - \frac{1}{6}\frac{b^2}{a^2+b^2} - \frac{1}{8}\frac{b^4}{(a^2+b^2)^2}\right\}\frac{b}{H}\right.$$
$$\left. - \left\{\frac{1}{3}\frac{a^2}{a^2+b^2} + \frac{1}{20}\frac{a^4}{(a^2+b^2)^2}\right\}\frac{a}{H} - 0.09\frac{b^2}{H^2} + \frac{3}{7}\left(\frac{a}{H}\right)^3 - \frac{1}{3}\frac{b^3}{H^3}\right] \quad \ldots \ldots (10)$$

If we put $a = b$ in these expressions, substitute common logarithms and call $\left(\frac{b}{H}\right)^3 = \left(\frac{b}{H}\right)^2$ we have

$$L = \frac{1.53\,H^3}{(d+t)^2}\left[1 - \frac{2b}{H} - 1.9\left(\frac{b}{H}\right)^2 + 3.7\left(\frac{b}{H}\right)^3\right] \quad \ldots \ldots \ldots (11)$$

and

$$F = \gamma \frac{4.58\,H}{(d+t)^2}\left[1 - \left\{1.12 - 3.18\log\frac{b}{H}\right\}\frac{b}{H} + 0.24\left(\frac{b}{H}\right)^2 + 0.23\left(\frac{b}{H}\right)^3\right] \quad \ldots \ldots (12)$$

These expressions hold with sufficient accuracy when H is greater than $3\,b$.

In practice we have L, not H, given, since we know the desired resistance of the coil and the diameter of the wire of which it is composed. Hence it is necessary to revert the series (11). This may be done by regarding (11) as a cubic equation and solving for H. In this way we obtain

$$H = 0.868\,L^{\frac{1}{3}}(d+t)^{\frac{2}{3}}\left[1 + 1.4\left(\frac{b}{L^{\frac{1}{3}}(d+t)^{\frac{2}{3}}}\right)^2 - 0.92\left(\frac{b}{L^{\frac{1}{3}}(d+t)^{\frac{2}{3}}}\right)^3 + 1.3\left(\frac{b}{L^{\frac{1}{3}}(d+t)^{\frac{2}{3}}}\right)^5\right] + \frac{2b}{3} \ldots (13)$$

Expressions (12) and (13) enable one to compute the radius H and the force F exerted at the centre for a coil of given resistance composed of one size of wire of known diameter. For from the resistance the value of L may be obtained by the use of wire tables. Substituting this value in (13) gives the value of H to be used in equation (12) to determine F.

When coils containing several sizes of wire are under investigation such a computation as has just been described suffices for the inmost coil. Substituting the value of H, thus obtained in equation (11), along with the values of d and t corresponding to the wire to be used in the next layer, the length of wire of this diameter which would replace the inmost coil is found. The force which this wire would exert is next calculated by equation (12). Now, adding the length which would be required to wind a coil of the dimensions of the inmost coil with the wire to be used in the second to that length which is actually to be used in the second layer, and substituting in equation (13), we obtain the extreme radius of the second coil, and from equation (12) we may compute the force which a coil wound to this radius with wire of uniform diameter would exert. Subtracting from this value the force already determined which would result from winding the inmost coil with wire of the same size as used in the second layer, we have the force actually exerted by the second layer. A similar procedure is to be followed with other layers.

In order to determine what size of wire should be used in the coils, and also how the force varies with the resistance, a series of such computations were performed for coils composed of a single size of wire and others composed of three sections. The computations included coils of 5, 25, 50, 250, and 500 ohms resistance, and the value of L and d made use of were those given in the Smithsonian Physical Tables for copper wire. The wire used was procured of A. F. Moore, of Philadelphia, Pa., for which the value of t was 0.0015 inch or 0.004 cm. The result of these computations appear in Table 32.

TABLE 32.—*Relative efficiency of galvanometer coils.*

Total resistance.	Division of resistance between sections.	No. of coil.	Diameters of wire in sections.			Gain of diameter by insulation.			Length of wire in sections.			External radii of sections.			Force at center from sections.			Total force.
Ohms.			d_1	d_2	d_3	t_1	t_2	t_3	L_1	L_2	L_3	R_1	R_2	R_3	F_1	F_2	F_3	F.
			cm.	cm.	cm.	cm.	cm.	cm.	cm.	cm.	cm.	cm.	cm.	cm.				
5	(¹)	1	0.0101	0.0202	0.0405	0.0038	0.0038	0.0038	81	328	1,318	0.223	0.598	1.297	298	321	301	*920
		2	.0101	.0255	.0644	.0038	.0038	.0038	81	521	3,332	.223	.753	2.257	298	359	290	*947
		3	.0127	.0255	.0511	.0038	.0038	.0038	130	521	2,095	.282	.762	1.827	338	327	254	920
		4	.0160	.0405	.1020	.0038	.0038	.0038	206	1,318	8,423	.466	1.279	3.467	377	341	182	899
25	(¹)	5	0.0050	0.0080	0.0127	0.0058	0.0038	0.0038	101	256	648	0.306	0.409	0.611	418	442	530	1,390
		6	.0050	.0101	.0202	.0058	.0038	.0038	101	407	1,639	.306	.480	.955	418	563	654	1,635
		7	.0050	.0127	.0321	.0058	.0038	.0038	101	648	4,144	.306	.583	1.608	418	675	741	1,834
		8	.0080	.0160	.0321	.0038	.0038	.0038	256	1,031	4,144	.383	.741	1.632	671	670	605	*1,946
		9	.0101	.0202	.0405	.0038	.0038	.0038	407	1,639	6,589	.453	.952	2.161	754	653	564	*1,971
	(²)	10	.0050			.0058			305			.282						794
		11	.0101			.0038			1,222			.620						1,251
		12	.0160			.0038			3,092			1.00						1,467
		13	.0255			.0038			7,817			1.71						1,407
50	(¹)	14	0.0050	0.0080	0.0127	0.0058	0.0038	0.0038	203	513	1,297	0.350	0.484	0.745	634	630	728	1,992
		15	.0050	.0101	.0202	.0058	.0038	.0038	203	815	3,278	.350	.577	1.182	634	797	873	2,304
		16	.0050	.0127	.0321	.0058	.0038	.0038	203	1,297	8,288	.350	.709	2.006	634	935	894	2,463
		17	.0080	.0160	.0321	.0038	.0038	.0038	513	2,062	8,288	.452	.910	2.034	991	912	784	*2,688
		18	.0101	.0202	.0405	.0038	.0038	.0038	815	3,278	13,178	.555	1.17	2.703	1,005	898	717	*2,710
	(²)	19	.0050			.0058			610			.451						1,161
		20	.0101			.0038			2,445			.756						1,908
		21	.0160			.0038			6,185			1.24						1,997
		22	.0255			.0038			15,635			2.13						1,863
250	5:8:12	23	0.0050	0.0080	0.0127	0.0058	0.0038	0.0038	1,017	2,567	6,483	0.463	0.740	1.205	1,871	1,517	1,452	4,840
	12:8:5	24	.0050	.0101	.0202	.0038	.0038	.0038	1,017	4,075	16,392	.463	.906	1.962	1,871	1,796	1,645	*5,314
	(¹)	25	.0050	.0127	.0321	.0058	.0038	.0038	1,017	6,483	41,442	.463	1.140	3.376	1,871	2,017	1,617	*5,505
		26	.0080	.0160	.0321	.0038	.0038	.0038	2,567	10,308	41,442	.698	1.496	3.426	2,310	1,761	1,409	*5,480
		27	.0101	.0202	.0405	.0038	.0038	.0038	4,075	16,392	65,892	.879	1.957	4.560	2,433	1,676	1,272	5,381
		28	.0050	.0101	.0202	.0038	.0038	.0038	*610	3,912	23,604	.408	.886	2.184	1,413	1,895	2,035	5,343
		29	.0050	.0101	.0202	.0038	.0038	.0038	1,464	3,912	9,835	.509	.906	1.696	2,295	1,630	1,220	5,145
	(²)	30	.0050			.0058	.0038	.0038	3,056			.621						3,271
		31	.0101			.0038	.0038	.0038	12,225			1.231						4,007
		32	.0160			.0038	.0038	.0038	30,925			2.067						3,923
		33	.0255			.0038	.0038	.0038	78,175			3.597						3,485
500	(¹)	34	0.0050	0.0080	0.0127	0.0038	0.0038	0.0038	2,033	5,133	12,967	0.555	0.908	1.496	2,731	2,027	*1,905	6,663
		35	.0050	.0101	.0202	.0038	.0038	.0038	2,033	8,150	32,783	.555	1.118	2.451	2,731	2,396	2,126	*7,253
		36	.0050	.0127	.0321	.0038	.0038	.0038	2,033	12,967	82,883	.555	1.417	4.235	2,731	2,677	2,068	*7,476
		37	.0080	.0160	.0321	.0038	.0038	.0038	5,133	20,617	82,883	.855	1.865	4.298	3,241	2,292	1,880	*7,413
		38	.0101	.0202	.0405	.0038	.0038	.0038	8,150	32,783	131,783	1.085	2.444	5.738	3,343	2,161	1,619	7,123
	(²)	39	.0050			.0058			6,100			.756						4,751
		40	.0101			.0038			24,450			1.530						5,398
		41	.0160			.0038			61,850			2.584						5,162
		42	.0255			.0038			156,350			4.512						4,512

¹ Three sections of equal resistance.   ² All in single section.
* Numbers distinguished by an asterisk show the best arrangements of coils found.

From these results it appears: First, that a considerable advantage is gained by using several sizes of wire; second, that the resistance may be advantageously distributed equally between the sections; third, that the force varies nearly proportionally to the square root of the resistance,[1] and finally, that giving due weight to the advantage of having the coils as small as possible, so as to decrease the length, weight, and moment of inertia of the needle, the best 25-ohm coil is that designated as No. 8, composed of Nos. 40, 34, and 28 B. & S. gauge wire. Accordingly, the coils were thus made.

THE NEEDLE SYSTEM.

Two needles have been tried with this galvanometer, one of 6.5 mgs. weight and the other weighing only 2.4 mgs. Each has six flat needles of about 1.2 mm. length. The disposition of these six needles is as follows: Two groups of three each are fastened with shellac to a glass stem with their centers 38 mm. apart, so that one needle at each end of the stem is on the same side of the stem as the mirror at its center, and the other pair at each end are on the other side of the stem and a little above and below the single needle in front. This gives a very well-balanced system and a very open arrangement of the magnets.

In preparing the magnets for the lighter system tungsten steel was used. The weight of the six was altogether 1.6 mgs., and that of the stem, mirror, and shellac 0.8 mgs. A selected silvered fragment of microscope cover glass was used for the mirror. It was found that the mirror gave rather poor definition for bolographic work and that this very light needle was more unsteady than allowable, so that all the bolographs and other records up to April 15, 1899, were taken with the needle of 6.5 mgs. weight, constructed exactly similar, except that the magnets weighed in all 3.8 mgs. and that a mirror of 2 mgs. weight, by Brashear, was employed. No damping vane was needed for either needle.

From observations made with the first needle at 2 seconds single swing and assuming the proportionality between deflections and squares of time of swing,[2] it was computed that a current of 0.000,000,000,005 ampere would give a deflection of 1 mm. on a scale of 1 meter with a time of single swing 10 seconds when the coils were connected in two series of two each in parallel. Observations with the heavier needle, at a time of swing of 3.5 seconds, gave a value similarly computed of 0.000,000,000,023.

---

[1] It appeared of some interest to know more exactly how the force varied with the resistance. Accordingly, the expression $F = C R^x$ was assumed and its constants were determined from the data $F = 947$, $R=5$; $F=1,971$, $R=25$; $F=2,710$, $R=50$; $F=5,505$, $R=250$; $F=7,476$, $R=500$, which are the best values given by the table. As thus determined we have $F = 457 R^{0.46}$, showing that the exponent is between one-half and two-fifths. From this formula we have the following values of F, corresponding to the five values of R, above given: 944, 1,955, 2,670, 5,525, and 7,544.

[2] While this proportionality does not hold in practice owing to the excessive damping of this very light needle by the resistance of the air to its magnets and mirror, yet if the air were exhausted from the case it would hold; and it is probable that an even smaller "constant" would then be obtained than that given above, for even at 2 seconds single swing the damping of the air is considerable, so that observation gives too great a time of swing.

SUMMARY.

For bolometric purposes the resistance of the bolometer strips should be small in comparison with that of the balancing coils.

The resistance of the galvanometer under these circumstances should be between one and two times that of the bolometer strips, but is more exactly specified by the expression given on page 246.

Where the deflection due to the galvanometer coils is nearly proportional to the square root of their resistance, as with the form of coils here employed, it is immaterial whether the best galvanometer resistance is secured by coils all in a series, all in parallel, or two in series, two in parallel.

The efficiency of coils of equal resistance varies largely with the sizes of wire used in their construction.

For a galvanometer of about 25 ohms resistance the coils may well be wound in three sections of equal resistance, using No. 40, No. 34, and No. 28 B. & S. gauge wire, gaining in diameter by the insulation 0.004 cm.

The sensitiveness of a galvanometer increases rapidly with decreasing weight of the needle, providing that the most advantageous distribution of weights is adopted.

The galvanometer constructed after these indications has justified expectations of great sensitiveness.

APPENDIX.

THE DETERMINATION OF WAVE LENGTHS IN THE INFRA-RED SPECTRUM OF ROCK SALT.

It would be very desirable if, when we knew the deviation of any line in a prismatic spectrum, we could by the use of some simple rule determine the corresponding wave-length, as we can do in the case of grating spectra. Unfortunately, the connection between wave-length and prismatic deviation is so obscure that no such simple formula as supposed above exists. Yet, from time to time there have been given, as the result of theoretical investigations, interpolation formulæ which professed to establish a connection between the wave-length and index of refraction through the employment of several constant coefficients to be determined for any particular prism by experiments with known wave-lengths.

DISPERSION FORMULÆ.

At the time the writer commenced his observations at Allegheny the formula of this kind most generally employed was Cauchy's, which is usually written:

$$n = A + \frac{B}{\lambda^2} + \frac{C}{\lambda^4}$$

where n is the index of refraction, λ the corresponding wave-length, and A, B, and C constants. If in this formula we assume λ to become infinite, then the terms $\frac{B}{\lambda^2}$ and $\frac{C}{\lambda^4}$ disappear and n becomes equal to A, which is its minimum value. The writer's earlier researches,[1] however, showed that for prisms, both of glass and rock salt, much smaller values of n were to be actually observed than were allowable according to this formula, so that, however well it might represent the facts in the visible spectrum, it failed completely for the long wave-lengths beyond the red. Among other dispersion formulæ current at that period, Redtenbacher's, Wüllner's, and Briot's were found by the writer[2] to imperfectly express the results of measure-

---
[1] S. P. Langley, American Journal of Science, third series, vol. 27, p. 169, 1884.
[2] Ibid., vol. 32, p. 83, 1886.

ments in the infra-red when their constants were determined from visual observations. It appeared that the formulæ of Wüllner and Briot would nevertheless fairly well serve for interpolation in the infra-red, provided their constants were determined from observations in that region.

More recently the dispersion formula of Ketteler,[1] deduced from theoretical considerations more general than those on which Briot's formula is based, has become the most accepted among physicists. The formula of Ketteler in its most extended form is capable of representing a very wide range of dispersion phenomena, but for the case of substances like rock salt, which retains a very high degree of transparency uniformly through a long series of wave-lengths, the formula may be much simplified, and takes the following form:

$$n^2 = a^2 + \frac{M_2}{\lambda^2 - \lambda_2^2}\lambda_2^2 - \frac{M_1}{\lambda_1^2 - \lambda^2} \quad \ldots \ldots \ldots \ldots \quad (1)$$

or still more simply,

$$n^2 = b^2 + \frac{M}{\lambda^2 - \lambda_2^2} - k\lambda^2 \quad \ldots \ldots \ldots \ldots \quad (2)$$

in which n is the index of refraction corresponding to the wave-length λ, and a^2, b^2, M_1, M_2, k, λ_1^2, and λ_2^2 are constants for whose significance the reader is referred to the article of Ketteler just cited. It need only be said here that λ_2^2 and λ_1^2 are positive quantities, of which λ_2^2 is less than λ^2 and λ_1^2 greater than λ^2.

We will next review the available experimental data for fixing the constants for rock salt in this equation, and for rigorously testing its validity.

OBSERVATIONS AT ALLEGHENY.

The investigation of the writer, carried on at Allegheny in 1885, has already been described in this memoir (see p. 16). With the apparatus there employed it was possible to proceed only as far as 5.3μ, corresponding to nine times the wave-length of D_2. In the following table are given the values obtained up to this point:

---
[1] E. Ketteler, Annalen der Physik und Chemie, neue Folge, vol. 30, p. 299, 1887.

TABLE 33.—*Dispersion of rock salt.*

[Observations of S. P. Langley.[1]]

Number.	1 Temperature of observation.	2 Mean deviation reduced for 60° prism.			3 Index of refraction. n	4 Wave-length. λ	5 Probable error of wave-length. $\delta\lambda$
	°	°	′	″		μ	μ
1	24	43	17	15	1.56833	0.39687	
2	24	41	54	09	1.55323	0.48614	
3	24	41	35	13	1.54975	0.51838	
4	24	41	05	00	1.54418	0.58901	
5	24	41	04	48	1.54414	0.58961	
6	24	40	45	13	1.54051	0.65630	
7	24	40	24	41	1.53670	0.75940	
8	24–26	39	49	18	1.5301	$2\lambda D_1$ 1.1780	0.002
9	24–26	39	34	06	1.5272	$3\lambda D_1$ 1.7670	0.005
10	26–(–1.1)	39	24	18	1.5254	$4\lambda D_1$ 2.3560	0.009
11	19–26.4	39	18	12	1.5243	$5\lambda D_1$ 2.9451	0.013
12	14–15	39	09	54	1.5227	$6\lambda D_1$ 3.5341	0.019
13	15–23.8	39	03	42	1.5215	$7\lambda D_1$ 4.1231	0.029
14	15	38	56	06	1.5201	$8\lambda D_1$ 4.7121	0.043
15	15.4	38	48	06	1.5186	$9\lambda D_1$ 5.3011	0.065

NOTE.—The wave-lengths are on Rowland's scale. No correction was applied to reduce the results in the invisible spectrum to a standard temperature from lack of data.

[1] S. P. Langley, American Journal of Science, third series, vol. 32, p. 98, 1886.

Upon plotting these results of observation it was found that up to the longest wave-length investigated the curve of refractive indices and wave-lengths appeared to be very slightly convex toward the axis of abscissæ (wave-lengths), so that it seemed reasonable to suppose[1] that wave-length values extrapolated by means of the tangent at any point would be safe approximations in the sense of being smaller than the true values, unless the character of the curve should change in a way which there was then no ground to expect.[2]

It has been shown by recent investigations, first, I think, by those of H. Rubens,[3] and, almost simultaneously by F. Paschen,[4] that there is a point of inflection in the curve at a wave-length of about 5μ, so that beyond 5μ the curve is concave toward the axis of abscissæ, and that therefore the contingency referred to above actually exists, namely, that a change in the character of the curvature occurs near the end of the

---

[1] See American Journal of Science, third series, vol. 32, p. 103, 1886.
[2] The writer has regretted that some physicists, despite the carefully guarded language in which the probable wave-length of certain points in the extreme infra-red was surmised by him, have thought him to attach such weight to a straight line extrapolation as to have published tables (giving the index of refraction so calculated) headed "Langley Extrapolation," and even have censured him for believing the curve to proceed in a straight line beyond 5μ, whereas he distinctly stated that the curve appeared still slightly convex to the axis, and used a tangent in his approximations only to obtain a minimum value of the wave-length.
[3] H. Rubens, Annalen der Physik and Chemie, neue Folge, vol. 53, p. 267, 1894.
[4] F. Paschen, Annalen der Physic und Chemie, neue Folge, vol. 53, p. 337, 1894.

observations of the writer. His observations, then, are confirmed, but the extrapolated wave-length values inferred from them are rendered too high.

MORE RECENT INVESTIGATIONS.

An excellent résumé of the several investigations of Rubens and Snow, of Paschen, and of others, touching the question of dispersion now under consideration, was published by Professor Keeler, then of Allegheny Observatory, in a recent number of the Astrophysical Journal,[1] and to that article, as well as to the original memoirs cited by Professor Keeler, the reader's attention is directed. It will suffice for the present purpose to include the latest results of these investigators with a short description of the methods which they have employed.

Both Rubens and Paschen appear to regard as of highest weight among their determinations of the dispersion and wave-length in the rock-salt spectrum certain measurements made indirectly by comparing the dispersion of rock salt with that of fluorite. It becomes necessary, therefore, to first discuss the properties of fluorite. In what follows I shall only mention the results of admittedly the highest weight.

Rubens and Snow, in 1892, published an article[2] in which the dispersion curve for fluorite was followed to a wave-length of 8μ by an interference method, which was found subsequently to be inaccurate for the longer wave-lengths, but which, according to Rubens,[3] furnished the most accurate results obtained up to 1895 for wave-lengths less than 2.5μ.

Carvallo,[4] by another ingenious interference method, obtained results which agree very closely with those of Rubens and Snow, but extend only as far as 1.8μ.

Rubens, in 1894,[5] published new dispersion and wave-length data for fluorite extending to 9μ, using a method somewhat similar to that employed by the writer in 1885. But as Rubens has more recently decided[6] that these values are less trustworthy than those of Paschen, shortly to be mentioned, no more need here be said of them. In this paper, however, is included the comparison of the dispersion of rock salt with that of fluorite, which, combined with part of the results of Rubens and Snow on the wave-lengths in the fluorite spectrum as already mentioned, and with those of Paschen shortly to be alluded to, has furnished the dispersion and wave-length values for rock salt which will be given below.

Paschen's contribution to the dispersion curve for rock salt is found in two articles published in 1894.[7] The first of these contains the determination of the dispersion and

---

[1] J. E. Keeler, Astrophysical Journal, vol. 3, p. 63, 1895.
[2] H. Rubens and B. W. Snow, Annalen der Physik und Chemie, neue Folge, vol. 46, p. 529, 1892.
[3] H. Rubens, Annalen der Physik und Chemie, neue Folge, vol. 53, p. 476, 1895.
[4] M. E. Carvallo, Académie des Sciences, Comptes rendus, vol. 116, p. 1189, 1893.
[5] H. Rubens, Annalen der Physik und Chemie, neue Folge, vol. 53, p. 267, 1894.
[6] Ibid., p. 476, 1895.
[7] F. Paschen, Annalen der Physik und Chemie neue Folge, vol. 53, pp. 301 and 337.

wave-length for fluorite between wave-lengths 0.88μ and 9.4μ. The method pursued was substantially that of the writer, which has already been referred to in this memoir (see p. 222). Changes were made in several particulars deserving of mention. First. The source of energy was a platinum strip coated with oxide of iron and heated to incandescence by means of an electric current from a storage battery. A spectrum particularly rich in infra-red rays was thus obtained. Second. Silvered mirrors were uniformly made use of in place of lenses. Third. The effective dispersion of the grating was diminished to only three times that of the prism by the expedients of employing two concave mirrors in the grating spectroscope, as well as in the prismatic, and of increasing the slit width of the grating spectroscope when the bolometric measures were made, so as to diminish the purity of its spectrum.

In Paschen's results the temperature of the prism is recorded, a quantity which it is surprising to see so often neglected. The probable errors of the measured deviations and refractive indices are given, and from them the corresponding values for the wave-lengths, assuming the deviations correct. The probable error in wave-length varies irregularly between 0.0011μ and 0.0074μ.

The second of the articles of Paschen above referred to is entitled "The dispersion of rock salt," and gives the wave-length of certain absorption bands in the infrared spectrum as deduced by comparing his earlier work on the emission spectra of gases[1] with his determination of the wave-lengths in the fluorite prismatic spectrum. The position of these bands in the rock-salt prismatic spectrum is taken from the result of Julius.[2] As these absorption bands are not of great sharpness the comparison gives, of course, only an approximation to the truth, but the results are as follows:

TABLE 34.—*Dispersion of rock salt.*

[Combined results of Paschen and Julius.]

Wave-length.	Refractive index.	Wave-length.	Refractive index.
μ		μ	
2.831	1.5250	6.70	1.5116?
2.831	1.5249	6.717	1.5118?
2.985	1.5242	7.690	1.5086?
4.403	1.5210	7.94	1.5051?
4.403	1.5209	8.188	1.5061
4.729	1.5198	8.443	1.5041?
5.322	1.5174?	8.660	1.5032
6.061	1.5147	9.098	1.5012
6.128	1.5162?	9.455	1.4993
6.456	1.5139	9.76	1.4974

[1] F. Paschen, Annalen der Physik und Chemie, neue Folge, vol. 50, p. 409, 1893; vol. 51, p. 1, 1894; vol. 52, p. 209, 1894.

[2] Verhandlungen des Vereines zur Beförderung des Gewerbfleisses, p. 232, 1893, and also Licht- und Wärmestrahlung verbrannter Gase, Preisschrift, 1890.

Rubens, in an article published in 1895,[1] accepts the work of Paschen on the dispersion of fluorite as more accurate than his own for wave-lengths greater than 2.5μ, while retaining that of himself and Snow for wave-lengths less than 2.5μ, and revises his wave-lengths for rock salt accordingly, using his determination of the relative dispersion of rock salt and fluorite already referred to. The table so obtained is given below.

TABLE 35.—*Dispersion of rock salt.*

[Combined results of Rubens, Rubens and Snow, and Paschen.]

Wave-length.	Deviation for salt prism of 60° 2′.	Index of refraction.	Wave-length.	Deviation for salt prism of 60° 2′.	Index of refraction.
μ	° ′		μ	° ′	
0.434	42 37	1.5607	4.01	39 06	1.5216
.485	41 56	1.5531	4.65	38 56	1.5197
.589	41 07	1.5441	5.22	38 47	1.5180
.656	40 47	1.5404	5.79	38 36	1.5159
.840	40 15	1.5345	6.78	38 16	1.5121
1.281	39 47	1.5291	7.22	38 06	1.5102
1.761	39 36	1.5271	7.59	37 57	1.5085
2.35	39 27	1.5255	8.04	37 46	1.5064
3.34	39 15	1.5233	8.67	37 23	1.5030

Before closing this discussion, attention should be drawn to the newer determination of the dispersion of fluorite by Paschen,[2] which apparently exceeds in accuracy his former one, already referred to.

Dispersion of fluorite prism. Angle, $59° 59' 15''$.

[F. Paschen, Ann. Phys. Chem., 56, 765, 1895.]

μ	$\dfrac{m}{n}$	δ	Probable error of δ.	Mean temperature.	Number measures.	n observed.	n calculated.	Δn
		° ′ ″	″	°				
0.32525						1.44987	1.44979	+0.00008
.34015						775	776	− 01
.34655						697	698	01
.36009						535	547	12
.39681						214	217	03
.41012					(³)	121	119	+ 02
.48607						.43713	.43712	01
.58932						392	388	04
.65618						257	253	04
.68671						200	203	− 03
.71836						157	157	00
.76040						101	104	03

[1] H. Rubens, Annalen der Physik und Chemie, neue Folge, vol. 54, p. 476, 1895.
[2] F. Paschen, Annalen der Physik und Chemie, neue Folge, vol. 56, p. 762, 1895.
[3] Sarasin.

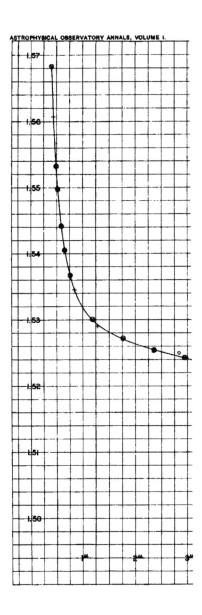

Dispersion of fluorite prism. Angle, 29° 59′ 15″—Continued.

[E. Paschen, Ann. Phys. Chem., 56, 765, 1895.]

μ	$\frac{m}{n}$	δ			Probable error of δ.	Mean temperature.	Number measures.	n observed.	n calculated.	Δn	
		°	′	″	″	°					
.8840	3/2	31	15	44.4	2.20	16.2	13	1.42982	.42982		00
1.1786	2/1		6	10.2	3.28	16.8	15	787	789		02
1.3756	7/3		1	26.4	2.27	16.9	9	690	690		00
.4733	5/2	30	58	58.8	3.81	17.0	11	641	644		03
.5715	8/3		56	48.0	3.65	18.0	8	596	598		02
.7680	3/1		52	27.6	2.49	17.2	17	507	506	+	01
.9153	13/4		49	0.6	1.63	17.3	4	437	434		03
1.9644	10/3		47	51.0	3.22	15.6	9	413	409		04
2.0626	7/2		45	12.6	2.43	17.1	9	359	359		00
.1608	11/3		42	42.0	5.78	16.2	9	308	308		00
.3573	4/1		37	23.4	1.83	16.9	22	199	199		00
.5537	13/3		31	58.2	5.12	17.7	19	088	084		04
(¹){ .6519	9/2		28	26.4	2.08	16.6	7	016	023	—	07
2.9466	5/1		19	9.0	2.33	16.7	11	.41826	.41829		03
3.2413	11/2		8	46.2	4.80	15.6	12	612	612		00
.5359	6/1	29	57	26.4	5.07	15.8	16	379	383		04
3.8306	13/2		44	52.8	5.09	17.6	16	120	125		05
4.1252	7/1		32	0.6	5.02	16.8	8	.40855	.40851	+	04
4.7146	8/1		2	12.6	3.84	17.1	7	238	233		05
5.3036	9/1	28	28	9.0	6.35	16.5	6	.39529	.39525		04
5.8932	10/1	27	49	23.4	9.04	15.4	9	.38719	.38721	—	02
6.4825	11/1		6	33.0	12.92	15.6	13	.37819	.37817	+	02
7.0718	12/1	26	18	36.6	22.61	14.9	9	.36805	.36807	—	02
7.6612	13/1	25	25	50.4	14.47	17.7	12	..35680	.35685		05
8.2505								1.34444	.34446		02
8.8398								(²){ .33079	.33081	·	02
9.4291								1.31612	.31581	+	31

¹ Paschen, new values. ² Paschen.

$$n^2 = A^2 + \frac{M_2}{\lambda^2 - \lambda_2^2} - k\lambda^2 - h\lambda^4$$

$A^2 = 2.03882$
$M_2 = 0.00621828$ $k = 0.00319987$
$\lambda_2^2 = 0.007706$ $h = 0.0000029160$

These results are reprinted in the annexed table, and have been made use of in the comparison between the dispersion of fluorite and rock salt included in Part II, Chapter I. If the writer's experience is a trustworthy guide, rock salt rather than fluorite is the substance most easy to procure, and, except for the deterioration of its surfaces (a difficulty not very serious), of greater value than fluorite for infra-red investigations. These observations are the more reinforced by the recent publications of Rubens and Nichols[1], and of Rubens and Trowbridge[2], in which data on the dispersion of rock salt are given to a wave-length of 18μ, and its absorption is found to be scarcely perceptible as far down as to a wave-length of 12μ, and not extremely prejudicial at

---
[1] H. Rubens und E. F. Nichols, Annalen der Physik und Chemie, neue Folge, vol. 60, p. 418, 1897.
[2] H. Rubens und A. Trowbridge, Annalen der Physik und Chemie, neue Folge, vol. 60, p. 724, 1897.

considerably greater wave-lengths, while fluorite appears to absorb almost completely at 11μ, and to be far from transparent at 8μ.

Constants for rock salt in Ketteler's dispersion formula.

Returning now, in closing, to the Ketteler dispersion formula given on page 254, let us see what values of the constants best fit the results thus far obtained on the dispersion of rock salt.

Ketteler[1] has calculated the following constants in applying formula (2) to the values of the writer already given.

$$b^2 = 2.32833$$
$$M_2 = 0.018496$$
$$\lambda_2^2 = 0.01621$$
$$k = 0.0008580$$

The greatest divergence between the observed and calculated values of n was found to be -32 units in the fifth place of decimals, and occurred for the value $n = 1.52270 \, (\pm 20)$.

Rubens, using formula (2), calculates the following constants to represent the values already given, as obtained from experimental data of Rubens and Snow, Rubens, and Paschen:

$$b^2 = 2.3285$$
$$M_2 = 0.018496$$
$$\lambda_2^2 = 0.01621$$
$$k = 0.000920$$

These values, as will be seen, are nearly identical with those given by Ketteler. The greatest deviation between observed and computed values of n, as computed by Rubens for the results just mentioned is one unit in the fourth place of decimals, which frequently occurs.

Rubens and Trowbridge[2] find the more exact formula (1) necessary to represent their data for rock-salt spectra of very long wave-lengths. They give constants as follows:

$$a^2 = 5.1790$$
$$M_2 = 0.018496$$
$$\lambda_2^2 = 0.01621$$
$$M_1 = 8977.0$$
$$\lambda_1^2 = 3149.3$$

Comparing observed and computed values of wave-lengths, the greatest deviation is 0.12μ, which occurs at $\lambda = 11.88\mu$.

It may, then, be said that the five constant formula of Ketteler represents the dispersion of rock salt within the limits of experimental error between wave-lengths of 0.4μ and 18μ.

---

[1] E. Ketteler, Annalen der Physik und Chemie, neue Folge, vol. 31, p. 322, 1887.
[2] H. Rubens und A. Trowbridge, Annalen der Physik und Chemie, neue Folge, vol. 60, p. 733, 1897.

THE CONSTANTS FOR ROCK SALT IN KETTELER'S DISPERSION FORMULA.

[From observations of the Smithsonian Astrophysical Observatory.]

Inasmuch as the formula of Ketteler has been found to follow the dispersion curves of rock salt, fluorite, and other substances through long ranges within the limits of error of determinations hitherto made, it appeared desirable to see if the extremely accurate results obtained at this observatory for rock salt, and already given in Part II, could be satisfactorily expressed by this formula. To do this the constants given by Rubens and Trowbridge, and already cited, were used as preliminary values and were corrected by the method of least squares from the nearly sixty observed values of n and λ determined experimentally at this observatory. The resulting formula to represent the dispersion of rock salt is as follows:

$$n^2 = a^2 + \frac{M_2}{\lambda^2 - \lambda_2^2} - \frac{M_1}{\lambda_1^2 - \lambda^2}$$

The values determined for the five constants are these:

$a^2 = 5.174714$ $M_1 = 8949.520$
$M_2 = 0.0183744$ $\lambda_1^2 = 3145.695$
$\lambda_2^2 = 0.015841$

In the following table is given a comparison between the values of refractive indexes experimentally determined for nearly sixty different wave-lengths and the corresponding refractive indexes computed by the aid of the above formula:

Application of Ketteler's formula to results on the dispersion of rock salt obtained at the Smithsonian Astrophysical Observatory.

Wave-length.	Observed refractive index.	Probable error of n.	Computed refractive index.	$n - n_1$	Wave-length.	Observed refractive index.	Probable error of n.	Computed refractive index.	$n - n_1$
μ	n		n_1		μ	n		n_1	
0.4861	1.553339	0.000009	1.553333	+0.000006	.5710	1.545546	0.000009	1.545525	+0.000021
.4920	2646	9	2647	— 1	.5758	5194	9	5189	+ 5
.4937	2456	9	2454	+ 2	.5786	5002	9	4997	+ 5
.4983	1940	9	1943	— 3	.5860	4505	9	4505	± 0
.5173	0006	9	.549984	+ 22	.5893	4273	9	4291	— 18
.5184	.549862	9	9878	— 16	.6105	3000	9	3004	— 4
.5273	9081	9	9042	+ 39	.6400	1405	9	1428	— 23
.5372	8175	9	8165	+ 10	.6563	0635	9	0651	— 16
.5530	6855	9	6868	— 13	.6874	.539807	9	.539822	— 15
.5660	5800	9	5884	— 84	.7190	8150	9	8151	— 1

Application of Ketteler's formula to results on the dispersion of rock salt obtained at the Smithsonian Astrophysical Observatory—Continued.

Wave-length.	Observed refractive index.	Probable error of n.	Computed refractive index.	$n-n_1$		Wave-length.	Observed refractive index.	Probable error of n.	Computed refractive index.	$n-n_1$	
μ	n		n_1			μ	n		n_1		
0.7604	1.536818	0.000009	1.536834	—	0.000016	1.7670	1.527377	0.000011	1.527349	+	0.000028
[1].7992	5691	20	5783	—	92	2.0736	6487	9	6467	+	20
.8424	4778	16	4780	—	2	.1824	6213	9	6193	+	20
.8835	3952	9	3958	—	6	.2464	6058	9	6038	+	20
.9033	3613	10	3602	+	11	.3560	5785	9	5778	+	7
.9724	2532	11	2519	+	13	[2].6505	4897				
.9916	2278	9	2257	+	21	[2].8080	4566	17			
1.0084	2057	9	2039	+	18	[2].9450	4359				
.0368	1695	10	1693	+	2	3.1104	4008	17	4084	—	76
.0540	1521	10	1497	+	24	.2736	3712	11	3712	±	0
.0810	1234	12	1206	+	28	.3696	3481	9	3490	—	9
.1058	0979	9	0979	±	0	.6288	2856	9	2874	—	18
.1420	0633	10	0618	+	15	.8192	2372	10	2405	—	33
.1780	0312	9	0309	+	3	4.1230	1564	18	1621	—	57
.2016	0139	10	0120	+	19	[2].7120	.519789	27	.519971	—	82
.2604	.529699	9	.529690	+	9	[3]5.3009	7874	42	8130	—	256
.3126	9368	12	9350	+	18	[3].8900	5497	42	6093	—	596
.4874	8452	10	8421	+	31	[3]6.4790	3413	69	3825	—	412
.5552	8144	9	8125	+	19	[4]13.96	.4373		.43653	+	770
.6368	7813	9	7803	+	10	[4]22.3	.340		.34000	±	000
.6848	7638	9	7628	+	10						

[1] The observations preceding this were obtained by comparing bolographs of the visible spectrum with Rowland's map. Those following are from n and λ experiments with grating and prism spectroscopes combined.
[2] Observations thus marked are unsatisfactory, and were not employed in the least square reduction.
[3] Observations thus marked were weighted in consideration of their large probable errors.
[4] The two observations thus marked are taken from results of Rubens and Trowbridge, and were not used in the least square reductions.

The small differences between the observed and computed values of n are, it will be seen, of the same order of magnitude as the probable error of the observations. Hence we may accept the formula given above as representing correctly the dispersion of rock salt.

THE MINUTE STRUCTURE OF THE ABSORPTION BAND ω_1 AS BOLOGRAPHICALLY OBSERVED IN THE SPECTRUM OF A 60° FLINT GLASS PRISM.

The following absorption lines were obtained by one observer, F. E. F., as the result of a comparison and measurement of four bolographs of ω_1 taken with the glass prism in October of 1899. Some of these bolographs are reproduced in Pl. XXII and are, it will be noted, far different from bolographs of the same region taken with a rock-salt prism owing to the great difference in the dispersion of rock salt and glass. The observer, C. G. A., examined the plates compared, though not to the extent of actually measuring them on the comparator, and he fully coincides in ascribing "reality" to the lines here given. Those lines marked with a query are not to be considered as absolutely certain without further evidence, but only as in a high degree probable. Letters a, b, c, d, under the caption "Intensity of line," have the same significance as in the tables of Part I.

Absorption lines in ω_1 bolographically determined in the spectrum of a 60° glass prism.

Number of line.	Intensity of line.	Measurement by F. E. F. on bolographs of—				Mean.	Mean intensity.	Distance from Φ.	Refractive index.	Wave-length.
		Oct. 13, 1899, 10ʰ 55ᵐ.	Oct. 13, 1899, 11ʰ 15ᵐ.	Oct. 13, 1899, 10ʰ 35ᵐ.	Oct. 14, 1899, 10ʰ 45ᵐ.					
		cm.	cm.	cm.	cm.	cm.		cm.	n.	μ.
1	c c c	0.538	....	0.550	0.566	0.551	c	63.100	1.584967	1.9745
2	d	v.	v.	0.497	v.	0.497	d?	63.154	1.584957	1.9750
3	d d d	0.376	0.364	0.402	v.?	0.381	d?	63.270	1.584936	1.9766
4	d d d c	0.296	0.304	0.251	0.316	0.292	d	63.359	1.584920	1.9777
5	d d d d	0.168	0.166	0.134	0.174	0.160	d }b	63.491	1.584897	1.9796
6	c c c c	0.000	0.000	0.000	0.000	0.000	c }b	63.651	1.584869	1.9819
7	c c c b	0.226	0.224	0.223	0.216	0.222	b	63.873	1.584829	1.9848
8	d d d	0.262	0.266	0.271	(¹)	0.266	d	63.917	1.584822	1.9853
9	d? d d?	n. v.	0.467	0.469	0.484	0.473	d?	64.124	1.584785	1.98825
10	d d d d	0.619	0.610	0.600	0.608	0.609	d	64.260	1.584761	1.99005
11	d d? d? d	0.705	0.706	0.667	0.682	0.690	d	64.341	1.584746	1.9911
12	d? d? d	0.778	....	0.756	0.787	0.774	d?	64.425	1.584732	1.9921
13	d d d d	1.083	1.120	1.077	1.072	1.089	d²	64.740	1.584675	1.9965
14	c d? c d	1.155	1.182	1.163	1.134	1.158	d	64.809	1.584664	1.9974
15	d d?d d?	1.240	1.225	1.209	1.184	1.215	d	64.886	1.584653	1.9980
16	c d c d	1.308	1.295	1.311	1.261	1.294	d	64.945	1.584640	1.9993
17	d? d? d d	1.385	1.358	1.377	1.340	1.365	d	65.016	1.584627	2.0102
18	d? d d? d	1.468	1.406	1.427	1.400	1.425	d	65.076	1.584617	2.0009
19	d? d d d	1.542	1.596	1.563	1.534	1.559	d ª	65.210	1.584593	2.0028

v. Visible but not measured.   n. v. Not visible.

¹ Distinctly single.   ² First in bottom of first depression of ω_1.   ³ Precedes maximum; possibly double.

Absorption lines in ω_1 bolographically determined in the spectrum of a 60° glass prism—Continued.

Number of line.	Intensity of line.	Measurement by F. E. F. on bolographs of—				Mean.	Mean intensity.	Distance from Φ.	Refractive index.	Wave length.
		Oct. 12, 1899, 10ʰ 55ᵐ.	Oct. 13, 1899, 11ʰ 15ᵐ.	Oct. 13, 1899, 10ʰ 35ᵐ.	Oct. 14, 1899, 10ʰ 45ᵐ.					
		cm.	cm.	cm.	cm.	cm.		cm.	n	μ.
20	d d ?	1.684	....	p. v.	1.661	1.672	d	65.322	1.584573	2.0043
21	d ? d d ?	....	1.728	1.727	1.718	1.724	d ? }b¹ ω₁	65.375	1.584564	2.0049
22	d d d	1.759	1.772	v. ?	1.794	1.775	d ?	65.425	1.584554	2.0057
23	d d ? d c	1.852	{1.816 / 1.870}	1.842	1.871	1.855	d a	65.506	1.584540	2.0067
24	d d c	1.944	1.924	1.916	v.	1.928	d	65.577	1.584528	2.0076
25	d ? d ? c	v.	2.038	2.039	2.032	2.036	d	65.687	1.584508	2.0090
26	d d c ? c	2.180	2.188	2.186	2.165	2.180	d	65.831	1.584482	2.0110
27	J ? d d d	2.248	2.254	2.284	2.296	2.270	d	65.921	1.584467	2.0122
28	d d ? d d ?	2.355	2.404	2.370	2.384	2.378	d ? }b	66.029	1.584447	2.0139
29	d d d	2.423	2.478	....	2.455	2.452	d ?	66.108	1.584435	2.0148
30	d d d d	2.642	2.680	2.617	2.611	2.650	d ?	66.301	1.584399	2.0174
31	d d ? d ? d	2.693	2.733	2.698	2.698	2.705	d ?	66.356	1.584390	2.0182
32	d d ? d	2.777	2.782	2.789	....	2.783	d	66.434	1.584375	2.0193
33	d d d d	3.042	3.060	3.060	3.059	3.055	d₁}b	66.706	1.584327	2.0230
34	d d d	3.110	3.108	3.121	?	3.113	d ?	66.764	1.584317	2.0238
35	d d d ? d ?	3.201	3.195	3.233	3.230	3.215	d ?	66.866	1.584298	2.0250
36	c d c c	3.406	3.456	3.465	3.468	3.452	c	67.103	1.584257	2.0283
37	d d d	3.918	....	3.940	3.950	3.936	d₁}c	67.587	1.584170	2.0347
38	d d d	3.998	4.016	3.993	....	4.002	d ?	67.658	1.584159	2.0355
39	c c d c	4.300	4.327	4.333	4.268	4.307	c	67.958	1.584105	2.0396

v. Visible but not measured.

¹ Seems different in second bolograph—2 in place of 1; not included in mean.   ² Apparently double—bottom second depression of ω_1.

The reader will not fail to note in examining Pl. XXII what a marked similarity is shown between the lines ω_1 and ω_2 and the line A shown on Pl. XXI A. It is significant to observe also that ω_1 like A, seems to be considerably influenced in form by the altitude of the sun, which indicates its terrestrial origin. This point is well shown in the series of curves in Pl. XXII.

INDEX.

	Page.
Abney, spectrum map of	9, 200
Absorption lines:	
Identification of	71, 74, 130
"Reality" of	72
Absorption, selective	11, 215
Absorption, variations in	205, 264
Seasonal	207
Irregular	207
Accidental deflections in records	105
Appropriation, Congressional	3
Assistants, names and titles of	2
Astrophysics, object of	2
Battery, for bolometer	63
Becquerel, spectrum map of	9
Bolographic method:	
Description of	22, 69
Errors inherent in	119
Bolographs:	
Comparison of	129
Measurements on	137
Notes accompanying	133
Ordinates of	159, 210
Reduction of measurements on	171
Bolometer:	
Accuracy of	238
Case	49, 51
Circuit, best resistances of	246
Description of	47
Invention of	10
Preparation of	48
Cauchy, dispersion formula	8, 253
Clock, driving	25, 56
Clouds, effect on bolographs	111
Collimation, discussion of	77, 82
Comparator	64
Measurements, accuracy of	118
Composite line spectra from bolographs	73
"Cylindrics"	37, 116
Diffraction, energy lost by	79
Dispersion formulæ	253
"Drift" of galvanometer	12, 22, 107
Error, sources of	86, 122, 125
Fluorite, dispersion of	219, 221, 258
Galvanometer:	
Coils, construction of	248
Description of	17, 24, 30, 59, 244
Needle, construction of	251
Sensitiveness	17, 25, 30, 60, 251
Suspension	62
Grating, constants of	225
Herschel, Sir J	8

INDEX.

	Page.
Higgs, spectrum map of	200
Identification of absorption lines	71, 74, 130
Infra-red solar spectrum:	
Illustrations of	127
Mapping of	126
Julius, W. H., method of suspension	38, 62
Kayser and Runge, formula for spectrum series	19
Ketteler, dispersion formula	254, 260, 261
Lamansky, M. S.	8
Latitude and longitude of observatory	4
Lens, cylindric	67, 116
Lewis, E. P	20
Light, cheapest form of	3
Mechanism, accuracy of	112
Melloni	7
Minimum deviation	88, 92
Mirror, cylindric	82
Mount Whitney	13, 23, 127, 200
Nichols, E. F.	21
Observatory:	
Description of	40
Early annals of	3
Founding of	1
Paschen, F	18, 120, 222, 256, 258
Photographic room	28
Plate carrier	58
Prism	45, 66
Aberration of	78
Angle determination of	92, 173
Concavity of faces	89
Resolving power of	76, 82
Radiation from terrestrial sources	240
Rayleigh, Lord	76, 78
Refractive index of A	172
Resolution:	
Bolometric	82
Visual	76
Rheostat, special for bolometer	38, 50, 51
Rock salt	14, 28
Dispersion of	16, 219, 222, 233, 236, 255, 261
Apparatus for determining	224
Method of determining	226
Prisms as standards of dispersion	236
Thermal conductivity of	102
Rubens, H	18, 256, 259
Shop, instrument making	4
Siderostat, Grubb	23, 45, 65
Slit	46, 65
Snow, B. W.	20
Spectrobolometer	30, 43, 46
Spectrum, infra-red solar	5, 15, 23, 198, 205
Tables of infra-red solar absorption lines	177, 188, 198
Temperature and deviation	96, 100, 101
Temperature, changes errors from	26, 94, 104
Temperature control, automatic	41, 95
Very, F. W	18
Wave-lengths of infra-red solar absorption lines	175, 198
Welsbach mantles, radiation of	240, 242
Wind, remarkable penetration of	108